PERGAMON INTERNATIONAL LIBRARY
of Science, Technology, Engineering and Social Studies

*The 1000-volume original paperback library in aid of education,
industrial training and the enjoyment of leisure*

Publisher: Robert Maxwell, M.C.

Chemistry for the
Engineering and Applied Sciences

Second edition of
Chemistry for the Applied Sciences

THE PERGAMON TEXTBOOK
INSPECTION COPY SERVICE

An inspection copy of any book published in the Pergamon International Library
will gladly be sent to academic staff without obligation for their consideration for
course adoption or recommendation. Copies may be retained for a period of 60
days from receipt and returned if not suitable. When a particular title is adopted or
recommended for adoption for class use and the recommendation results in a sale
of 12 or more copies, the inspection copy may be retained with our compliments.
The Publishers will be pleased to receive suggestions for revised editions and new
titles to be published in this important International Library.

Other Titles of Interest

Books

COOK and CULLEN
Chemical Plant and its Operation
(including Safety and Health Aspects)
Second (SI) edition

COULSON and RICHARDSON
Chemical Engineering (in 5 Vols,)

FORDHAM
High Explosives and Propellants, Second edition

FRANCIS and PETERS
Fuels and Fuel Technology, Second (SI) edition

GREEN, HARTLEY and WEST
Chemicals for Crop Protection and Pest Control

Journals*

Progress in Analytical Atomic Spectroscopy

* Free specimen copy available on request

Chemistry for the
Engineering and Applied Sciences

Second edition of
Chemistry for the Applied Sciences

by

W. STEEDMAN,
B.Sc., Ph.D., C.Chem., MRIC,
Senior Lecturer, Heriot-Watt University, Edinburgh

R. B. SNADDEN,
B.Sc., Ph.D., C.Chem., MRIC,
Lecturer, Heriot-Watt University, Edinburgh

I. H. ANDERSON,
B.Sc., Ph.D., C.Chem., FRIC,
*Senior Education Officer,
Scottish Council for Technical Education (SCOTEC)*

PERGAMON PRESS
Oxford · New York · Toronto · Sydney · Paris · Frankfurt

U.K.	Pergamon Press Ltd., Headington Hill Hall, Oxford OX3 0BW, England
U.S.A.	Pergamon Press Inc., Maxwell House, Fairview Park, Elmsford, New York 10523, U.S.A.
CANADA	Pergamon of Canada, Suite 104, 150 Consumers Road, Willowdale, Ontario M2J 1P9, Canada
AUSTRALIA	Pergamon Press (Aust.) Pty. Ltd., P.O. Box 544, Potts Point, N.S.W. 2011, Australia
FRANCE	Pergamon Press SARL, 24 rue des Ecoles, 75240 Paris, Cedex 05, France
FEDERAL REPUBLIC OF GERMANY	Pergamon Press GmbH, 6242 Kronberg-Taunus, Pferdstrasse 1, Federal Republic of Germany

First edition 1970
Second edition 1980

British Library Cataloguing in Publication Data
Steedman, William, b. 1934
Chemistry for the engineering and applied sciences. — 2nd ed.
1. Chemistry
I. Title II. Snadden, Robert Bruce
III. Anderson, Ian Howe
IV. Chemistry for the applied sciences
540′.2′462 QD33 78-41325

ISBN 0-08-022851-8 Hardcover
ISBN 0-08-022852-6 Flexicover

Printed in Hungary by Franklin Printing House

Contents

Contents

Preface

AT A time when the disadvantages of over-specialization in science and technology are becoming more clearly recognizable there is little need to defend the practice of teaching chemistry to prospective graduates in engineering and other applied sciences. The desirability of interpreting the behaviour of a large variety of materials in terms of their physical and chemical structures is acknowledged by the emergence of materials science as a discipline in itself. It is our experience that such an interpretation is possible only if a student has some reasonable acquaintance with chemical principles. The first edition of this book attempted to set out these principles and to examine as far as was possible within its scope their significance in the applied sciences. In the writing of this revision the basic philosophy remains unaltered although a number of additions have been incorporated. These include an extension of the chapter on "Radioactivity" (reflecting the increased importance of nuclear power) and the insertion of a chapter on "Colloids" (reflecting the industrial importance of this often sadly neglected topic). Also now included at the end of each chapter is a selection of "Exercises" which the reader is strongly recommended to tackle himself.

Since the first edition of this book was printed, a concerted international effort has been mounted to adopt a single coherent set of units in science and technology, referred to as the SI system (short for *Système Internationale d'Unités*). However, during what is still a transitional period, familiarity is required with both the old and new unit systems, and consequently in this edition certain quantities are still quoted in non-SI units. Thus while the joule almost entirely replaces the calorie and the British thermal unit, we have preferred to express concentration in moles per litre rather than moles per cubic decimetre. Atmospheres and millimetres of mercury have been retained for pressure, but when necessary newtons per square metre are employed. Similarly, temperatures are quoted in both degrees Celsius and Kelvin units. Readers who are unfamiliar with SI units will find the short appendix on SI units helpful.

The book divides naturally into two parts, the first part being mainly general physical and inorganic chemistry, the second organic chemistry. The early chapters (1–6) contain accounts of fundamental topics concerned with chemical bonding and structure. In Chapters 7–19 these concepts are developed and their relevance to a number of topics in applied science is examined. Chapters 20–24 constitute a short, non-mechanistic introduction to organic chemistry, permitting in Chapters 25–29 a study of a number of important organic materials. Throughout, the accent is on the application of chemical principles rather than process descriptions. The book is designed to meet the varied needs of undergraduates studying within the broad area of engineering and the applied sciences. We fully recognize that such students are likely to have widely varying chemical backgrounds, but we are confident that much of the material will stimulate the interest of the student with good entry qualifications in chemistry, and at the same time provide sufficient fundamental material to allow those with lower attainments in chemistry to acquire a sound grounding in the subject.

W. STEEDMAN
R. B. SNADDEN
I. H. ANDERSON

CHAPTER 1

Atomic Structure and Bonding

1.1. Atomic theory

The idea that all matter is composed of minute discrete particles, or atoms, has existed for hundreds of years, but it was not until the nineteenth century that this idea became a quantitative theory in accord with known experimental facts. It was at this time that Dalton advanced his theory which postulated that:

(i) matter is composed of small particles called atoms;

(ii) atoms can be neither created nor destroyed;

(iii) atoms of one particular element are all exactly alike, but different from the atoms of all other elements;

(iv) chemical combination takes place between whole numbers of atoms in definite numerical proportions.

Although we now know that not all of these ideas are strictly correct, they are the foundations on which modern atomic theory has been built. While a detailed study of the structure of the atom lies more in the realm of physics, some knowledge is essential for a reasonable understanding of how atoms bond together.

1.2. Subatomic particles

It is now accepted that the atom is not the ultimate particle, but is composed of even smaller entities. A fairly large number of subatomic particles have been discovered to date, but only three are considered to be stable and only these are of direct interest to the chemist. These particles are the *proton, neutron* and *electron*.

Protons and neutrons have about the same mass but differ in that a proton carries unit positive charge while the neutron is electrically neutral. The mass of an electron is about 1/1850 that of a proton and the electron carries unit negative charge. It can be seen therefore that practically all of the mass of an atom is accounted for by protons and neutrons, and since atoms are electrically neutral the number of protons and electrons in any given atom must be equal. The next problem we must consider is how these individual particles are arranged within the actual atom.

1.3.   The nuclear atom

Whereas it was once believed that atoms had a homogeneous structure, it is now considered that the atom consists of a small positively charged *nucleus* containing the neutrons and protons, and that this is surrounded by a relatively diffuse arrangement of electrons. The diameter of the nucleus is of the order of 1.0×10^{-12} cm, while that of the atom as a whole is about 1.0×10^{-8} cm. The nucleus is held together by attractive forces operating between protons and neutrons, and there is a tendency for these particles to exist in equal numbers within the nucleus. The structure and stability of an atomic nucleus are considered in more detail in Chapter 2.

Every atom is characterized by its *atomic number* (Z), which is the number of protons contained in its nucleus. Hydrogen, the simplest element, which has a nucleus consisting of only one proton, has an atomic number $Z = 1$. Helium, the next simplest element, whose nucleus is composed of two protons and two neutrons, has an atomic number $Z = 2$.

So far we have a picture fairly similar to that described by Dalton, with atoms characterized by their atomic number and all atoms of any particular element having the same mass. We find, however, that this is not always the case. Hydrogen, for example, can exist in three different forms: hydrogen with a nucleus containing one proton, hydrogen with a nucleus composed of one proton and one neutron (*deuterium* or "heavy" *hydrogen*), and hydrogen with a nucleus containing one proton and two neutrons (*tritium*). These atoms of hydrogen, containing the same number of protons but different numbers of neutrons, are called *isotopes* of hydrogen and obviously the mass of each isotope differs from that of the others.

Many of the elements can exist in two or more isotopic forms, and this has in the past created some difficulties in the definition of atomic mass. Since the mass of an atom is extremely small it is convenient to define a relative atomic mass scale, choosing one element as a reference. Prior to 1961 the reference element was oxygen which was given a mass value of 16 atomic mass units (a.m.u.). Since naturally occurring oxygen contains three isotopes, each containing eight protons together with eight, nine and ten neutrons respectively, a certain amount of ambiguity existed because chemists ascribed a mass of 16 a.m.u. to the average mixture of isotopes, whilst physicists referred the scale to the single isotope containing eight protons and eight neutrons. Now confusion is avoided by assigning 12 a.m.u. to the isotope of carbon containing six protons and six neutrons.

On this scale the relative masses of protons, neutrons and electrons are 1·00728, 1·00867 and 0·000549 a.m.u. respectively. It would be wrong, however, to assume that the atomic mass of any element can be calculated exactly simply by adding together the relative masses of its constituent parts. In the formation of a nucleus from individual protons and neutrons, a small proportion of the total mass is converted into energy. The consequences of this phenomenon are discussed more fully in Chapter 2.

1.4. Electronic structure

In general terms the structure of the nucleus does not have any marked effect on chemical behaviour. Thus we find only very slight differences in chemical reactivity between the various isotopes of oxygen. It is true that differences are rather more pronounced when we consider the chemical behaviour of hydrogen and deuterium, because in this case the addition of only one neutron to the nucleus doubles the mass of the atom. This is the exception rather than the rule, and it is the electronic structure of atoms we must investigate if we wish to understand their chemical behaviour.

Modern theories of atomic structure are highly mathematical rather than pictorial, and although they are more useful than the older theories they are correspondingly more difficult to understand. One of the basic postulates of modern theory (Heisenberg's Uncertainty Principle) is that it is impossible to define both the position and the energy of an electron simultaneously, since the process of attempting to measure the location or momentum of an electron will itself have some effect on its position or energy. The behaviour of an electron is therefore described in terms of the *probability* of finding an electron at any particular point in space. We can regard an electron in two different ways, either as a discrete particle which will spend most of its time in regions of high probability or alternatively as a cloud of negative charge with a high density in high probability regions and a low density in low probability areas. For example, we could picture a hydrogen atom as in Fig. 1.1, where the single electron of the hydrogen atom is represented as a diffuse cloud of negative charge.

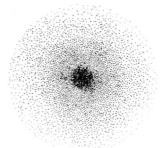

FIG. 1.1. The hydrogen atom.

This diagram describes two aspects of the behaviour of the electron: first that the electron is most likely to be found in the regions close to the nucleus where the density of shading is greatest, and, secondly, that no direction is favoured over any other, in other words the electron is equally likely to exist on one side of the nucleus as the other. A description of the probability of finding an electron at any given point such as we have just discussed is called an *orbital*.

Any electron can be characterized by a set of four *quantum numbers*, the *principal* quantum number (*n*), the *azimuthal* quantum number (*l*), the *magnetic* quantum number (*m*) and the *spin* quantum number (*s*). Each of these quantum numbers can have only certain values assigned to it.

The principal quantum number can have any positive integral value.

$$n = 1, 2, 3, \ldots$$

and describes the average distance of the electron from the nucleus. This is the most important factor in determining the energy of the electron; a low value of *n* implies a stable low energy electron, while a high value of *n* signifies an electron of high energy.

The azimuthal quantum number is determined in part by the value of *n*, the allowed values being

$$l = 0, 1, 2, \ldots (n-1)$$

Thus for a value of $n = 3$ only three values of *l* are possible, viz. 0, 1 and 2. The value of the azimuthal quantum number describes the shape of the orbital, and generally the value of *l* is represented by a letter rather than by a number:

Numerical value of *l*	0	1	2	3
Representative letter	*s*	*p*	*d*	*f*

The magnetic quantum number can assume any integral value within the limits.

$$m = 0, \pm 1, \pm 2 \ldots \pm l$$

While the value of *l* describes the shape of the orbital, the value of *m* describes the orientation of the orbital relative to the nucleus.

The spin quantum number describes the direction of spin of the electron, which may be envisaged as spinning about its own axis in either a clockwise or anti-clockwise direction. There are therefore only two possible values for *s*, $+\frac{1}{2}$ or $-\frac{1}{2}$.

Any electron then can be characterized by quoting values for these four quantum numbers, and in particular the energy of the electron is determined predominantly by the values of *n* and *l*. For our purposes we can assume that the energies of orbitals with the same values of n and l but different values of m and s are the same, and only n and l values need be specified.

Another basic postulate of modern theory is that *no two electrons on the same atom can have the same quantum number values*. With this in mind we can use these definitions of the various quantum numbers to see how electrons arrange themselves around the nucleus. For a value of $n = 1$, only two electrons can exist, having the quantum numbers:

n	l	m	s
1	0	0	$+\frac{1}{2}$
1	0	0	$-\frac{1}{2}$

Similarly, for $n = 2$ we find a group of eight electrons:

n	l	m	s
2	0	0	$+\frac{1}{2}$
2	0	0	$-\frac{1}{2}$
2	1	-1	$+\frac{1}{2}$
2	1	-1	$-\frac{1}{2}$
2	1	0	$+\frac{1}{2}$
2	1	0	$-\frac{1}{2}$
2	1	$+1$	$+\frac{1}{2}$
2	1	$+1$	$-\frac{1}{2}$

For $n = 3$ there are eighteen possible combinations of quantum numbers and therefore eighteen electrons can exist with this value for the principal quantum number. The full significance of this sequence of numbers will be discussed again in Chapter 3.

We have already learned that the energy of an electron is described by the values of the principal and azimuthal quantum numbers. Figure 1.2 shows how the relative energy varies with the quantum number value, each circle in the diagram representing an orbital capable of containing two electrons. If we now accept that the electrons on any atom tend to adopt the configuration of lowest energy we can build up a picture of the electronic structure of an atom.

Hydrogen, the simplest atom with an atomic number $Z = 1$, has one electron which will be held in the $1s$ orbital, this being the orbital of lowest energy as indicated in Fig. 1.2. The electronic configuration of this atom would then be represented $1s^1$.

The atom with atomic number $Z = 2$ is helium, which has two electrons to be arranged around the nucleus. By reference to Fig. 1.2 we can see that both these electrons can be accommodated in the $1s$ orbital.

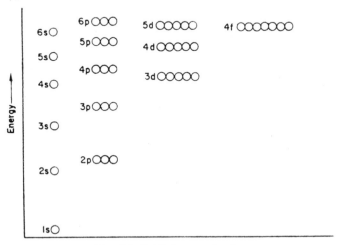

FIG. 1.2. Orbital energy levels.

Hence the electronic configuration of helium can be represented $1s^2$. At this point it can be seen that the $1s$ orbital now holds its maximum quota of electrons, and since helium is known to be chemically unreactive we can say that this situation is associated with a high degree of stability.

The lithium atom has three electrons only two of which can occupy the low energy $1s$ orbital, leaving the third electron to be accommodated in the $2s$ orbital. Thus the configuration will be represented $1s^2.2s^1$. Again by reference to Fig. 1.2 it can be seen that eight electrons can be accommodated in the group of orbitals having a principal quantum number $n = 2$, before this group is completely filled. As is shown in Table 1.1 this situation arises with the neon atom which has the configuration $1s^2 2s^2 2p^6$. Neon, like helium, is highly inert chemically, and again we note that a high degree of stability is associated with the state where all the orbitals of a group hold their maximum number of electrons.

TABLE 1.1. ELECTRONIC

Z	Element	1 s	2 s	2 p	3 s	3 p	3 d
1	H	1					
2	He	2					
3	Li	2	1				
4	Be	2	2				
5	B	2	2	1			
6	C	2	2	2			
7	N	2	2	3			
8	O	2	2	4			
9	F	2	2	5			
10	Ne	2	2	6			
11	Na	2	2	6	1		
12	Mg	2	2	6	2		
13	Al	2	2	6	2	1	
14	Si	2	2	6	2	2	
15	P	2	2	6	2	3	
16	S	2	2	6	2	4	
17	Cl	2	2	6	2	5	
18	Ar	2	2	6	2	6	
19	K	2	2	6	2	6	
20	Ca	2	2	6	2	6	
21	Sc	2	2	6	2	6	1
22	Ti	2	2	6	2	6	2
23	V	2	2	6	2	6	3
24	Cr	2	2	6	2	6	5
25	Mn	2	2	6	2	6	5
26	Fe	2	2	6	2	6	6
27	Co	2	2	6	2	6	7
28	Ni	2	2	6	2	6	8
29	Cu	2	2	6	2	6	10
30	Zn	2	2	6	2	6	10
31	Ga	2	2	6	2	6	10
32	Ge	2	2	6	2	6	10
33	As	2	2	6	2	6	10
34	Se	2	2	6	2	6	10
35	Br	2	2	6	2	6	10
36	Kr	2	2	6	2	6	10
37	Rb	2	2	6	2	6	10
38	Sr	2	2	6	2	6	10
39	Y	2	2	6	2	6	10
40	Zr	2	2	6	2	6	10
41	Nb	2	2	6	2	6	10
42	No	2	2	6	2	6	10

STRUCTURE OF THE ATOMS

| | 4 | | | | 5 | | | | 6 | | | 7 |
s	p	d	f	s	p	d	f	s	p	d	f	s
1												
2												
2												
2												
2												
1												
2												
2												
2												
2												
1												
2												
2	1											
2	2											
2	3											
2	4											
2	5											
2	6											
2	6			1								
2	6			2								
2	6	1		2								
2	6	2		2								
2	6	4		1								
2	6	5		1								

(continued overleaf)

TABLE 1.1

Z	Element	1 s	2 s	2 p	3 s	3 p	3 d
43	Tc	2	2	6	2	6	10
44	Ru	2	2	6	2	6	10
45	Rh	2	2	6	2	6	10
46	Pd	2	2	6	2	6	10
47	Ag	2	2	6	2	6	10
48	Cd	2	2	6	2	6	10
49	In	2	2	6	2	6	10
50	Sn	2	2	6	2	6	10
51	Sb	2	2	6	2	6	10
52	Te	2	2	6	2	6	10
53	I	2	2	6	2	6	10
54	Xe	2	2	6	2	6	10
55	Cs	2	2	6	2	6	10
56	Ba	2	2	6	2	6	10
57	La	2	2	6	2	6	10
58	Ce	2	2	6	2	6	10
59	Pr	2	2	6	2	6	10
60	Nd	2	2	6	2	6	10
61	Pm	2	2	6	2	6	10
62	Sm	2	2	6	2	6	10
63	Eu	2	2	6	2	6	10
64	Gd	2	2	6	2	6	10
65	Tb	2	2	6	2	6	10
66	Dy	2	2	6	2	6	10
67	Ho	2	2	6	2	6	10
68	Er	2	2	6	2	6	10
69	Tm	2	2	6	2	6	10
70	Yb	2	2	6	2	6	10
71	Lu	2	2	6	2	6	10
72	Hf	2	2	6	2	6	10
73	Ta	2	2	6	2	6	10
74	W	2	2	6	2	6	10
75	Re	2	2	6	2	6	10
76	Os	2	2	6	2	6	10
77	Ir	2	2	6	2	6	10
78	Pt	2	2	6	2	6	10
79	Au	2	2	6	2	6	10
80	Hg	2	2	6	2	6	10
81	Tl	2	2	6	2	6	10
82	Pb	2	2	6	2	6	10
83	Bi	2	2	6	2	6	10
84	Po	2	2	6	2	6	10

(*cont.*)

4				5				6				7
s	p	d	f	s	p	d	f	s	p	d	f	s
2	6	6		1								
2	6	7		1								
2	6	8		1								
2	6	10										
2	6	10		1								
2	6	10		2								
2	6	10		2	1							
2	6	10		2	2							
2	6	10		2	3							
2	6	10		2	4							
2	6	10		2	5							
2	6	10		2	6							
2	6	10		2	6			1				
2	6	10		2	6			2				
2	6	10		2	6	1		2				
2	6	10	2	2	6			2				
2	6	10	3	2	6			2				
2	6	10	4	2	6			2				
2	6	10	5	2	6			2				
2	6	10	6	2	6			2				
2	6	10	7	2	6			2				
2	6	10	7	2	6	1		2				
2	6	10	9	2	6			2				
2	6	10	10	2	6			2				
2	6	10	11	2	6			2				
2	6	10	12	2	6			2				
2	6	10	13	2	6			2				
2	6	10	14	2	6			2				
2	6	10	14	2	6	1		2				
2	6	10	14	2	6	2		2				
2	6	10	14	2	6	3		2				
2	6	10	14	2	6	4		2				
2	6	10	14	2	6	5		2				
2	6	10	14	2	6	6		2				
2	6	10	14	2	6	9						
2	6	10	14	2	6	9		1				
2	6	10	14	2	6	10		1				
2	6	10	14	2	6	10		2				
2	6	10	14	2	6	10		2	1			
2	6	10	14	2	6	10		2	2			
2	6	10	14	2	6	10		2	3			
2	6	10	14	2	6	10		2	4			

(*cont. overleaf*)

2*

TABLE 1.1

Z	Element	1 s	2 s	2 p	3 s	3 p	3 d
85	At	2	2	6	2	6	10
86	Rn	2	2	6	2	6	10
87	Fr	2	2	6	2	6	10
88	Ra	2	2	6	2	6	10
89	Ac	2	2	6	2	6	10
90	Th	2	2	6	2	6	10
91	Pa	2	2	6	2	6	10
92	U	2	2	6	2	6	10
93	Np	2	2	6	2	6	10
94	Pu	2	2	6	2	6	10
95	Am	2	2	6	2	6	10
96	Cm	2	2	6	2	6	10
97	Bk	2	2	6	2	6	10
98	Cf	2	2	6	2	6	10
99	Es	2	2	6	2	6	10
100	Fm	2	2	6	2	6	10
101	Md	2	2	6	2	6	10
102	No	2	2	6	2	6	10
103	Lw	2	2	6	2	6	10

1.5. Chemical bonding

We have built up a picture of an isolated atom but have not as yet discussed how these atoms can combine with similar atoms or with atoms of a different element. We want to understand *why* different atoms combine to form a new molecule and also *how* the bonding is achieved. The answer to the first question is that any system will tend to adjust itself in order to attain the most stable possible state, i.e. the state of lowest energy. If it is assumed that electronic rearrangements will readily occur provided that the end result is a more stable arrangement, then the formation of bonds between different atoms is readily understood. It has already been stated in the previous section that an arrangement of electrons where a group of orbitals holds the maximum possible number of electrons is one of high stability. It is from this viewpoint, viz. that electronic rearrangements will occur to attain these configurations of high stability, that we shall attempt to account for the formation of chemical bonds.

(*cont.*)

4				5				6				7
s	p	d	f	s	p	d	f	s	p	d	f	s
2	6	10	14	2	6	10		2	5			
2	6	10	14	2	6	10		2	6			
2	6	10	14	2	6	10		2	6			1
2	6	10	14	2	6	10		2	6			2
2	6	10	14	2	6	10		2	6	1		2
2	6	10	14	2	6	10		2	6	2		2
2	6	10	14	2	6	10	2	2	6	1		2
2	6	10	14	2	6	10	3	2	6	1		2
2	6	10	14	2	6	10	4	2	6	1		2
2	6	10	14	2	6	10	6	2	6			2
2	6	10	14	2	6	10	7	2	6			2
2	6	10	14	2	6	10	7	2	6	1		2
2	6	10	14	2	6	10	8	2	6	1		2
2	6	10	14	2	6	10	10	2	6			2
2	6	10	14	2	6	10	11	2	6			2
2	6	10	14	2	6	10	12	2	6			2
2	6	10	14	2	6	10	13	2	6			2
2	6	10	14	2	6	10	13	2	6	1		2
2	6	10	14	2	6	10	14	2	6	1		2

The answer to how bonding is achieved leads to a classification of bond types under two main headings depending on what type of electron rearrangement takes place. Chemical bonds can be described either as *ionic*, where electrons are completely transferred from one atom to another, or as *covalent*, where electrons are shared between atoms. It must be emphasized, however, that most chemical bonds are intermediate between these two extreme types although in practice they are usually classed as one or the other.

1.6. Ionic bonds

A good example of the formation of an ionic bond is the formation of sodium chloride from sodium and chlorine atoms. By reference to Table 1.1 we find that the electronic configuration of the sodium atom is

$$1s^2 \quad 2s^2 2p^6 \quad 3s^1$$

We know that the configurations $1s^2$ and $2s^2 2p^6$ are stable closed systems where the groups of orbitals contain their maximum number of electrons, and that the single electron in the $3s$ orbital is a fairly high energy electron. This electron is obviously in a different category from the others and it is found that this electron can be easily removed from the atom leaving a *sodium ion* (Na^+) which is positively charged and whose electronic structure is a known stable configuration.

Similarly, we know that the electronic configuration of the chlorine atom is

$$1s^2 \quad 2s^2 2p^6 \quad 3s^2 3p^5$$

Here again we find two closed stable configurations together with an outer shell of seven electrons. It can readily be seen that if the chlorine atom could acquire an extra electron to form a negatively charged *chloride ion* (Cl^-), again a stable electronic configuration would be achieved. Thus by the transfer of one electron from a sodium atom to a chlorine atom two *ions* are produced, each with a stable arrangement of electrons, and since these ions are of opposite electrical charge they attract each other to form an ionic bond. By considering only the outer electrons, the bond formation may be represented

$$Na^{\circ} + \cdot \ddot{\underset{\cdot\cdot}{C}}l\colon \rightarrow Na^+ \colon \ddot{\underset{\cdot\cdot}{C}}l\colon^-$$

It is important to remember that in the formation of a bond of this type there must be a balance between the electrons lost and the electrons gained. If we consider the formation of another compound, calcium chloride, we find a slightly different situation. Calcium has the configuration:

$$1s^2 \quad 2s^2 2p^6 \quad 3s^2 3p^6 \quad 4s^2$$

Here *two* electrons must be transferred from the atom in order to attain a stable configuration, hence *two* chlorine atoms must be involved to maintain the electron balance.

$$\colon\ddot{\underset{\cdot\cdot}{C}}l \cdot + \overset{\cdot\cdot}{C}a + \cdot\ddot{\underset{\cdot\cdot}{C}}l\colon \rightarrow \colon\ddot{\underset{\cdot\cdot}{C}}l\colon^- Ca^{2+} \colon\ddot{\underset{\cdot\cdot}{C}}l\colon^-$$

The term valency is used to describe the ability of atoms to bond together. From the above examples sodium is said to have a valency of one, and calcium a valency of two.

1.7. Covalent bonds

It can easily be seen that only a limited number of elements will be capable of forming bonds by the total transfer of electrons. Since negatively charged electrons are attracted to the positively charged nucleus by electrostatic forces, there will obviously be a limit to the number of electrons which can be removed from any particular atom, because the removal of an electron increases the residual positive charge by which the others are retained. In general it is energetically not feasible to transfer more than two electrons. For this reason the majority of compounds are formed by *covalent* bonding, i.e. by the *sharing* of electrons.

The simplest example of this type of bonding is the formation of a hydrogen molecule from two atoms of hydrogen. We have already learned that the hydrogen atom has only a single $1s$ electron and that the $1s$ orbital is capable of holding two electrons. Although it would be possible to describe bond formation between two hydrogen atoms formally as an ionic bond:

$$H^{\bullet} + H^{\circ} \rightarrow H^{+}{}_{\bullet}^{\bullet}H^{-}$$

it seems unreasonable to suppose that one hydrogen atom should be capable of abstracting an electron from another exactly similar hydrogen atom. Instead, the bond is described as the sharing of the two electrons between the two atoms:

$$H^{\bullet} + H^{\circ} \rightarrow H^{\bullet}_{\bullet}H$$

Just as the hydrogen atom can be represented as in Fig. 1.1 the hydrogen molecule can be pictured as in Fig. 1.3 where the electron clouds of the separate hydrogen atoms have merged to produce a region of high electron density between the two atoms, thus forming a covalent bond.

A slightly more complicated example is the formation of the hydrogen chloride molecule. Here the hydrogen atom has one electron and the

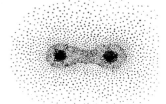

Fɪɢ. 1.3. The hydrogen molecule.

chlorine atom has seven outer electrons, hence covalent bonding can occur to give the hydrogen atom its full complement of two electrons and the chlorine atom a stable arrangement of eight electrons:

$$\text{H}\cdot + \cdot \overset{\circ\circ}{\underset{\circ\circ}{\text{C}}}\text{l}_\circ^\circ \rightarrow \text{H}\overset{\circ\circ}{\underset{\circ\circ}{\text{Cl}}}\overset{\circ\circ}{\underset{\circ\circ}{}}$$

Although the H—Cl bond is covalent it differs in one important respect from the covalent H—H bond. In the hydrogen molecule the pair of electrons forming the bond is shared equally between the two atoms, whereas in the hydrogen chloride molecule this is not the case. Because of differences in internal atomic structure atoms of different elements vary in their capacity to attract electrons. An element that tends to acquire rather than lose electrons in its reactions is said to be *electronegative*, while an element which tends to lose electrons is described as *electropositive*. In the above example the chlorine atom is more electronegative than the hydrogen, thus the electron pair forming the bond will have a greater probability of being found near the chlorine atom, which will therefore tend to be negatively charged with respect to the hydrogen. Such a bond is described as *polar*, and will be expected to influence the properties of the material.

A molecule can obviously contain more than two atoms. For example, in a water molecule the oxygen atom, which has six outer electrons, forms two covalent bonds with hydrogen atoms in order to attain a stable configuration:

$$\text{H}\cdot + \cdot \overset{\circ\circ}{\underset{\circ\circ}{\text{O}}}\circ + \text{H}\cdot \rightarrow \text{H}\overset{\circ\circ}{\underset{\circ\circ}{\text{O}}}\text{H}$$

As was the case in ionic bonding the valency of a particular atom is determined by its electronic structure, and in the examples we have considered the valency of hydrogen and of chlorine is one, while that of oxygen is two.

If there is an insufficient number of electrons to attain a stable configuration by the sharing of one electron pair, it is possible for two atoms to share two pairs to form a *double* bond. An example of this is the carbon dioxide molecule, where the carbon atom has four valency electrons and the oxygen atoms each have six:

$$\overset{\circ\circ}{\underset{\circ\circ}{\text{O}}}\!\circ + \overset{\bullet\bullet}{\underset{\bullet}{\text{C}}} + \circ\!\overset{\bullet\bullet}{\underset{\circ\circ}{\text{O}}}\!\circ \rightarrow \overset{\bullet\bullet}{\underset{\circ\circ}{\text{O}}}\!{\scriptstyle{\circ\atop\circ}}\!{\scriptstyle\bullet}\text{C}\!{\scriptstyle\bullet}\!{\scriptstyle{\circ\atop\circ}}\!\overset{\bullet\bullet}{\underset{}{\text{O}}}$$

Similarly the formation of a *triple* bond may occur when three electron pairs must be shared between two atoms to attain a stable configuration. This occurs, for example, in the nitrogen molecule, where each nitrogen atom possesses only five valency electrons.

$$\overset{\circ}{\underset{\circ}{\text{N}}}\!\circ + \bullet\overset{\bullet}{\underset{\bullet}{\text{N}}}{\scriptstyle\bullet} \rightarrow {\scriptstyle\circ}\text{N}{\scriptstyle{\circ\circ\atop\bullet\bullet}}\text{N}{\scriptstyle\bullet}$$

In all examples considered up to this point a covalent bond has been formed by the sharing of a pair of electrons, one electron being donated by each atom. There are occasions where the two electrons required for bond formation are donated by only one of the atoms concerned. An example of this is the formation of the ammonium ion (NH_4^+) which may be regarded as the combination of a hydrogen ion (H^+) with a molecule of ammonia:

$$\text{H} \quad : \overset{\bullet\bullet}{\underset{\bullet\bullet}{\text{N}}} : + \text{H}^+ \rightarrow \left[\text{H} \quad : \overset{\bullet\bullet}{\underset{\bullet\bullet}{\text{N}}} : \text{H} \right]^+$$

with H above and below the N on both sides.

Here the ammonia molecule contains three covalent N—H bonds, leaving a *lone pair* of electrons not involved in bond formation. This lone pair of electrons is available for sharing with the hydrogen ion to form a bond which is usually described as a *co-ordinate covalent bond*.

1.8. Hybridization

Since a large part of this book deals with the chemistry of carbon compounds, it is worth while at this stage to study in rather more detail the bonding involved in such compounds.

The electronic configuration of carbon is

$$1s^2 \quad 2s^2 2p^2$$

i.e. there exists a stable closed system of two electrons, leaving four valency electrons.

In Section 1.4 we learned that the shape of an orbital was described by the value of the azimuthal quantum number and that an s orbital was spherically symmetrical as shown in Fig. 1.1. There are three p orbitals each of which can hold two electrons, and these dumb-bell-shaped orbitals are arranged mutually at right angles to one another as is shown in Fig. 1.4.

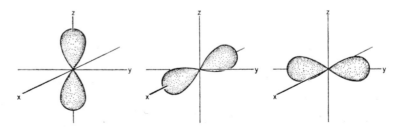

FIG. 1.4. p orbitals.

We now have a situation where a single carbon atom is pictured as having two valency electrons in a spherical s orbital, and two in two of the three p orbitals. On the other hand, since it is found from experiment that when carbon forms a compound such as methane the four bonds are equivalent, we must modify our picture to obtain four equivalent orbitals around the carbon atom. This is achieved by postulating a process of *hybridization* where by "blending" one s orbital and three p orbitals we obtain four sp^3 orbitals directed tetrahedrally (i.e. mutually at an angle of 109°28') as shown in Fig. 1.5(a). Since the formation of a

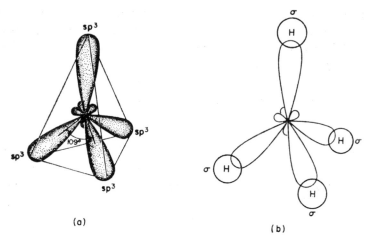

(a)

(b)

FIG. 1.5. *sp³* orbitals and the methane molecule.

covalent bond can be pictured as the overlapping or merging of the electron orbitals, we would picture the molecule of methane as in Fig. 1.5(b) with the central carbon atom surrounded tetrahedrally by the four hydrogen atoms.

When a carbon–carbon double bond is formed, as is the case in ethylene (CH_2=CH_2), a different type of hybridization occurs. Here one *s* orbital and two *p* orbitals are blended to give three equivalent *sp²* hybrids, mutually at an angle of 120°, leaving one *p* orbital unchanged (Fig. 1.6(a)).

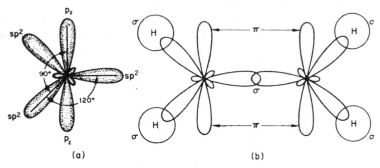

(a)

(b)

FIG. 1.6. *sp²* orbitals and the ethylene molecule.

Again bond formation has involved the overlapping of orbitals to give a flat molecule where each carbon atom forms three normal covalent bonds (σ bonds) and a second carbon–carbon bond (π bond) is formed by the merging of the unhybridized p orbitals above and below the plane of the molecule in Fig. 1.6(b). It can be seen that this second carbon–carbon bond is much more diffuse than the others in the molecule and is found to be more susceptible to chemical attack as will be explained in Chapter 20.

An analogous situation is found in acetylene ($CH\equiv\!CH$) where a triple bond is formed. In this case only one p orbital is combined with the s orbital to give two sp orbitals and the remaining two p orbitals are left unchanged as shown in Fig. 1.7(a). If bond formation is again pictured as

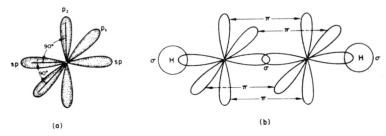

(a)   (b)

FIG. 1.7. *sp* orbitals and the acetylene molecule.

the overlapping of orbitals it can be seen that each carbon atom will form two normal σ bonds, while at the same time two diffuse π bonds are formed between the carbon atoms as illustrated in Fig. 1.7(b).

If we now consider a molecule containing *conjugated* double bonds, i.e. alternate single and double bonds, we find a further complication. A molecule of this type is *benzene* whose formula may be written

By analogy with ethylene we might expect that *sp²* hybridization would occur giving a planar cyclic molecule containing three well-defined double bonds. This description does not explain the observed chemistry of benzene, as you will learn in Chapter 22.

We must picture the molecule as in Fig. 1.8. Instead of having three separate electron clouds forming three well-defined, separate double

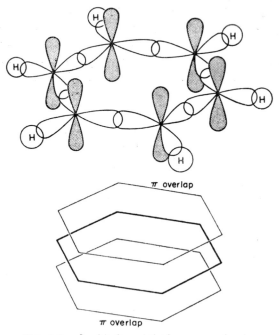

FIG. 1.8. *sp²* orbitals and the benzene molecule.

bonds, we allow those isolated clouds to merge to form a continuous electron cloud above and below the plane of the molecule. This implies that the electrons forming the double bonds are not localized between any two carbon atoms but are free to move around the molecule, imparting a certain amount of double bond character to each carbon–carbon bond. The electrons are said to be *delocalized*. Any molecule where delocalization of electrons is possible is found to be far more stable than would otherwise be expected.

1.9.   The hydrogen bond

It is convenient to introduce at this point an unusual feature of certain compounds containing hydrogen. It is found that where molecules contain hydrogen bonded to a highly electronegative element, e.g. —OH, —NH, —FH, unexpected forces of attraction exist which may considerably affect the properties of the material. If we compare the boiling points of the hydrogen halides (Table 1.2) we see that for the latter members of the group the boiling points increase with increasing molecular weight. Hydrogen fluoride, however, has an abnormally high boiling point, showing that the molecules of this material must be held together in some unusual way.

TABLE 1.2. BOILING POINTS OF THE
HYDROGEN HALIDES

Compound	HF	HCl	HBr	HI
Boiling point (°C)	19·4	− 83·1	− 68·7	− 36·7

Since a hydrogen atom has only a single $1s$ electron it can form only one normal covalent bond. Nevertheless, it can at the same time form a special type of bond called the *hydrogen bond*, which is much weaker, usually by a factor of about twenty, than the normal covalent type. The bond may be envisaged as being mainly electrostatic in character, the attraction being between the electropositive hydrogen and a lone pair of electrons on the electronegative element involved.

The abnormally high boiling point of hydrogen fluoride can now be understood in terms of the individual molecules being linked together by hydrogen bond "bridges" which must be broken before vaporization can occur.

Since the elements chlorine, bromine and iodine are much less electro-negative than fluorine, hydrogen bonding is negligible in the hydrides of these elements.

Hydrogen bonding involving oxygen atoms is of considerable importance in organic chemistry and drastically modifies the physical properties of many organic compounds. The lower alcohols, for example, have properties very different from closely related compounds of similar molecular weight (Table 1.3) because of the effect of hydrogen bond bridging between molecules:

$$O—H---O—H---O—H---O—H$$
$$|\qquad|\qquad|\qquad|$$
$$R\qquad R\qquad R\qquad R$$

TABLE 1.3. PHYSICAL PROPERTIES OF SOME LOWER ALCOHOLS AND HYDROCARBONS

Compound	Molecular weight	Melting point (°C)	Boiling point (°C)
CH_3OH	32	$-97\cdot8$	$64\cdot7$
CH_3CH_3	30	-172	$-89\cdot0$
CH_3CH_2OH	46	$-117\cdot6$	$78\cdot5$
$CH_3CH_2CH_3$	44	$-189\cdot9$	$-42\cdot2$

Exercises

1.1. Define the following terms:

(i) principal quantum number;
(ii) magnetic quantum number;
(iii) atomic number;
(iv) neutron;
(v) hybridization:
(vi) hydrogen bond.

1.2. Using the rules for permitted values of the four quantum numbers, construct a table indicating the 18 possible combinations of quantum numbers for the value of $n = 3$.

1.3. Referring to the orbital energy levels in Fig. 1.2 deduce the electronic configurations of the elements having atomic numbers:

$$Z = 8, \quad 11, \quad 16, \quad 26, \quad 31, \quad 35$$

and express the answer using s, p, d, f notation.

1.4. Summarize the methods whereby electronic rearrangements may occur between atoms to form stable chemical bonds.

1.5. Describe the main differences between the Dalton and the present-day model of the atom.

CHAPTER 2

Radioactivity

WE HAVE already learned that for the chemist it is the *electronic structure* of the atoms which governs chemical behaviour. In normal chemical reactions the atomic nuclei remain unchanged and the formation of new chemical bonds is ascribed to a rearrangement of electrons in the system. There are certain types of reaction, however, where changes take place within the nucleus itself. These reactions are accompanied by very large energy changes, and elements which undergo such reactions spontaneously are described as *radioactive*.

Natural radioactivity was first discovered in 1896 by Becquerel, who observed that uranium salts emitted some kind of high energy radiation which had the effect of blackening a photographic plate. Soon after this, Pierre and Marie Curie isolated a new radioactive element, radium, and established that the radiation was an atomic phenomenon, characteristic of the element involved.

2.1. Nuclear stability

We learned in Chapter 1 that the nucleus of an atom is composed essentially of protons and neutrons, and that these particles are held together in a very small volume by strong attractive forces. At the same time it is clear that a number of positively charged protons held in a small volume must repel one another, and this repulsive force must increase as the number of protons increases. It is found that as the number of protons in a nucleus increases, so the number of neutrons present increases to overcome this repulsive force between the protons; that is, as the atomic number increases, so does the neutron: proton ($n : p$) ratio. This is

3

25

shown graphically in Fig. 2.1, where the number of neutrons is plotted against the number of protons for naturally occurring stable nuclei.

FIG. 2.1. $n : p$ ratios in stable nuclei.

We see that the $n : p$ ratio must fall within a certain range for the nucleus to be stable, and that this ratio increases as the atomic number increases. Furthermore, we find that there is no stable ratio for elements having an atomic number greater than 83. If, therefore, a nucleus has a structure outside this stable $n : p$ ratio range it will tend to undergo a spontaneous change in order to reach a more stable state.

2.2.   Types of radioactive decay

In terms of the graph given in Fig. 2.1 we see that three types of unstable nucleus can exist:

(i) a nucleus with too high a $n : p$ ratio;
(ii) a nucleus with too low a $n : p$ ratio;
(iii) a nucleus with an atomic number greater than 83.

A nucleus of the first type, where the ratio of neutrons to protons is too high, can reduce this ratio by emitting *beta radiation*. Beta radiation consists simply of a stream of electrons and is usually described by either the symbol

$$\beta^- \quad \text{or} \quad _{-1}e^0$$

e.g.
$$_6C^{14} \rightarrow {}_7N^{14} + {}_{-1}e^0$$

Note that the usual way of writing such an equation involves describing each species present in the form $_aX^b$, where X is the usual symbol for the element involved, the subscript a is the atomic number and the superscript b the mass number. We see that the above example involves the isotope of carbon containing six protons and eight neutrons. Reference to Fig. 2.1 shows that this $n : p$ ratio is too high for stability, therefore β emission occurs. It is formally assumed that one neutron in the unstable nucleus breaks up to yield a proton and an electron, which is then emitted:

$$\text{neutron} \rightleftharpoons \text{proton} + \text{electron}$$

Since the mass of the emitted electron is negligible, the overall effect of β radiation is that the mass number remains unchanged while the atomic number increases its value by one.

Another possible method of reducing the $n : p$ ratio is simply by the emission of unwanted neutrons from the unstable nucleus. While this reaction can occur, it is extremely uncommon.

Nuclei of the second type, where the $n : p$ ratio is too low, undergo nuclear reactions which decrease their nuclear charge. Thus whereas nuclei with a high $n : p$ ratio emit electrons, nuclei with low $n : p$ ratios emit *positrons* (particles having the same mass as an electron, but carrying a positive charge). A proton in the unstable nucleus is regarded as breaking up to form a neutron and a positron ($_1e^0$) which is then emitted.

$$\text{proton} \rightleftharpoons \text{neutron} + \text{positron}$$

e.g.
$$_{10}Ne^{19} \rightarrow {}_9F^{19} + {}_1e^0$$

In this example the unstable neon nucleus undergoes positron emission. Since the mass of a positron is negligible the mass number of the nucleus is unchanged, but the atomic number is decreased by one unit.

3*

Another method of increasing the $n:p$ ratio is by *electron capture.* Here one of the orbital electrons is captured by the nucleus, thus converting one of the nuclear protons to a neutron and thereby increasing the $n:p$ ratio to a more stable value,

e.g. $$_{18}A^{37} \xrightarrow[\text{capture}]{\text{electron}} {}_{17}Cl^{37}$$

The third type of unstable nucleus, where the atomic number is greater than 83, usually achieves stability by emitting *alpha particles* (written α or $_2He^4$), which are *dipositive helium atoms,*

e.g. $$_{92}U^{238} \rightarrow {}_{90}Th^{234} + {}_2He^4$$

Here we start with a uranium isotope which is unstable because of its high atomic number and consequently undergoes α emission to form the element thorium. This isotope is still unstable and emits β particles to form the element protactinium:

$$_{90}Th^{234} \rightarrow {}_{91}Pa^{234} + {}_{-1}e^0$$

This nucleus is still unstable and a series of nuclear reactions takes place, involving the emission of both α and β particles until eventually the stable lead isotope $_{82}Pb^{206}$ is obtained.

One further type of radiation must be described, namely *gamma rays* (γ rays). These do not consist of particles and are neither positively nor negatively charged, but are high energy electromagnetic waves, similar to X rays, and are often emitted when a nucleus undergoes a disintegration reaction.

2.3.   Rate of radioactive decay

The rate at which the above reactions occur is independent of external influences such as temperature or pressure, and depends on only two factors, viz. the type of nucleus involved and the number of atoms present. In other words, for any given radioactive element a definite fraction of the nuclei present will disintegrate per unit time, thus the greater the number of atoms present the more disintegrations will occur. The rate of decay of any species is described by the term *half-life* which is the time required for half the number of atoms present to disintegrate.

The half-life is a measure of the stability of any given element: the shorter the half-life the more unstable the nucleus. Since the stabilities of different radioactive elements vary enormously, the half-life can vary from millions of years to only a fraction of a second.

2.4. Artificial radioactivity

We have discussed how some naturally occurring elements may have nuclei which are unstable and which therefore spontaneously disintegrate to form a stable nucleus, at the same time emitting some sort of radiation. We must now consider what would happen if these reactions were reversible; i.e. the effect of bombarding a stable nucleus with high energy particles.

Such a situation was first observed by Rutherford in 1919 when he irradiated nitrogen with high energy α particles and found that an oxygen isotope was produced along with protons. The reaction may be represented

$$_7N^{14} + {_2He^4} \rightarrow {_8O^{17}} + {_1H^1}$$

Here the nitrogen nucleus has absorbed an α particle to form an unstable intermediate which then disintegrates to form oxygen. Other positively charged projectiles used are protons ($_1H^1$) and deuterons ($_1H^2$).

Although each of these projectiles can successfully induce nuclear reactions they all suffer from one disadvantage, viz. they are positively charged. Since the experiment involves the penetration of a positively charged nucleus by one of the bombarding particles, it can easily be seen that the particles will need to possess a very large amount of energy in order to overcome the repulsive force of the nucleus. The necessary energy can be achieved by means of *particle accelerators* where the particles can be made to move with very high velocities by subjecting them to large potential differences, but the process is fairly costly.

For this reason, the most useful projectile is the neutron which is not repelled by the nucleus and therefore does not require a high velocity, but can be absorbed at low speeds. If we bombard a light element such as aluminium with neutrons a variety of reactions can take place. For example, if high velocity neutrons are used we get reactions similar to the bombardment of nitrogen with α particles.

$$_{13}Al^{27} + {}_0n^1 \rightarrow {}_{11}Na^{24} + {}_2He^4$$
$$_{13}Al^{27} + {}_0n^1 \rightarrow {}_{12}Mg^{27} + {}_1H^1$$

In each case new particles are ejected from the nucleus and a different element is produced. Both the sodium and magnesium isotopes are in fact unstable and undergo β decay.

If, instead of fast neutrons, slow neutrons are used, a different type of reaction results.

$$_{13}Al^{27} + {}_0n^1 \rightarrow {}_{13}Al^{28}$$

$$_{12}Mg^{27} + {}_1H^1 \qquad {}_{13}Al^{28} + \gamma \qquad {}_{13}Al^{26} + 2{}_0n^1$$

Here the bombarding neutron is *captured* by the nucleus to form a new isotope which may eventually break down in a variety of ways. Note that in one case we obtain an extra neutron, a process known as *spallation*.

2.5. Energy changes

We learned at the beginning of this chapter that nuclear reactions involved very large amounts of energy, but we have not yet discussed the source of this energy. Normal chemical reactions do not usually involve energy changes much greater than 400 kJ mol^{-1}, whereas the energy changes involved in nuclear transformations may be a million times greater than this.

The source of this vast amount of energy is found in the Einstein relationship, $E = mc^2$, where E is the energy, m the mass and c the velocity of light. The law of conservation of mass, which states that mass is neither created nor destroyed, can be safely applied to normal chemical reactions where energy changes are small. If however we try to apply this law to nuclear changes, we find that the "law" appears to be invalid.

Thus we know that a deuteron (deuterium nucleus) is composed of one neutron and one proton, and we might expect that the mass of a deuteron would equal the mass of a proton (1·00783 a.m.u.) plus that of a neutron (1·00867 a.m.u.), viz. 2·01650 a.m.u. However, the experimental value for the mass of a deuteron is 2·01410 a.m.u., implying that there is a mass loss of 0·00240 units. This mass loss is explained by the fact that the Einstein relationship shows mass and energy to be equivalent.

In a nuclear reaction we should think then not in terms of conservation of mass, but of mass–energy, i.e.

$$\text{neutron} + \text{proton} = \text{deuteron} + \text{energy}$$

That the amount of energy involved is very large can easily be seen by calculating the energy equivalent of a mass of 1 gram:

$$
\begin{aligned}
E &= mc^2 \\
&= (1 \times 10^{-3})(3 \times 10^8)^2 \qquad &1\,\text{g} = 10^{-3}\,\text{kg} \\
&= 9 \times 10^{13}\,\text{J} \qquad &c = 3 \times 10^8\,\text{m s}^{-1}
\end{aligned}
$$

2.6. Radioactive tracers

The useful applications of atomic energy can broadly be classified under two main headings:

(i) the use of individual radioisotopes as "tracers";
(ii) the production of energy.

Let us consider first of all the former category. The property that concerns us here is the ease with which we can detect the presence of radioisotopes and this in turn depends on the ability of nuclear radiation to ionize matter. Perhaps the simplest type of detector is the one used by the French physicist Becquerel when he first detected such radiations. Becquerel noticed that certain salts of uranium could produce an image on a photographic plate. The principle involved is that the radiation interacts with the silver halide in the photographic emulsion and converts some of the silver ions to silver atoms. These serve as centres for development as described more fully in Section 9.8. Small badges containing photographic film are commonly used for determining the radiation dosage received by personnel working in an environment where radiation may be present.

Other types of detector are designed to produce a pulse of electrons in the presence of radiation. An example of this type is the Geiger–Müller counter which consists of a wire with a positive potential several hundred volts higher than the potential of a surrounding cylindrical shell. The potentials are held just below the values which would cause a spontaneous discharge between the wire and the shell. The presence of radiation which can ionize any molecules present will trigger a discharge which can then easily be measured.

It is easy to visualize the usefulness of this type of detector which produces an electronic signal, since it is a relatively simple matter to amplify that signal and to transfer it to some recording device. Also, when the detector provides a signal for each ionizing event then the equipment can be used not only to detect the presence of radiation, but to measure accurately the amount of radiation present.

We can now see an immense variety of uses of radioisotopes when they are used as "labels" which can be detected in minute quantities in situations where normal chemical analyses would be impossible. For example, they may be used in medical diagnosis. Radioisotopes such as iodine-131 (i.e. $_{53}I^{131}$) can be safely swallowed or injected and will become concentrated in particular organs of the body. Although present in very small amounts, the radioactivity can be measured fairly readily and hence help in the diagnosis of whether an organ is functioning normally or not.

Radioactive tracers are used in industry in a similar fashion. For example, they may be used to detect leaks in underground pipelines. By introducing a short-lived radioisotope into the pipe and raising the pressure, some radioactive liquid will be forced into the soil. After the pipe has been flushed out the radiation remaining in the soil in the vicinity of the leak can readily be detected.

Another industrial application is the use of radioisotopes to measure and control the thickness of materials produced in sheet form (e.g. paper or metal sheet) or to measure the thickness of a surface coating. Radiation from a radioactive source is directed at the material in question and the amount of radiation which penetrates is measured by a suitable detector. In this way the thickness of the material can be monitored constantly, and the electrical signal from the detector can even be used to adjust the control machinery so as to maintain a constant thickness of material passing the detector.

2.7. Radioactive dating

Yet another interesting example of the use of radioisotopes is the technique of radioactive dating. A knowledge of the rate of decay of radioactive isotopes makes it possible for us to determine the age of materials containing these isotopes. For example, a radioactive isotope of carbon with a half-life of 5700 years is present in the atmosphere in the form of carbon dioxide. This radioactive carbon is assimilated by all living matter which therefore contains a known amount of radioactive carbon. At death the absorption of radioactive carbon ceases and the amount present in the material steadily decreases due to radioactive decay. By measuring the activity remaining, the age of the material can be calculated.

The radiocarbon method is suitable only for the determination of ages of thousands of years. The ages of rocks millions of years old can, however, be determined by a similar method based on the decay of uranium to stable lead isotopes. Since the half-life of uranium is $4 \cdot 5 \times 10^9$ years, this method can be used to date rocks 500 million years old, although the errors involved may be quite considerable.

2.8. Power sources

We can now consider the second general type of application, namely the use of the energy produced by the radioisotopes. First of all there are a variety of uses where the amounts of energy involved are relatively small. One of the most useful radioisotopes of this kind is cobalt-60 which is made by bombarding inactive cobalt-59 with neutrons in the core of a nuclear reactor.

Cobalt-60 has a nucleus containing 27 protons and 33 neutrons, and reference to Fig. 2.1 shows that this $n : p$ ratio is too high for stability. One neutron will therefore break up to yield a proton and an electron which is then emitted, together with a pulse of high energy γ radiation:

$$_{27}Co^{60} \rightarrow \ _{28}Ni^{60} + \ _{-1}e^0 + \gamma$$

Cobalt-60 has a half-life of $5\frac{1}{4}$ years and the γ rays which it emits make it particularly valuable for radiotherapy and for radiation processing. Thus tumours and other growths can sometimes be destroyed by irradiation with high energy γ rays. Similarly prepackaged foodstuffs and medical equipment can be sterilized by irradiation. Any micro-organisms present are effectively destroyed and the products will remain sterile in storage until the seal is broken.

A somewhat different use of the energy emitted is the initiation of chemical reactions. Thus after polyethylene has been produced, its properties can be modified by inducing a secondary reaction which will cause *cross-links* to form between the long molecular chains (see Chapter 28). Similarly useful composite materials can be produced by impregnating timber or other porous materials with a polymer forming liquid and then irradiating the material with highly penetrating γ radiation. The radiation initiates the polymerization process throughout the porous material, producing a composite with very good mechanical properties and with a high resistance to chemical attack.

Yet another application of harnessing the energy emitted is the production of very small light sources and luminous paints. When small amounts of radioactive material are combined with a phosphor such as zinc sulphide, the radiation emitted interacts with the phosphor to produce a steady glow. If a more intense light source is required this can be achieved rather more indirectly by converting the heat energy released by radioactive materials to electricity which can then power a conventional light source. Although this method is rather costly, such power units are most suitable for remote installations such as navigation beacons.

2.9.   Nuclear fission

Although all the nuclear reactions we have discussed so far are potentially sources of large amounts of energy in practice these reactions can only be carried out on a fairly small scale. However, in 1939 a different type of nuclear reaction was discovered. It was found that when uranium was bombarded by neutrons the large nucleus split into fragments of approximately equal mass, at the same time liberating large amounts of energy and several other neutrons. In fact there are no unique products of this reaction, and the process of *nuclear fission* produces a variety of fragments with mass numbers ranging from approximately 70–160. One of the many fission reactions which can occur is:

$$_{92}U^{235} + {}_{0}n^{1} \rightarrow {}_{36}Kr^{90} + {}_{56}Ba^{144} + 3{}_{0}n^{1}$$

Many other modes of fission are possible, but on average approximately 3×10^{-11} J of energy and 2·5 neutrons are released in the fission of each atom of uranium-235.

The point of interest here is that the extra neutrons can be captured by other nuclei causing other fission reactions which in turn will produce still more neutrons. This type of reaction is self-maintaining and is described as a *chain reaction*. The two major applications of this type of chain reaction are (i) atomic weapons, and (ii) nuclear-powered generators.

The rate at which neutrons escape from a sample of uranium depends on the amount of material present and on its shape. If the effective surface area is large the neutrons can readily escape and reaction will effectively cease. If on the other hand the size of the sample is increased a point is reached where the rate of generation of neutrons exceeds the rate of loss, and the chain reaction develops. There is therefore a *critical mass* below which only a slow disintegration will occur but above which an accelerating chain reaction occurs. In the nuclear bomb a critical mass of uranium-235 is brought together and an uncontrolled chain reaction is established.

A more interesting use of nuclear fission is the harnessing of a controlled chain reaction to generate heat which can in turn be used to produce electricity. There are two basic types of nuclear power reactor: *thermal reactors* and *fast reactors*.

A typical thermal reactor is illustrated in Fig. 2.2 which shows the working of Britain's first nuclear power station at Calder Hall. The uranium fuel is held in tubes of a magnesium alloy (Magnox) with a low

FIG. 2.2. Thermal nuclear reactor.

neutron absorption, and these tubes are in turn held within blocks of graphite. The most abundant isotope of carbon (graphite) is $_6C^{12}$ which has a particularly stable nucleus and has no affinity for neutrons. The high energy neutrons released by the uranium enter the graphite and are slowed down by repeated collision with the carbon nuclei. The neutrons are not absorbed and diffuse back into the fuel rods where they can now be readily captured by a uranium nucleus and so sustain the nuclear fission process. The graphite, which slows down the high energy neutrons, is called a *moderator*, and the presence of a moderator of some kind is a feature of all thermal reactors.

We can see therefore that within the core of the reactor the fuel is arranged in such a way that a self-sustaining chain reaction can occur. We must now introduce a mechanism by which the rate of generation of neutrons can be controlled. This is achieved by inserting into the core control rods made of a material which has a high affinity for neutrons. One such material is boron which on capturing a neutron splits up to form other stable nuclei:

$$_5B^{10} + _0n^1 \rightarrow _2He^4 + _3Li^7$$

Whenever the neutron flux rises above a predetermined level the control rods are pushed further into the core thus absorbing some of the neutrons and controlling the chain reaction.

The reaction within the core produces large amounts of heat and of radiation. It is therefore necessary to remove the heat by the circulation of a suitable coolant, in this case CO_2 gas, and to surround the entire structure with a heavy concrete shield to prevent the escape of dangerous radiation.

There is now a large family of thermal reactors similar to that described above. The fuel is either natural uranium or uranium enriched with uranium-235, the moderator either graphite or heavy water (D_2O), and the coolant is either a gas or in some cases the moderator itself.

Fast reactors are so called because the neutrons are not slowed down by using a moderator. The fuel currently used is plutonium, produced initially as a byproduct from thermal reactors. The plutonium is sufficiently concentrated for the chain reaction to be sustained by fast neutrons. Excess neutrons are absorbed in uranium-238 both in the core and in a "blanket" around the core, thus generating more plutonium fuel than is used up in the production of heat. The coolant used in this type of reactor is liquid sodium.

2.10. Nuclear fusion

Although a great deal of energy is released from a nuclear fission reaction, the mass changes involved are relatively small. It was realized that if *nuclear fusion* could be achieved even greater amounts of energy would be released. Nuclear fusion involves the formation of a single nucleus from two smaller nuclei. For example, it is believed that the main source of the sun's energy arises from the conversion of hydrogen to helium by the following reaction sequence:

$$_1H^1 + {_1}H^1 \rightarrow {_1}H^2 + {_1}e^0$$
$$_1H^2 + {_1}H^1 \rightarrow {_2}He^3 + \gamma$$
$$_2He^3 + {_2}He^3 \rightarrow {_2}He^4 + 2{_1}H^1$$

The mass difference between helium and the four hydrogen atoms is 0·2759 a.m.u. and this appears as an equivalent amount of energy as explained in Section 2.5. It can be appreciated then that the production of energy by the nuclear fusion of light elements represents a vast resource for mankind, but we must also consider the technical problems involved and we shall see that they are far more difficult to solve than those for controlled fission reactions.

First of all let us consider the nature of the reaction we wish to bring about. We know that nuclei have positive charge, so nuclei which approach one another will be mutually repelled. For fusion to occur, however, the nuclei must somehow be brought together. The only practical way of achieving this is to raise the material to a very high temperature (in excess of 10^6 K) when the particles involved will possess sufficient energy to overcome the repulsive forces between them.

We now encounter the second main problem, which is how to contain the mixture of hydrogen isotopes at these extremely high temperatures for long enough for fusion to occur. Obviously steel and concrete vessels such as were used for fission reactions cannot be used here since they would vaporize at these temperatures. Fortunately there appears to be a possibility of using a property of the reaction material itself to give a solution to this problem. At the temperatures we are considering the kinetic energies of the atoms are much greater than the binding energies of many of the electrons. Therefore as the atoms collide with each other they tend to lose their electrons and form a material which eventually consists of positive ions and free electrons, all moving at very high velocities. Such a material is called a *plasma*.

We know that moving charged particles can be deflected by means of magnetic fields, and since a plasma consists of high velocity charged particles it is conceivable that it could be contained by a suitable arrange-

FIG. 2.3. Plasma "bottle".

ment of magnetic fields. Such an arrangement is illustrated in Fig. 2.3, where a doughnut-shaped magnetic field is produced by a series of coils. Charged particles within such a system are deflected along in the direction of the magnetic field and are contained within the ring. When such a current is induced in the plasma, a secondary magnetic field is produced which "pinches" the plasma in towards the core of the tube. The induced current both heats the plasma and contains it within a magnetic "bottle".

So far projects of this kind are only at the experimental stage and it has not proved possible to contain the plasma for long enough for fusion to occur.

Exercises

2.1. Explain each of the following terms:

(i) nuclear fission;
(ii) critical mass;
(iii) moderator;
(iv) radioactive dating;
(v) plasma;
(vi) beta emission.

2.2. Complete the following equations for radioactive disintegration reactions:

(i) $_{55}Cs^{140} \rightarrow {}_?Ba^{140} + _{-1}e^0$

(ii) $_?U^? + {}_0n^1 \rightarrow {}_{92}U^{239}$

(iii) $_{90}Th^{232} \rightarrow {}_{88}Ra^{228} + {}_?He^?$

(iv) $_{30}Zn^{68} + {}_0n^1 \rightarrow {}_{28}?^? + {}_2H^4$

[*Ans.*: 56; 92, 238; 2, 4; Ni, 65.]

2.3. The isotope $_{82}Pb^{214}$ radioactively decays to a stable lead isotope through a series of successive emissions of β, α, β, β, β, α. Construct a decay diagram and specify the atomic mass of the final lead isotope.

[*Ans.*: $_{82}Pb^{206}$.]

2.4. Summarize the principal differences between a thermal reactor and a fast breeder reactor.

CHAPTER 3

The Periodic Table

3.1. The periodic table

Very early in the history of chemistry it was realized that the ever-increasing knowledge of chemical behaviour must be systematized. Many attempted this task, but it was not until the latter half of the nineteenth century that much success was achieved when Mendeleev and Meyer independently classified the known elements in order of increasing atomic weight and observed a significant pattern of chemical behaviour. It is a refinement of this idea that leads to the modern version of the periodic classification of elements.

Even now new attempts are being made to present the periodic table in a more convenient form, but by far the most usual way is that represented in Fig. 3.1. Here the elements are arranged in order of increasing atomic number and are arranged horizontally in *periods* and vertically in *groups*.

3.2. Electronic structure

While we must remember that the periodic classification of elements was achieved by considering the chemical properties of the elements, it is perhaps more satisfactory to consider how the above classification is a direct consequence of the electronic structure of atoms which we discussed in Chapter 1.

The first period contains only two elements, hydrogen and helium, whose electronic configurations are $1s^1$ and $1s^2$ respectively. You will recall that because of the rules governing the energies of electrons it is impossible to have more than two electrons with a principal quantum

1 H																	2 He
3 Li	4 Be											5 B	6 C	7 N	8 O	9 F	10 Ne
11 Na	12 Mg											13 Al	14 Si	15 P	16 S	17 Cl	18 Ar
19 K	20 Ca	21 Sc	22 Ti	23 V	24 Cr	25 Mn	26 Fe	27 Co	28 Ni	29 Cu	30 Zn	31 Ga	32 Ge	33 As	34 Se	35 Br	36 Kr
37 Rb	38 Sr	39 Y	40 Zr	41 Nb	42 Mo	43 Tc	44 Ru	45 Rh	46 Pd	47 Ag	48 Cd	49 In	50 Sn	51 Sb	52 Te	53 I	54 Xe
55 Cs	56 Ba	57 *La	72 Hf	73 Ta	74 W	75 Re	76 Os	77 Ir	78 Pt	79 Au	80 Hg	81 Tl	82 Pb	83 Bi	84 Po	85 At	86 Rn
87 Fr	88 Ra	89 †Ac															

*Lanthanide series		58 Ce	59 Pr	60 Nd	61 Pm	62 Sm	63 Eu	64 Gd	65 Tb	66 Dy	67 Ho	68 Er	69 Tm	70 Yb	71 Lu
†Actinide series		90 Th	91 Pa	92 U	93 Np	94 Pu	95 Am	96 Cm	97 Bk	98 Cf	99 Es	100 Fm	101 Md	102 No	103 Lw

Fig. 3.1. Periodic table of the elements.

number of unity. Hence the first period corresponds to filling up the $1s$ atomic orbital.

The second period contains eight elements from lithium ($Z = 3$) to neon ($Z = 10$). The outer electrons, i.e. the electrons which principally govern chemical behaviour, are characterized by having a principal quantum number equal to two. Thus, as we proceed from the element lithium, electrons are progressively added to the $2s$ and $2p$ orbitals, and, as was explained in Chapter 1, these orbitals are completely filled when eight electrons have been added, i.e. at the element neon.

The third period corresponds very closely to the second. Again the period contains eight elements which in this case correspond to the progressive addition of electrons to the $3s$ and $3p$ orbitals, these being completely filled at the element argon.

Consideration of Fig. 1.2 shows that at the beginning of the fourth period two possibilities exist—electrons may enter the $4s$ orbital or the $3d$ orbitals, since these are approximately of equal energy. In fact, we find that for potassium and calcium the electrons enter the $4s$ orbital. Commencing at the element scandium, however, electrons begin to fill the available $3d$ orbitals giving rise to the *first transition series* of elements. The addition of electrons to the $3d$ orbitals continues up to the element zinc where the $3d$ orbitals are completely filled. The remaining six elements of this period, from gallium to krypton, correspond to the completion of the $4p$ orbitals.

The fifth period containing the eighteen elements from rubidium to xenon is analogous to the preceding period, the orbitals concerned being the $5s$, $4d$ and $5p$. Again we find a series of eight elements similar to the early periods interrupted by a *second transition series* of ten elements.

The sixth and seventh periods are built up in a similar fashion, except that in these periods an *inner transition series* of fourteen elements is inserted after the first element of the normal transition series. These groups of fourteen elements correspond to the filling of f orbitals, but an account of the chemistry of these elements is outside the scope of this text.

3.3.   Atomic radii

We have seen in the previous section that the electronic structure of the elements is a periodic function of their atomic numbers, i.e. a pattern exists which repeats itself in a more or less regular fashion. It would seem reasonable to assume that other atomic properties may exhibit periodicity.

A particularly important atomic property is that of size. The *atomic radius* cannot be defined exactly because of the diffuse nature of the electronic structure. Also, the atomic radius varies slightly depending on the precise nature of the bonds holding it to its neighbours. We can, however, define the *covalent atomic radius* as being half the distance between the nuclei of atoms in a suitable molecule and this may be used to compare the relative sizes of atoms.

In Fig. 3.2 the atomic radii are plotted against atomic number and we can readily see that atomic size is a periodic function of atomic number. Two main features may be distinguished. First, if we consider how the atomic radius varies along any given period, we see that the radii progressively decrease as we move along the period only to rise sharply at

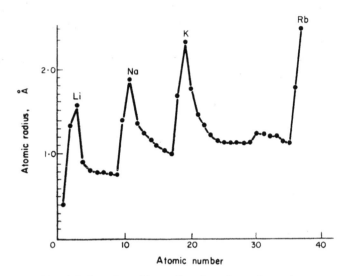

Fig. 3.2. Atomic radii as a function of atomic number.

the last member of that particular period. This is so because, although throughout the period extra electrons are being added, at the same time the nuclear charge is being increased. The overall effect is that the electrons are attracted more strongly to the nucleus, the greater the nuclear charge. Second, we can consider the variation of atomic radii within a given group. In this case we see a different trend, i.e. the atomic radius increases with atomic number. The reason for this is that the increase in nuclear charge is more than cancelled out by the "screening" effect of the extra shells of electrons added. The outer electrons are therefore less firmly bound to the nucleus as we descend any group in the periodic table.

3.4.   Electronegativity

A useful concept for explaining some aspects of the behaviour of molecules is that of *electronegativity*, which is defined as the power of an atom to attract electrons to itself. Since the bonding in any molecule depends on the behaviour of the electrons, it is important to have some knowledge of the probable effect of any particular atom on the other electrons in the molecule.

Numerical values for the electronegativities of the elements have been obtained and Fig. 3.3 shows that again a periodic variation occurs.

Fig. 3.3. Electronegativity as a function of atomic number.

We see that the values range from 0·7 for caesium and francium, indicating a very low affinity for electrons, to 4·0 for fluorine, the most electronegative element, which has a high capacity for attracting electrons. As was the case with atomic radii we find two general trends. The electronegativity increases with atomic number within a given period and, in general, decreases with increasing atomic number within a given group.

3.5. Chemical behaviour—non-transition elements

We learned in Chapter 1 how chemical bonds were formed by the rearrangement of electrons to give the most stable configuration. We can now see how the periodic table helps us to classify the behaviour of elements and compounds whose properties are largely determined by the nature of their chemical bonds.

If we neglect for the moment the transitional elements, we see in the groups of the periodic table a *valency classification*. The group I elements all have a single valency electron and all are strongly electropositive. These elements all have a valency of one and very readily form ionic bonds because of their lack of affinity for electrons. Most of the compounds of these elements are ionic in nature and show characteristic properties. Ionic compounds in general are hard solids with high melting and boiling points. In the solid state they act as insulators but in the molten state or in solution are good conductors of electricity. These properties are a consequence of the nature of the forces acting to hold the structure together in the solid state (see Chapter 6). All of these elements are soft, low melting metals and are chemically extremely reactive.

All the group II elements have two valency electrons and are strongly electropositive. These elements have a valency of two and generally form ionic compounds. Again all the elements are fairly soft metals and are highly reactive.

The elements of group III each have three valency electrons and, with the exception of boron, are only weakly electronegative. These elements generally form trivalent compounds whose bonds have only a small percentage of ionic character. This is so because very large amounts of energy are required to remove three electrons from an atom. Thus covalent compounds are formed which have in general much lower melting and boiling points than ionic materials. With the exception of boron the elements are metals, and boron itself does exhibit some metallic properties.

The chemical behaviour of group IV and V elements is more varied than that in any other group. In this part of the periodic table we find elements of intermediate electronegativity which range from non-metals in the early members of each group to typical metals in the later members. Carbon, silicon and germanium form covalent compounds, exhibiting a valency of four. Tin and lead, on the other hand, whilst forming some tetravalent compounds, also form ionic salts containing divalent ions. Group IV elements are relatively unreactive, although paradoxically carbon probably forms more compounds than any other element. This is because carbon has a facility for forming complicated chain and ring structures to a much greater extent than other elements as is described in greater detail in Chapter 20.

The overall trend in group V is similar to that in group IV, with nitrogen and phosphorus behaving as non-metals, bismuth being typically metallic and arsenic and antimony showing some metallic properties. Apart from bismuth the elements react to form covalent compounds. The group valency can be either three or five, and although nitrogen exhibits only a valency of three the other members of the group form compounds of both types, e.g. PCl_3 and PCl_5.

The group VI elements with the exception of polonium are in general fairly strongly electronegative and are classed as non-metals although tellurium does exhibit some metallic properties. The group valency is two, the bonding being either ionic or covalent depending on the nature of the other element. Thus with weakly electronegative elements typical ionic compounds containing the X^{2-} ion are formed, while in combination with, e.g. hydrogen, the compounds are covalent. Oxygen apart, the group VI elements can exhibit valencies of four and six.

The group VII elements, whose electronic configurations are only one short of the inert gas structure, are strongly electronegative and highly reactive. Fluorine, for example, attacks most other elements even at room temperature. The halogens normally exhibit a valency of one, the type of bond depending on the other element present. The halides of electropositive metals are typical ionic compounds whereas their compounds with hydrogen are covalent.

The group VIII elements, commonly described as the *inert gases*, show a great reluctance to take part in chemical reactions. In recent years a small number of inert gas compounds have been successfully prepared, showing that these elements are not quite so unreactive as was once believed.

3.6. Chemical behaviour—transition elements

We have yet to consider the transition elements, i.e. those elements with inner d orbitals incompletely filled, and we shall for the moment include the group zinc, cadmium and mercury under this general heading. The properties of these elements, which are all metals, are quite distinctive and obviously different from those of the typical non-transitional metals. These differences are a direct consequence of the rather unusual electronic configurations of these elements.

The first striking characteristic of the transition metals is the similarity of chemical behaviour horizontally along the period. The reason for this is that although electrons are added as we move along the series, they are accommodated in the incomplete inner d orbitals, leaving the important outer electronic configuration unchanged. Since, to a large extent, it is the outer electrons which determine chemical behaviour, we find adjacent elements behaving very similarly.

Another distinguishing feature of these metals is that they have a variable valency. While it is true that some non-transitional elements also have this ability, it is far more pronounced in the transition series. Titanium, for example, forms three different chlorides, $TiCl_2$, $TiCl_3$ and $TiCl_4$. The most common valency encountered is that of two (e.g. $FeCl_2$, $CoCl_2$) since this involves bonds using the two $4s$ outer electrons, but as there is little difference in energy between the $4s$ and $3d$ electrons, these inner electrons readily take part in bond formation as well.

Yet another characteristic of this series of metals is that they have a strong tendency to form *co-ordination compounds* containing complex ions, e.g. $K_4Fe(CN)_6$. Again this is due to the incompletely filled d orbitals which can be used to accommodate pairs of electrons from other chemical groups and so form chemical bonds. This property can be one of great importance. Although gold is a very unreactive metal and is insoluble in common acids it can be dissolved in *aqua regia* (3 volumes concentrated hydrochloric acid $+1$ volume concentrated nitric acid) because of the formation of the complex ion $[AuCl_4]^-$.

Finally, we find that the majority of transition metal salts are highly coloured. Again this is not a property exclusive to transition metal compounds but is fairly characteristic and is due to transitions involving unpaired *d* electrons. Compounds of zinc and scandium, whose ions do not contain unpaired *d* electrons, are colourless.

Exercises

3.1. Summarize the evidence for periodicity.

3.2. Define the following terms:

    (i) periods;
    (ii) groups;
    (iii) first transition series;
    (iv) inner transition series;
    (v) covalent atomic radius;
    (vi) electronegativity.

3.3. Make a copy of the outline structure of the periodic table similar to that on page 41, but omitting all details of the actual elements.
    On your diagram indicate roughly:

    (i) the area of maximum electronegativity;
    (ii) the area of minimum electronegativity;
    (iii) the metallic elements;
    (iv) the non-metallic elements;
    (v) the transition elements;
    (vi) the inert gases;
    (vii) for group I and II elements, the direction in which the atomic size increases;
    (viii) for group I and II elements, the direction in which the melting point increases (see Chapter 9 for data).

CHAPTER 4

The Gaseous State

4.1. The kinetic theory of matter

There are basically three states of matter—the solid, liquid and gaseous states. It is not unreasonable to suppose that the observed properties of a particular state are likely to be related to and indeed are a consequence of the way in which the individual units behave, be they molecules, atoms or ions.

To this end, let us imagine the changes that occur at the molecular level in the gradual transition of solid to liquid to gas. In water, for example, the basic unit is the molecule H_2O. X-ray examination reveals that the water molecules in ice are positioned in space in a highly ordered fashion. A still more detailed analysis reveals that these molecules are not stationary but are vibrating about a mean position. As the temperature is raised the vibrations become more energetic until at 0°C melting of the ice occurs, and liquid water is formed. This sudden change of state that occurs is due to the vibrations becoming so violent that the molecules, although still exerting an attractive influence on each other, are now free to alter their position. We can express this alternatively by saying that molecules in the liquid state have a greater *degree of freedom* and are more disordered than molecules in the solid.

As the temperature is increased beyond 0°C the molecules, in absorbing the energy supplied to them as heat, move about still more rapidly; the system becomes even more disordered, until at 100°C the molecules are now sufficiently energetic to overcome the attractive forces and the liquid boils, producing a gas, viz. steam. In this state the molecules exert only a very weak mutual attraction and as a result are free to move about

anywhere within the container; the continual collisions occurring between the molecules and the walls of the container create the effect we
describe as pressure. Clearly, the gaseous state is much less ordered than
the liquid state with the molecules having a much greater degree of
freedom.

We must now consider whether the theory is consistent with experimental observation; like any theory it must be put to the test.

TEST I. *The absolute zero of temperature.* In describing what we mean
by a kinetic theory of matter we have identified heat with molecular
motion for we have presumed that the effect of increasing the temperature increases this motion and produces a consequent increase in
disorder; i.e. we can say that not only is temperature a measure of the
"heat content" of a system, but it is indeed a measure of the average kinetic
energy of the molecules. We might well feel tempted at this point to pose
the question—to what value may the temperature be reduced?

It follows from our previous remarks that if heat is removed from a
system its temperature will fall, molecular motion will become less
vigorous and a more ordered state will result. Inevitably a temperature
must be reached when the system assumes a state of maximum order.[†]
Thus, on the basis of the kinetic theory, *an absolute zero of temperature*
is predicted; Lord Kelvin deduced that for *any* substance, this temperature is $-273\cdot15°C$. Although this temperature has never been reached,
experiments have been performed where, for a short period of time, the
temperature has been lowered to $-273\cdot1499°C$.

It is convenient to introduce an *absolute temperature scale* which has
as its zero point the absolute zero of temperature. Any temperature on
this scale is indicated by the symbol T and is expressed in K (Kelvin).
Thus, if the size of the absolute degree is made the same as the centi-
grade degree it follows that

$$T\,(\text{K}) = t\,(°\text{C}) + 273\ (\text{approx.})$$

[†] At first sight we might reasonably expect a condition of maximum order to be
associated with a complete cessation of molecular motion. However, modern physics
requires that molecular motion can never cease entirely.

We can now be more definitive in describing the relationship between the kinetic energy of gas molecules and the temperature by stating that the average kinetic energy E is directly proportional to the absolute temperature T,

i.e. $$E \propto T$$

TEST II. *Brownian movement.* The behaviour of very small pollen grains suspended in water provides a striking confirmation of the kinetic theory. These tiny particles, when examined under a microscope, are observed to be in constant motion, moving about in a completely unpredictable fashion. This chaotic motion, referred to as the *Brownian movement* (after the botanist Robert Brown who first observed the phenomenon), can only be attributed to the fact that the molecules of the liquid in which the pollen particles are suspended are moving about in a similarly chaotic, random manner. Thus, at one instant of time, a small pollen particle may be struck by, say, three water molecules in a manner as shown in Fig. 4.1(a). The net result of this molecular bombardment would be to cause it to move off in a new direction. Over a period of time, the particle would experience many changes of direction.

A similar effect can be seen if smoke is examined under a microscope. In this case the denser solid particles experience the Brownian movement and for this reason do not settle out under the influence of gravity.

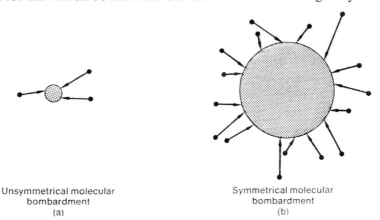

Unsymmetrical molecular
bombardment
(a)

Symmetrical molecular
bombardment
(b)

FIG. 4.1. Brownian movement.

As the particle size increases the particle motion becomes less violent and indeed for large particles the Brownian movement is completely absent. That this is so is due to the fact that the large particles are more likely to suffer a symmetrical molecular bombardment in which molecular collisions neutralize each other (Fig. 4.1(b)).

4.2.   Gas laws

Charles' law

The pressure exerted by a gas when measured under the condition of constant volume is found to vary linearly with its temperature (°C). This relationship, expressed graphically in Fig. 4.2, implies that if the gas

FIG. 4.2. Variation of gas pressure with temperature.

continued to behave in the manner indicated by the unbroken line, then its pressure would become zero at a temperature of -273°C. Now in the light of the kinetic theory, pressure is the result of molecular collisions with the walls of the container; zero pressure must therefore mean a total absence of molecular motion and the temperature corresponding to this state must be an absolute zero of temperature. Thus, *for a fixed amount of gas under the condition of constant volume the pressure is directly proportional to the absolute temperature*; this statement is one form of Charles' law,

i.e. $P \propto T$ (fixed amount of gas at constant volume)

This result is entirely consistent with the kinetic theory, for as the temperature is raised, molecular velocity will increase and molecular collisions with the container walls will be both more frequent and more vigorous, giving rise to a greater pressure.

An alternative form of Charles' law states that *for a fixed amount of gas under the condition of constant pressure the volume is directly proportional to the absolute temperature,*

i.e. $V \propto T$ (fixed amount of gas at constant pressure)

Boyle's law

The relationship between gas pressure and volume was investigated by Robert Boyle, who was able to show that, *at constant temperature, the pressure exerted by a given amount of gas varies inversely with its volume* (Fig. 4.3),

i.e. $P \propto 1/V$ (fixed amount of gas at constant temperature)

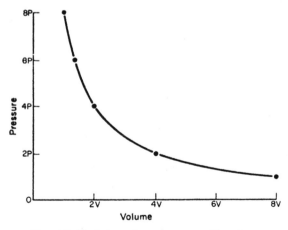

FIG. 4.3. Variation of gas pressure with volume.

This relationship, referred to as Boyle's law, is exactly what we would expect on the basis of our theory. If the temperature remains constant the average velocity of the molecules likewise does not change, and if the volume of the container is halved, the number of molecules striking a unit area of surface per second must double, i.e. the pressure must double.

Avogadro's law

Avogadro postulated that *equal volumes of all gases measured under the same conditions of pressure and temperature contain an equal number of molecules.* This postulate has since been verified,

i.e. $V \propto$ no. of molecules (constant pressure and temperature)

The importance of this relationship is that it suggests the possibility of counting molecules by measuring the volume under specified conditions of pressure and temperature. However, before this can be done we must introduce a new quantity, the *mole*, defined as *that amount of substance which contains the same number of particles (atoms, molecules, ions, etc.) as are contained in 12 g of carbon-12.* This number has been experimentally determined to be $6 \cdot 022 \times 10^{23}$ and is referred to as the *Avogadro number (L)*, i.e. one mole of any substance contains $6 \cdot 022 \times 10^{23}$ particles.

We can now modify Avogadro's law to read

$V \propto$ no. of moles (n) (constant pressure and temperature)

The molar mass (M) of a substance is simply the mass of a mole and because of the way in which the mole is defined it follows that the molar mass, if expressed in g mol^{-1}, is numerically equal to the molecular weight M_r, itself a dimensionless quantity.[†] Thus if we consider one mole of *any* gas (i.e. $6 \cdot 022 \times 10^{23}$ molecules) under standard conditions of pressure and temperature (1 atm and 0°C) it is found to occupy a volume of 22·414 l,

i.e.   molar volume $= 22 \cdot 414$ l at 1 atm and 0°C

[†] The molecular weight of a compound is the sum of the atomic weights in the molecular formula. It is therefore the weight of a molecule relative to one-twelfth the weight of an atom of carbon-12.

Graham's law

Gases have the ability to diffuse through porous materials. Graham in 1883 studied this aspect of gaseous behaviour and showed that *the rate of diffusion was inversely proportional to the square root of the molecular weight of the gas* (Graham's law),

i.e. $$\text{rate of diffusion} \propto \sqrt{\frac{1}{M_r}}$$

where M_r = molecular weight.

On the basis of the kinetic theory, it seems reasonable to suppose that the rate of diffusion will be determined by the average speed of the molecules. Consider therefore two gases A and B, each at the same temperature and therefore each being associated with the same amount of kinetic energy per molecule. Now if c_A and c_B are the respective average velocities, with m_A and m_B, the respective molecular masses, then it follows that for the two gases at the same temperature

$$\tfrac{1}{2}m_A c_A^2 = \tfrac{1}{2}m_B c_B^2 \quad (\text{kinetic energy} = \tfrac{1}{2}mc^2)$$

$$\therefore \qquad \frac{c_A}{c_B} = \sqrt{\frac{m_B}{m_A}}$$

The molar mass (M) of a gas is the weight of one mole of the gas, i.e. the weight of the Avogadro number (L) of molecules.

$$\therefore \qquad M_A = \text{molar mass of gas A} = Lm_A$$

and $$M_B = \text{molar mass of gas B} = Lm_B$$

$$\therefore \qquad \frac{c_A}{c_B} = \sqrt{\frac{M_B}{M_A}}$$

Provided the molar mass is expressed in units of g mol^{-1} it will be numerically equal to molecular weight (M_r) which is itself a dimensionless quantity,

i.e. $$\text{average molecular velocity} \propto \sqrt{\frac{1}{M_r}}$$

$$\therefore \qquad \text{rate of diffusion} \propto \sqrt{\frac{1}{M_r}}$$

4.3.   Ideal gas equation of state

The laws of Boyle, Charles, and Avogadro are found to be only truly valid under conditions of low pressure and high temperature. Under these conditions the gas is said to behave ideally and is described as an *ideal gas*. Under the conditions for which the laws are not accurately obeyed the gas is said to be non-ideal. For the moment we shall restrict ourselves to the consideration of an ideal gas, and show how the three laws can be combined to produce an equation of state which displays the relationship between the four variables P, V, T and n.

Boyle's law $V \propto \dfrac{1}{P}$ (constant n and T)

Charles' law $V \propto T$ (constant n and P)

Avogadro's law $V \propto n$ (constant P and T)

Combination of the above yields

$$V \propto \frac{nT}{P}$$

∴ $$PV = nRT$$

where R is a proportionality constant whose value is the same for all gases. It is called the *gas constant*.

The numerical value of R may be determined by inserting into the ideal gas equation of state known values of P, V, T and n. Thus under standard conditions of pressure and temperature, i.e. at 1 atm and 0°C (STP conditions), one mole of gas occupies a volume of 22·414 l. Considerable care must be exercised in this substitution to ensure that all quantities are expressed in their appropriate SI units.

Hence

$$1 \text{ atm} = 1 \cdot 01325 \times 10^5 \text{ N m}^{-2}$$
$$1 \text{ l} \quad = 1 \text{ dm}^3 = 10^{-3} \text{ m}^3$$
$$0°\text{C} \quad = 273 \cdot 15 \text{ K}$$

$$\therefore \quad R = \frac{1{\cdot}01325\times10^5\ (\text{N m}^{-2})\times22{\cdot}414\times10^{-3}\ (\text{m}^3)}{273{\cdot}15\ (\text{K})\times1\ (\text{mol})}$$

$$= 8{\cdot}3143\ \text{N m K}^{-1}\text{mol}^{-1}$$

$$= 8{\cdot}3143\ \text{J K}^{-1}\text{mol}^{-1} \qquad\qquad 1\,\text{N m} = 1\,\text{J}$$

Clearly R is seen to have units of energy.

Prior to the adoption of SI units (Appendix) the volume and pressure of a gas were expressed in units of litres and atmospheres respectively. The choice of such units produced a value of R equal to 0·082 1 atm K^{-1} mol^{-1}. This value is now regarded as obsolete and its use is not recommended except in a certain type of equilibrium calculation (Section 14.3).

The ideal gas equation is one whose use in physical chemistry is considerable. The following example illustrates some of the simple manipulations which must be performed in the use of the equation.

EXAMPLE 4.1. *Calculate the number of molecules remaining in an electronic vacuum tube of capacity 180 cm³ which has been evacuated and sealed off at 25°C to a pressure of* $1{\cdot}0\times10^{-7}$ *mmHg.*

The fact that R is expressed in $J K^{-1}$ mol^{-1} requires us always to express pressure and volume in N m^{-2} and m³ respectively.

$$V = 180\ \text{cm}^3 = 180\times10^{-6}\ \text{m}^3$$

$$P = 1{\cdot}0\times10^7\ \text{mmHg} = \frac{1{\cdot}0\times10^7\times1{\cdot}013\times10^5}{760}\ \text{N m}^{-2} = 1{\cdot}33\times10^{-5}\,\text{N m}^{-2}$$

$$T = 298\ \text{K}$$

$$\therefore\quad n = \frac{PV}{RT} = \frac{1{\cdot}33\times10^{-5}\,(\text{N m}^{-2})\times180\times10^{-6}\,(\text{m}^3)}{8{\cdot}314\,(\text{J K}^{-1}\text{mol}^{-1})\times298\,(\text{K})} = 9{\cdot}7\times10^{-13}\ \text{mol}$$

$$\therefore\quad \text{no. of molecules} = \text{no. of moles}\times\text{Avogadro no.}$$

$$= 9{\cdot}7\times10^{-13}\times6{\cdot}022\times10^{23}$$

$$= 5{\cdot}8\times10^{11}$$

4.4. Gas mixtures

Not infrequently the chemist is faced with a problem involving not one single gas but a gas mixture. In order to tackle such a problem it is necessary to know what contribution will be made to the total pressure by each gas in the mixture, relative to the amount of that particular constituent present. This contribution is termed the partial pressure of the gas and is defined as being *the pressure it would exert if it alone occupied the entire volume of the container at the specified temperature.*

Consider now a mixture of gases A and B for which the measured (i.e. total) pressure is *P*. If now we imagine gas B to be removed the pressure within the container will be that due to gas A, i.e. the partial pressure of A.

From the ideal-gas equation

$$P_A = \frac{n_A RT}{V}$$

where P_A = partial pressure of A and n_A = no. of moles of A. Similarly, the pressure within the container upon the imaginary removal of gas A is the partial pressure of B,

i.e.
$$P_B = \frac{n_B RT}{V}$$

where P_B = partial pressure of B, n_B = no. of moles of B.

For the gas mixture:

$$P = \frac{n_T RT}{V}$$

where n_T = total moles present = $n_A + n_B$.

$$\therefore \qquad P_A = \frac{n_A P}{n_T} \quad \text{and} \quad P_B = \frac{n_B P}{n_T}$$

The fractions n_A/n_T and n_B/n_T are described as being the *mole fractions* of A and B respectively. The symbol x is the one most commonly used to represent a mole fraction.

Thus

$$\frac{n_A}{n_T} = x_A \quad \text{and} \quad \frac{n_B}{n_T} = x_B$$

$\therefore$

$$P_A = x_A P \quad \text{and} \quad P_B = x_B P$$

It is now possible to see that the contribution made by each gas in a mixture to the total pressure is determined by the relative amount of the gas present. Thus, if the gases A and B are present in the mole ratio of 1 : 3 respectively, then

$$P_A = \frac{P}{4} \quad \text{and} \quad P_B = \frac{3P}{4}$$

The relationship between the measured total pressure and the respective partial pressure can also be rapidly deduced, for

$$P_A = \frac{n_A P}{n_T} \quad \text{and} \quad P_B = \frac{n_B P}{n_T}$$

$\therefore$

$$P_A + P_B = \left(\frac{n_A}{n_T} + \frac{n_B}{n_T}\right)P$$

$$= \left(\frac{n_A + n_B}{n_T}\right)P$$

$$= P$$

Thus, *the total pressure of a mixture of gases is the sum of the individual partial pressures of the gases in the mixture.* This statement is *Dalton's law of partial pressures.*

EXAMPLE 4.2. *A volume 0·54 l of argon measured at 72 cmHg and 15°C together with 2·32 g of neon (at. wt. 20) are placed in a 2-l flask. What will be the partial pressures of the two gases when the gas mixture is brought to a temperature of* 100°C?

5*

Method I

$$\text{No. of moles of argon} = \frac{PV}{RT} = \frac{72 \times 1 \cdot 013 \times 10^5 \times 0 \cdot 54 \times 10^{-3}}{76 \times 8 \cdot 314 \times 288}$$

$$= 0 \cdot 0216$$

$$\text{No. of moles of neon} = \frac{\text{wt. of neon}}{\text{molar mass of neon}} = \frac{2 \cdot 32 \text{ g}}{20 \text{ g mol}^{-1}}$$

$$= 0 \cdot 116$$

$$\therefore \qquad \text{Total no. of moles} = 0 \cdot 1376$$

$$PV = nRT$$

$$\therefore \qquad P = \frac{0 \cdot 1376 \times 8 \cdot 314 \times 373}{2 \times 10^{-3}}$$

$$= 2 \cdot 134 \times 10^5 \text{ N m}^{-2}$$

$$\therefore \quad P_{\text{argon}} = x_{\text{argon}} P = \frac{0 \cdot 0216}{0 \cdot 1376} \times 2 \cdot 134 \times 10^5 = 0 \cdot 3350 \times 10^5 \text{ N m}^{-2}$$

$$\therefore \quad P_{\text{neon}} = x_{\text{neon}} P = \frac{0 \cdot 116}{0 \cdot 1376} \times 2 \cdot 134 \times 10^5 = 1 \cdot 799 \times 10^5 \text{ N m}^{-2}$$

Method II

In this alternative method, use is made of the relationships

$$P_{\text{argon}} = \frac{n_{\text{argon}} RT}{V} \qquad \text{and} \qquad P_{\text{neon}} = \frac{n_{\text{neon}} RT}{V}$$

$$= \frac{0 \cdot 0216 \times 8 \cdot 314 \times 373}{2 \times 10^{-3}} \qquad\qquad = \frac{0 \cdot 116 \times 8 \cdot 314 \times 373}{2 \times 10^{-3}}$$

$$= 0 \cdot 3350 \times 10^5 \text{ N m}^{-2} \qquad\qquad = 1 \cdot 799 \times 10^5 \text{ N m}^{-2}$$

4.5. Energy and speed distributions

The molecules in a sample of gas do not all possess the same kinetic energy nor therefore the same speed. $\left(\text{Kinetic energy per molecule} = \frac{1}{2}mc^2.\right)$ There is in reality a *distribution of both energy and speed*. This distribution arises as a direct consequence of the chaotic motion of the molecules. Let us suppose that it was possible to introduce into a container gas molecules all of which initially were travelling at the same speed and therefore possessing the same kinetic energy. This situation would soon alter, for some molecules would, during collision, lose kinetic energy thereby reducing their speed while others would experience a gain in kinetic energy with a consequent increase in speed. Although during a collision between two molecules, there is likely to be an exchange of energy (one molecule gaining in kinetic energy at the expense of the other) the total kinetic energy is conserved. Such a collision is described as being perfectly elastic. If such were not the case then, as a result of continual collision, all the molecules would experience a gradual decrease in speed and would inevitably settle out to the foot of the container; that this does not happen is due to the elastic character of the collisionsl

The manner in which the molecules distribute themselves over al possible speeds is shown in Fig. 4.4(a). A somewhat different shape of curve is given for the corresponding energy distribution (Fig. 4.4(b)).

Fig. 4.4. Speed and energy distributions.

Examination of Fig. 4.4(a) reveals that

(i) very few molecules have either extremely low or high speeds; and

(ii) there is one speed which is more probable than any other, this speed corresponding to the maximum in the curve.

Similar deductions apply to Fig. 4.4(b).

Any increase in temperature will increase both the average kinetic energy and the speed of the molecules and although the general shape of the above curves remains the same, there is a displacement of the maxima towards higher speeds and higher energies.

One final point of interest with regard to the above diagrams is to consider the significance of various areas under the curve. In Fig. 4.4(b), the area A, relative to the total area under the curve, represents the fraction of molecules having energy values between E_1 and E_2. Of greater interest is the fraction of molecules which have energies in excess of a particular value. Thus, area B is a measure of the fraction of molecules having energy values between E_3 and infinity, i.e. having energies in excess of E_3.

When we come to consider evaporation (Chapter 5) and reaction rates (Chapter 13) we shall learn that it is these highly energetic molecules which to a large extent control these processes.

4.6.   Real gases

An ideal gas is one in which (a) the molecules are regarded as point masses, and (b) attraction between the molecules is ignored. In reality the molecules of a gas *do* experience a mutual attraction and *do* occupy

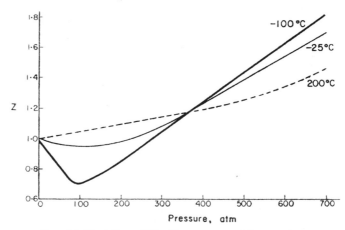

Fig. 4.5. Variation of Z with pressure and temperature.

space, small though it may be. As a result, the ideal generalization $PV = nRT$ is hardly valid when the behaviour of gases is to be studied accurately. Deviations from ideal behaviour are conveniently stated in terms of the *compressibility factor*, which is defined by

$$Z = PV/nRT$$

For an ideal gas, $Z = 1$, and any deviation from ideal behaviour will be measured by the departure of the value of Z from unity. In Fig. 4.5, Z is plotted against the pressure for the gas nitrogen at various temperatures. Clearly, the extent of the deviations from ideality is dependent upon both the pressure and temperature. Before any attempt is made to account for this pattern of behaviour some knowledge of the nature of attractive forces is required.

4.7. Molecular interactions

Interactions occurring between molecules are electrostatic in origin and may be either repulsive or attractive. The repulsive interaction principally arises through the overlap of electron clouds and only becomes appreciable at very close intermolecular separations; attractive interactions are similarly dependent on the distance of separation and only when the latter is extremely small does the repulsive effect exceed its attractive counterpart.

Attractive interactions are set up between all molecules, polar or non-polar, and are referred to as *van der Waals* interactions. The origin of this interaction for polar molecules is easy to appreciate for within such a molecule the centres of positive and negative charge do not coincide, and this condition leads to a situation in which the preferred configuration is one of attraction, with the positive end of one molecule tending to align itself with the negative end of a second and so on.

The origin of attractive forces between non-polar molecules is less clear. While such molecules possess no permanent dipole moment, there will very probably be at any instant of time a separation of the centres of positive and negative charge. This *instantaneous* dipole, arising through the continual movement of the electrons, can induce an electric dipole in a nearby molecule and thereby create an attractive interaction. At some later moment in time there will be a new electron distribution but again an overall attraction results through the interaction of the instantaneous dipole of one molecule with the induced dipole of a neighbouring molecule.

4.8. The van der Waals treatment of a real gas

The first attempt to improve the simplified model of the ideal gas was made by the Dutch scientist van der Waals in 1873. In the ideal gas, V is identified with the volume of the container; in reality, it represents the *space* in which the molecules are free to move, and account must be taken of the fact that some of this space is unavailable for this purpose, is in fact "excluded", due to the finite volume of the molecules. Hence, the space in which the molecules are free to move corresponds to a volume somewhat less than that of the container, and V in the ideal gas equation must be replaced by $V-b$, where V is the molar volume and b, the excluded volume, corresponds to the effective volume of all the molecules in one mole. Thus, the first modification of the ideal gas model results in the equation

$$P(V-b) = RT \quad \text{for one mole of gas}$$

An examination of this equation reveals the nature of the deviation produced by the finite volume of the molecules, for

$$\frac{PV}{RT} = Z = 1 + \frac{Pb}{RT}$$

Hence, the deviation produced as a result of this volume effect causes the value of Z to increase its value in excess of unity, and the greater the pressure, the greater will be the value of Z.

The second correction made to the ideal gas equation results from a consideration of the intermolecular attractive forces. The way in which these forces produce deviation from ideal behaviour is illustrated in the following diagram (Fig. 4.6). A molecule within the bulk of the gas experiences no resultant attractive force because the field of force surrounding this molecule is spherically symmetrical. This symmetry is absent

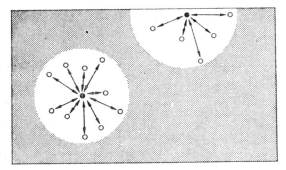

FIG. 4.6. Molecular attractions in the bulk of the gas and at the container wall.

when the molecule is about to strike the wall of the container, and such a molecule will consequently be under the influence of a net inward pull. This effect will decrease the pressure below that expected for an ideal gas and it therefore becomes necessary to add to the observed pressure a correction term in order to obtain that pressure an ideal gas would exert under the same conditions. It can be shown that the force of attraction is proportional to the square of the density and thus inversely proportional to the square of the volume,

i.e.

$$\text{correction term} \propto \frac{1}{V^2} = \frac{a}{V^2}$$

where V is again the molar volume and a is the constant of proportionality whose magnitude gives a measure of the strength of the intermolecular forces. Neglecting for the moment the correction previously made to V in the ideal gas equation, the effect of molecular attraction would alter the latter to

$$\left(P + \frac{a}{V^2}\right)V = RT \quad \text{for one mole of gas}$$

Hence

$$\frac{PV}{RT} = Z = 1 - \frac{a}{RTV}$$

Thus, the deviation produced in consequence of this effect causes the value of Z to decrease below unity, and as the pressure is increased, the

volume will decrease and the value of Z will fall still more below that of
unity. This is readily understandable because as the pressure is increased
the molecules are separated from each other by smaller distances and the
attractive forces increase in strength—as a result, the net inward pull
suffered by molecules about to strike the surface is increased.

It can now be appreciated that the two effects producing deviation
from ideal behaviour act in opposition to each other with one tending to
increase, the other to decrease the value of Z. Whichever effect predominates will depend on the gas, the temperature and the pressure.

The equation which corrects for both the size of the molecules and
their mutual attraction is called the *van der Waals* equation of state, viz.

$$\left(P + \frac{a}{V^2}\right)(V - b) = RT \quad \text{for one mole of gas}$$

We can now explain in semi-quantitative terms the various features of
the graphs in Fig. 4.5. Consider the behaviour of nitrogen at $-100°C$.
Within the region of pressure in which $Z < 1$, the effect of molecular
attraction predominates, whilst at higher pressures, where $Z > 1$, the
influence of the excluded volume is the more marked. At exceedingly low
pressures, both the correction terms a/V^2 and b are very small compared
with P and V, in which case the ideal gas equation is obeyed.

The reason that the deviations caused by van der Waals forces depend
on temperature and indeed (considering this effect alone) are more pronounced at low rather than high temperatures can be explained as
follows. It will be recalled that heat is identified with molecular motion—
as a result of this, the attractive forces are more effectively opposed by
fast moving molecules than by molecules travelling on the average at
lower velocities.

Also, it is an observed fact that heavy molecules deviate more than
light molecules under the same conditions of temperature and pressure;
this is so because all molecules have, at the same temperature, approximately the same kinetic energy. Consequently, heavy molecules will have
lower velocities than light molecules and the forces of attraction will have
a greater influence on them.

4.9. The critical state of a gas

The forces of attraction existing between molecules, which in the gaseous state give rise to deviations from ideal behaviour, are responsible for the existence of a liquid state, which indeed can be thought of as being an extremely imperfect gas. To bring about the conversion of gas to liquid we must select conditions which will increase the magnitude of the attractive forces. In this context the work of Andrews, carried out in 1869, is of considerable interest. He established the pressure–volume–temperature relationships of carbon dioxide, as a result of which the conditions essential for the liquefaction process were revealed. He showed that unless the temperature of the carbon dioxide was below a certain critical value, sensibly called the *critical temperature* (T_c), liquefaction of the gas was impossible irrespective of what pressure was applied.

Further investigation has shown that this phenomenon is quite general, the limiting temperature above which liquefaction cannot occur depending on the gas being considered. Below this temperature,[†] the kinetic energy of the molecules is sufficient to allow the forces of attraction to produce liquid when an appropriate pressure is applied. Clearly gases having high a values will also have high T_c values, and in view of the fact that the normal boiling point T_b is approximately related to the critical temperature by the equation

$$T_b \approx \tfrac{2}{3} T_c$$

it follows that the normal boiling point is a measure of attractive forces acting between the molecules.

4.10. Liquefaction of gases

By the early part of the nineteenth century, a large number of gases had been successfully liquefied, but attempts to liquefy hydrogen, oxygen, nitrogen and carbon monoxide (all of which are characterized by low a values) failed. The consensus of opinion at that time was that these gases

[†] A gaseous substance below the critical temperature is described as a vapour since it can be condensed to a liquid by pressure alone.

could not be liquefied and were consequently given the name *permanent gases*. The work of Andrews strongly suggested that the failure to liquefy the so-called permanent gases was due to the fact that the temperatures reached were insufficiently low. His contribution acted as a spur to other workers to develop new techniques for obtaining low temperatures.

At present, two general principles (or a combination of the two) are employed to liquefy gases on an industrial scale. The first of these makes use of the *Joule–Thomson effect*. Joule and Thomson in 1854 showed that when a gas under pressure is allowed to pass through a valve or throttle into a region of lower pressure, a cooling effect is observed. This change in temperature, known as the Joule–Thomson effect, is a result of the gas not being ideal, and arises from the energy required to overcome the forces of attraction between the molecules in moving them further apart. If the apparatus is effectively insulated from its surroundings,[†] the only available source of energy is that due to the thermal motion of the molecules. Thus the kinetic energy of the molecules will decrease and this must result in a fall in temperature. At ordinary temperatures this cooling effect is, for most gases, extremely small, and as the temperature is raised it becomes even smaller until at one particular temperature characteristic of the gas there is no change in temperature at all. The gas is then said to be at its *inversion temperature*. Above this the gas, on expansion through a valve, experiences a heating effect. This rise in temperature is related to the shapes of the high temperature isotherms in Fig. 4.5, where Z continually increases with an increase in pressure due to the effect of the excluded volume predominating over that of molecular attraction.

Hydrogen, helium and neon are the only gases not below their inversion points at ordinary temperatures. Hence, precooling before expansion would be necessary in order to liquefy those three gases by the Joule–Thomson effect.

Linde process. Carl von Linde in 1895 successfully developed a technique for the liquefaction of air. In this process (Fig. 4.7) air, previously freed of carbon dioxide by passage over caustic soda, is compressed to

---

[†] An expansion in which heat is allowed neither to enter nor leave the system is called an *adiabatic expansion*.

FIG. 4.7. Linde process for liquefaction of air.

about 200 atm, and passed through cooling coils in order that the heat generated during the compression be removed. The gas stream then passes through the heat exchanger from which it is expanded to about 1 atm through the expansion valve where it is cooled by the Joule–Thomson effect. The gas is then returned to the compressor via the heat exchanger, thus causing the temperature of the incoming compressed gas to be lowered prior to its expansion. The repetition of this cycle results in the temperature of the expanding gas dropping sufficiently to cause partial liquefaction.

The second of the two principles involves cooling by the performance of external work. Consider a gas under pressure enclosed in a cylinder fitted with a piston. When the pressure exerted on the piston is reduced to a value below that of the gas, the latter will expand and cause the piston to move up the cylinder. This performance of work by the gas requires the expenditure of energy. If the expansion is carried out under adiabatic conditions, the performance of mechanical work can only be achieved at the expense of the kinetic energy of the molecules, hence the temperature of the gas drops.

Claude process. The Claude process for the liquefaction of air employs a combination of the two principles discussed. Air is compressed to 200 atm, the heat generated being removed by cold water. It then passes to an expansion engine where, in performing work, it experiences a very considerable drop in temperature for the reason just outlined. If this cycle were repeated several times, liquefaction of the air would result. However, such a process in itself would be highly inefficient because of the formation of liquid in the expansion engine. In practice, therefore, this method of cooling is used only to produce a sufficiently low temperature for liquefaction to occur by the Joule–Thomson effect. Hence, the air from the compressor is divided into two streams (Fig. 4.8). One goes to the expansion engine where it is cooled. It is returned to the compressor via

Fig. 4.8. Claude process for liquefaction of air.

the heat exchanger where it precools the second stream of air, the temperature of which is further reduced by its passage through the expansion valve at which point partial liquefaction results. The Claude process is economically more attractive than the Linde process, for the energy released by the gas during its expansion is recoverable. No such recovery is possible in the Linde process; it has, however, the attraction of having no moving parts—this is very much an advantage when we bear in mind that the possibility of mechanical failure is greater at lower temperatures.

Exercises

4.1. 10 moles of an ideal gas has an initial pressure of 8 atm and a temperature of 300 K. The gas is first expanded at 300 K until the volume has doubled. Next the temperature is raised to 450 K at constant volume. Finally, an additional 5 moles of gas is added at constant volume. What is the final pressure of the gas?

[*Ans.*: 9 atm.]

4.2. Tyre pressures of a stationary car are 26·0 lb in^{-2} at 18°C. After a journey the air in the tyres rises to 31°C. Calculate the new tyre pressure. (Assume no expansion of the tyres.)

[*Ans.*: 27·2 lb in^{-2}.]

4.3. If the density of nitrous oxide gas is 0·986 g dm^{-3} at 0°C and 0·500 atm, what is its molar mass? Is your result consistent with the formula N_2O?

[*Ans.*: 44·2 g mol^{-1}.]

4.4. The composition of air is approximately 80% N_2 and 20% O_2 by volume. Calculate the density of air at 20°C and 1 atm pressure.

[*Ans.*: 1·2 g dm^{-3}.]

4.5. A balloon having a capacity of 2000 m^3 is filled with helium at 20°C and 1 atm pressure. What is the maximum load it can lift at ground level? Density of air at 20°C and 1 atm = 1·2 g dm^{-3}. (Recall Archimedes' principle.)

[*Ans.*: 2·1×10^3 kg.]

4.6. The average speed of an O_2 molecule in air at 28°C is calculated to be 1080 miles per hour. What is the average speed of a molecule of N_2 in air at 27°C?

[*Ans.*: 1155 miles per hour.]

4.7. At what temperature is the average speed of an O_2 molecule equal to 2160 miles per hour?

[*Ans.*: 1200 K.]

CHAPTER 5

The Liquid State

A LIQUID can be conveniently thought of as being a continuation of the gas phase into a region of smaller volume and higher molecular attraction. The attractive forces are sufficiently powerful to prevent the liquid from filling the whole of the containing vessel (as would a gas) but too weak to allow the liquid to have any characteristic shape (as in the case of a solid).

Although the attractive forces in a liquid restrict molecular motion, they do not eliminate it. The molecules of the liquid are in continuous random motion, their velocities and kinetic energies following the same distribution pattern as for molecules in the gaseous state.

5.1.   Surface tension

In Chapter 4 we saw that a molecule in the gaseous state, about to strike the wall of the container, is under the influence of a net inward pull. Because of this, the observed pressure is less than that which an ideal gas would exert under the same conditions.

A similar situation arises at the liquid surface with the surface molecules being attracted inwards as a result of the imbalance in the intermolecular attractions at the surface (Fig. 5.1). Because of this purely surface effect, the surface molecules behave differently from those in the bulk of the liquid. The difference manifests itself in a number of interesting ways. For example, a drop of liquid tends to assume a shape which presents the smallest surface area for a given volume of liquid, viz. a sphere. In this way the number of surface molecules is minimized. Also, certain objects such as a razor blade or a needle can be made to float on the sur-

Fig. 5.1. Molecular attractions in the bulk of a liquid and at
the surface.

face of water even though they are very much more dense than water.
These phenomena demonstrate that the surface behaves very much
like an elastic membrane under tension, tending in the case of a liquid
drop to reduce its area to a minimum and, in the case of the razor blade,
to resist penetration of the surface. For this reason the term *surface
tension* is used to describe the effect.

Because surface tension is a consequence of intermolecular attractions,
it is not surprising that surface tension decreases with increase in temperature, for as the temperature rises, forces of attraction grow weaker.
Surface tension can also be reduced by adding "wetting agents". Thus
the solution of a soap in water lowers the surface tension of water, thereby
increasing its cleansing action (Chapter 29).

5.2.   Viscosity

Viscosity is the resistance to flow of a liquid. Thus water shows a greater
tendency to flow than syrup and we describe water as being less viscous
than syrup.

What explanation can be offered for this particular aspect of liquid
behaviour? The flow process is one which involves molecules sliding
past each other under the influence of some applied stress. The rate of
flow will depend upon (i) the magnitude of the stress, (ii) the shape of the
molecules, and (iii) the magnitude of the forces of intermolecular attraction. Thus, molecules approximately spherical in shape will more readily

slide past each other than, for example, long chain polymer molecules which can become entangled in each other. Also, molecules among which attractive forces are weak will flow more easily than molecules strongly bound to each other.

The viscosity of a liquid nearly always decreases with increase in temperature. This again can be thought of as a consequence of the way in which temperature influences forces of molecular attraction.

5.3. Evaporation

No matter how strong may be the attractive forces in a body of liquid, there will always be some molecules which will overcome these forces, leave the liquid and continue their existence in the gaseous state. Such a transformation of liquid into vapour is referred to as evaporation. The amount of energy in the form of heat that is required to bring about the evaporation of a stated amount of liquid (either 1 gram or 1 mole) is the *latent heat of vaporization,* the magnitude of this quantity being yet another measure of the strength of molecular attraction.

The mechanism by which liquids evaporate is explained on the basis of the kinetic theory. We have already remarked that molecules in the

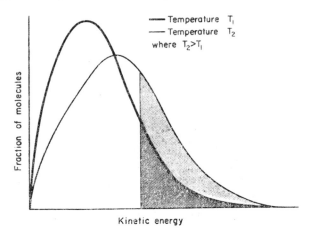

Fig. 5.2. Influence of temperature on the distribution of molecular energies.

liquid state have their energies distributed over a wide range of values as in Fig. 5.2. At any one instant of time some surface molecules will, through collision, be directed upwards and out of the liquid. The future of such a molecule then depends on how energetic it is at the moment of its departure, for only molecules which possess kinetic energy in excess of a particular value will be sufficiently energetic to overcome the attractive forces operative in the liquid state and hence leave the surface. Thus in Fig. 5.2, if E represents this critical value of kinetic energy, the shaded area is a measure of the fraction of molecules which possess kinetic energy in excess of this, and the unshaded area corresponds to the fraction of molecules which does not meet the energy requirements for evaporation. In short, two conditions must be satisfied if a molecule is to evaporate:

(i) It must occupy a position at the surface.
(ii) It must have kinetic energy in excess of a particular value.

5.4. Rate of evaporation

Extending the above argument, it follows that the rate of evaporation, expressed by the number of molecules leaving the liquid surface per minute, will be dependent upon two factors:

(i) *The amount of surface presented by the liquid*—the greater the surface area, the greater is the number of molecules occupying positions at the surface.

(ii) *The fraction of molecules having kinetic energy in excess of the critical amount.* This quantity will be strongly influenced by the temperature. Reference to Fig. 5.2 shows that a much larger fraction of molecules possesses the necessary kinetic energy required to escape from the influence of the attractive forces at the temperature T_2 than at the lower temperature T_1. As a result, the rate of evaporation increases with increase in temperature. Furthermore, for one particular temperature, this critical fraction will vary from liquid to liquid. For example, molecules of carbon tetrachloride experience only very weak mutual attraction (small van der Waals a value) and therefore the minimum kinetic energy that one such molecule requires in order to break free from the surface is small. As a result, the fraction of molecules which, at any one instant of time, meets the necessary energy requirements is high and so the rate of evaporation will likewise be high.

5.5. Evaporation and cooling

Consider a liquid evaporating in an open vessel, so that vapour molecules can diffuse into the atmosphere. If only the more energetic molecules are those lost by the liquid then it follows that the average kinetic energy of the remaining molecules must decrease. As heat is the kinetic energy of molecular motion the temperature of the liquid must therefore fall. *There is thus an attendant cooling effect accompanying evaporation*, the magnitude of which will depend on how effectively insulated is the liquid from its surroundings. If, for the moment, we imagine that no heat enters the system from the surroundings, then cooling would continue until a temperature was reached at which virtually no molecules would possess sufficient kinetic energy to break free of the surface. In this condition the evaporation rate would be zero. In practice this situation can never be realized completely, as there must always be at least a slight heat flow from the surroundings no matter how well insulated the vessel may be.

5.6. Evaporation in a closed container

Attention must now be drawn to the possibility of vapour molecules returning to the liquid state—the *condensation process*. The extent to which this process can superimpose itself on evaporation depends on whether the liquid is evaporating in an open or a closed vessel. In an open vessel, where vapour molecules are free to diffuse away from the liquid surface, condensation does not occur to an appreciable extent. However, in a closed vessel, where vapour molecules are not in a position to leave the system completely, the condensation process can no longer be ignored.

Consider therefore a vessel from which all air has been removed and into which there is injected liquid. Initially the pressure within the container will be zero. However, as a result of evaporation the pressure will increase as more and more molecules enter the space above the liquid. At any one instant of time, some of these molecules will through sheer chance collisions be directed back to the liquid surface. Thus the two opposing processes of evaporation and condensation proceed simultaneously.

If a condition of constant temperature is imposed on the system, the fraction of liquid molecules possessing kinetic energy in excess of the critical value will be constant and evaporation will therefore proceed at a constant rate (Fig. 5.3). On the other hand, the rate of condensation

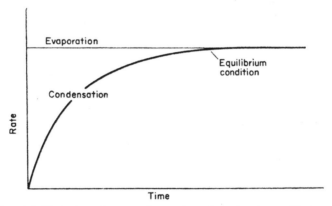

FIG. 5.3. Variation of evaporation and condensation rates with time.

will be proportional to the number of molecules occupying the space above the liquid. This number, initially zero, increases up to a limiting value at which point the number of molecules leaving the surface equals the number returning at any one instant of time,

i.e.   Rate of evaporation = Rate of condensation

Such a system is described as being in a condition of *dynamic equilibrium*, the term dynamic being used to emphasize the fact that the equilibrium is by no means static. In this equilibrium state, the pressure exerted by the vapour is referred to as the *vapour pressure of the liquid* and is defined as being *the pressure the vapour exerts when liquid and vapour coexist in a condition of dynamic equilibrium*. Its value is dependent on the nature of the liquid and upon the temperature.

5.7. Variation of vapour pressure with temperature: boiling

The fraction of molecules capable of leaving a liquid surface increases rapidly with temperature. The result, true for all liquids, is a rapid increase in the vapour pressure with increase in temperature.

When the vapour pressure of a liquid equals the external pressure, the liquid is said to boil. Thus, water boils at 100°C under a pressure of 1 atm because at this temperature the vapour pressure of water is 1 atm. If the external pressure is increased the water will no longer boil at 100°C but at some temperature in excess of this value. Similarly, if the external pressure is reduced below 1 atm, water can be made to boil at temperatures lower than 100°C.[†]

During boiling bubbles of vapour are formed. That these bubbles do not collapse in upon themselves is a result of the equality between the pressure within the bubble and the pressure acting on the bubble. When the liquid is at a temperature such that its vapour pressure is less than that of the confining atmosphere, bubble formation is impossible.

The rate of evaporation is markedly increased when a liquid is brought to the boil. We have seen that the evaporation rate is dependent upon the amount of surface presented by the liquid. Bubble formation during boiling results in a considerable increase in the area of the liquid–vapour surface. This increase allows more molecules the opportunity of breaking free from the liquid surface.

[†] Because the boiling point of a liquid varies with the external pressure, it is necessary, when quoting a boiling point, to state the pressure. When the latter is 1 atm the term "normal boiling point" is employed.

5.8. The Clapeyron equation

The quantitative relationship between vapour pressure and temperature is expressed by the Clapeyron equation:

$$\frac{dP}{dT} = \frac{L}{T \, \Delta V}$$

In this equation, the derivative dP/dT represents the rate at which the vapour pressure changes with temperature at the temperature T (K). The quantity L is the latent heat of vaporization per mole and, for the transformation from liquid to vapour, is always positive. The quantity ΔV represents a volume change and, for the transformation from liquid to vapour, is equal to the difference between the molar volumes of vapour and liquid,

i.e.
$$\Delta V = V_{vapour} - V_{liquid}$$

Now at temperatures less than the critical value

$$V_{vapour} > V_{liquid}$$

and as a result ΔV, like L, will always have a positive value.

If the change being considered is one from vapour to liquid, then L will have a negative value as a result of heat being in this case liberated in the transformation (Chapter 12). The sign of ΔV also changes for now

$$\Delta V = V_{liquid} - V_{vapour}$$

Thus, the sign of the derivative dP/dT is at all times positive irrespective of how one views the transformation. This deduction is consistent with the experimental observation that the vapour pressure of all liquids increases with an increase of temperature.

The manner in which vapour pressure changes with temperature is shown in Fig. 5.4(a). Points on the curve (called the *vapour pressure curve*) represent the conditions of pressure and temperature under which both liquid and vapour exist in equilibrium. It should be noted that the derivative dP/dT is given by the slope of the curve at any temperature T.

The Clapeyron equation is rarely used in the form given for the liquid–vapour equilibrium. A more useful equation can be developed if certain fairly justifiable assumptions are made. Thus, if the temperature is far removed from the critical value

$$\Delta V \approx V_{\text{vapour}}, \quad V_{\text{vapour}} \gg V_{\text{liquid}}$$

Further, if we suppose the vapour to behave ideally

$$V_{\text{vapour}} = \frac{RT}{P} \quad n = 1$$

Hence
$$\frac{dP}{dT} = \frac{LP}{RT^2}$$

∴
$$\frac{dP}{P} = \frac{L}{RT^2} \, dT$$

∴
$$d \ln P = \frac{L}{RT^2} \, dT$$

If we regard the latent heat of vaporization to be a constant, independent of temperature, then

$$\int_{P_1}^{P_2} d \ln P = \frac{L}{R} \int_{T_1}^{T_2} \frac{1}{T^2} \, dT$$

∴
$$\ln \frac{P_2}{P_1} = \frac{L}{R} \left(\frac{T_2 - T_1}{T_1 T_2} \right)$$

∴
$$\log \frac{P_2}{P_1} = \frac{L}{2 \cdot 303R} \left(\frac{T_2 - T_1}{T_1 T_2} \right)$$

This last equation is particularly useful for it allows us to calculate the vapour pressure of a liquid at a second temperature, knowing its vapour pressure at a given temperature. Furthermore, in view of the fact that a plot of vapour pressure against temperature is equivalent to a plot of external pressure against boiling point, it follows that the above equation can be used to calculate the boiling point of a liquid under a variety of pressures.

EXAMPLE 5.1. *It is desired to produce superheated steam at 120°C. Under what pressure must water be boiled to achieve this? Normal boiling point of water = 100°C. Latent heat of vaporization of water = 39·70 kJ mol⁻¹.*

Basically, the information we wish to obtain for the solution of this problem is the vapour pressure of water at 120°C; let P be this value. In order to obtain a value for P we must know the vapour pressure of water at some temperature other than 120°C. Now we are told that water has a normal boiling point of 100°C—this is the temperature at which boiling commences under a pressure of 1 atm. It follows from this that at 100°C, the vapour pressure of water is 1 atm.

Hence
$$\log \frac{P}{1} = \frac{39\,700\,(393-373)}{2\cdot303 \times 8\cdot314 \times 393 \times 373} \quad ^{\dagger}$$

∴
$$P = 1\cdot92 \text{ atm.}$$

Therefore to produce superheated steam at a temperature of 120°C water must be boiled under a confining pressure of 1·92 atm.

The value of the Clapeyron equation lies in the fact that it can be applied to *any* phase change. Thus, for the *sublimation process*, i.e. the process whereby solid passes directly into vapour without the intermediate formation of liquid, the relevant equations are of the same form as those already discussed. In this case

$$L = \text{latent heat of sublimation per mole,}$$
$$\Delta V = V_{\text{vapour}} - V_{\text{solid}} \approx V_{\text{vapour}},$$

and dP/dT = rate at which the vapour pressure of the solid changes with respect to temperature and is at all times positive.

The dependence of vapour pressure on temperature is shown in Fig. 5.4(b). The points on the curve (called the *sublimation curve*) represent the conditions of pressure and temperature under which both solid and vapour exists in equilibrium.

---

† In the above calculation it was unnecessary to convert pressures to $N\,m^{-2}$ because we were dealing with a ratio of pressures.

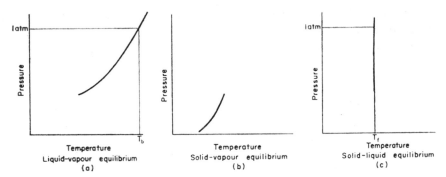

FIG. 5.4. Equilibrium conditions of pressure and temperature for
liquid–vapour, solid–vapour and solid–liquid equilibria.

In the case of the solid to liquid transformation

$$L = \text{latent heat of fusion per mole}$$
$$\Delta V = V_{\text{liquid}} - V_{\text{solid}}$$

and dT/dP = rate at which the freezing point changes with pressure.

Thus
$$\frac{dP}{dT} = \frac{L}{T(V_{\text{liquid}} - V_{\text{solid}})}$$

Now $L = lM;$ $V_{\text{liquid}} = v_{\text{liquid}} M;$ $V_{\text{solid}} = v_{\text{solid}} M$

where l is the latent heat of fusion per gram, M is the molar mass, and
v_{liquid} and v_{solid} are the *specific volumes* (cm³ g⁻¹) of the liquid and
solid respectively.

$$\therefore \qquad \frac{dP}{dT} = \frac{l}{T(v_{\text{liquid}} - v_{\text{solid}})}$$

Now for this particular change of state, ΔV is normally positive
because most solids expand when melted. If then one views the transformation as taking place from solid to liquid, L, ΔV and consequently
dP/dT are all positive, from which it follows that an increase in pressure
elevates the freezing point as shown in Fig. 5.4(c). Points on this line
(called the *freezing-point curve*) represent the conditions of pressure and
temperature under which solid and liquid exist together in equilibrium.

A few substances, of which water is the prime example, experience a contraction upon melting, i.e. ΔV is negative, and this has the effect of making dP/dT negative. Consequently for such substances, the freezing point is depressed by an increase of pressure. Furthermore, because the molar (or specific) volumes of solid and liquid do not differ appreciably from each other, the quantity ΔV is small, thus making dP/dT large. This explains why the slope of the line in Fig. 5.4(c) is very nearly vertical, and indicates that a large change in pressure is necessary to produce even a small change in the freezing point. (The normal freezing point is the temperature at which solid and liquid are in equilibrium under a pressure of 1 atm.)

EXAMPLE 5.2. *Calculate the temperature at which ice melts under a pressure of* 11 *atm. The specific volumes of ice and water are* 1·0907 *and* 1·0001 *cm*^{3} *g*$^{-1}$ *respectively at* 0°C *under* 1 *atm. The latent heat of fusion is* 335 *J g*$^{-1}$.

Both l and ΔV are virtually constant and independent of pressure. Also, as we have already seen, the freezing point T, although changing with pressure, does so only slightly, so that we can write

$$\int_{P_1}^{P_2} dP = \frac{l}{T \, \Delta v} \int_{T_1}^{T_2} dT$$

$$\therefore \quad P_2 - P_1 = \frac{l}{T \, \Delta v}(T_2 - T_1)$$

$$\therefore \quad \frac{\Delta P}{\Delta T} = \frac{l}{T \, \Delta v}$$

or more conveniently for our purpose

$$\Delta T = \frac{T \, \Delta v \, P}{l}$$

This type of problem is one in which extreme care must be exercised in the choice of units in order that there should be numerical equality between both sides of the equation. The safest approach, although arithmetically tedious, is to express every quantity in SI units. Thus

$$\Delta v = v_{\text{liquid}} - v_{\text{solid}} = -0.0906 \text{ cm}^3 \text{ g}^{-1} = -0.0906 \times 10^{-6} \text{ m}^3 \text{ g}^{-1}$$

$$T = 273 \text{ K} \qquad l = 335 \text{ J g}^{-1}$$

$$\Delta P = 11 - 1 = 10 \text{ atm} = 10.13 \times 10^5 \text{ N m}^{-2}$$

$$\therefore \quad \Delta T = \frac{-273 \text{ (K)} \times 0.0906 \times 10^{-6} \text{ (m}^3 \text{ g}^{-1}) \times 10.13 \times 10^5 \text{ (N m}^{-2})}{335 \text{ (J g}^{-1})}$$

$$= -0.075 \text{ (K N m J}^{-1})$$

$$= -0.075 \text{ K} \qquad\qquad\qquad 1 \text{ N m} = 1 \text{ J}$$

$\therefore$ Under a pressure of 11 atm, ice melts at $-0.075°C$.

The three curves shown in Fig. 5.4 can be combined to give a composite diagram (Fig. 5.5).

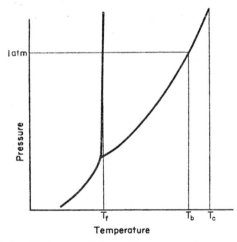

FIG. 5.5. Equilibrium conditions of pressure and temperature for liquid–vapour, solid–vapour and solid–liquid equilibria.

It will be noted that the three curves intersect at a point, referred to as the *triple point*, the significance of which is that it represents the *one* condition of pressure and temperature under which all three phases are in equilibrium with each other.

The critical temperature T_c is included to remind the reader that above this temperature the liquid cannot exist and so the vapour pressure curve has both an upper and a lower limit.

5.9.   Refrigeration

Any refrigerator works on the principle that an evaporating liquid, through absorption of heat, cools its surroundings as well as itself.

An ideal refrigerant should have the following characteristics:

(i) It must be a gas which can be easily liquefied at ordinary temperatures.

(ii) It should have a high latent heat of vaporization.

(iii) It should be non-inflammable, non-toxic and non-corrosive.

Refrigeration on an industrial scale normally involves the use of ammonia as refrigerant, its principal advantage being that it has a higher latent heat than any other liquid except water. One such refrigeration

FIG. 5.6. Refrigeration unit.

unit is shown in Fig. 5.6. Here, ammonia is first compressed, then passed through condenser coils cooled by a water spray where the heat generated in the compression is removed. Also absorbed by the water is the heat of vaporization released in the condenser coils as vapour liquefies. The ammonia, now in the liquid condition, is forced through an expansion valve from which it emerges as a fine spray; evaporation then occurs in the evaporator coils, resulting in a cooling of the brine.[†] The ammonia vapour is then returned to the compressor and the cycle repeated. The cold brine may now be circulated to parts of an industrial plant where cooling is necessary at some stage in a manufacturing process, or where cold storage of either product or raw materials is essential. In a commercial ice plant, large tanks filled with water are placed in the brine and withdrawn when the water freezes.

In domestic refrigeration the same basic principles are employed involving the same basic components, namely the compressor, condenser, evaporator and expansion valve. However, a few modifications are necessary. Firstly, air cooling is used to cool the compressed gas in the condenser. Secondly, ammonia is replaced by a less toxic refrigerant such as dichloro-difluoromethane which combines a fairly high value of latent heat of vaporization with a non-toxic nature.

5.10. The heat pump

Precisely the same components which go to make up a refrigerator are incorporated in what is referred to as a *heat pump*. The heat pump system differs from the refrigeration system in that the extracted heat, which in the latter is dissipated at the condenser, is in the former transferred from the condenser to a heating circuit, e.g. domestic central heating.

One such heat pump circuit is outlined in Fig. 5.7. From the liquid reservoir liquid at approximately room temperature passes through an expansion valve, emerging as a fine spray. This aids the vaporization which occurs under low pressure in the evaporator coils buried in the ground. Here the liquid, in addition to cooling itself, will, upon evaporating, absorb heat from its surroundings—in this case the cold ground. Emerging from the evaporator coils the vapour now at low pressure passes on to the compressor, where the high pressure is restored and the

---

[†] The "brine" is a calcium chloride solution which freezes at a temperature well below 0°C.

Fig. 5.7. Heat pump.

temperature is raised. Finally, in the condenser coils, the vapour condenses and in so doing releases its latent heat of vaporization. This, together with the heat generated in the compression stage, is then transferred to the central heating circuit. In this manner heat is transferred or "pumped" from outside to inside the house—hence the description "heat pump".

By itself this novel method of house heating is not at present an economic proposition. However, by the simple inclusion of a few valves, the system becomes capable of functioning in a dual role, viz. house heating in the winter—house cooling in the summer. As the demand for summer comfort increases (i.e. in countries which experience hot summers) the system can be expected to become more economically viable.

In house cooling, the system is made to function as a typical refrigerator. The change-over from heat pump to refrigerator is accomplished simply by reversing the flow of the gas. In this event, the outside coils become the condenser coils, whilst the inside coils become evaporator coils. In the latter, the liquid evaporates, absorbs its latent heat of vaporization from the surroundings (the house interior), then "pumps" this heat to the condenser. Here the vapour condenses and in the process loses its latent heat of vaporization to the ground.

Exercises

5.1. Use the following data to sketch the pressure–temperature phase diagram of carbon dioxide. Critical temperature = 31°C; critical pressure = 73 atm. Triple point temperature = −56°C; triple point pressure = 5·1 atm. Density of solid > density of liquid.

From the phase diagram deduce the phase changes that will occur when:

(i) CO_2 at   0°C and 1 atm → CO_2 at 0°C and 7 atm.

[*Ans.*: v → l.]

(ii) CO_2 at −100°C and 5 atm → CO_2 at 0°C and 5 atm.

[*Ans.*: s → v.]

(iii) CO_2 at −100°C and 10 atm → CO_2 at 0°C and 10 atm.

[*Ans.*: s → l → v.]

5.2. Show that over a limited temperature range the variation of vapour pressure with temperature can be expressed by an equation of the form

$$\log P = \frac{A}{T} + B$$

How does a knowledge of A permit the determination of latent heat of vaporization?

$$\left[Ans.: A = -\frac{L}{2 \cdot 303 R} . \right]$$

5.3. Liquid sodium is used as a coolant in nuclear reactors. Its vapour pressure P, between 700°C and 900°C, is given by the equation

$$\log P/\text{mmHg} = -\frac{5215}{T} + 7 \cdot 36$$

Determine the normal boiling point and the latent heat of vaporization of liquid sodium.

[*Ans.*: 891°C; 99·9 kJ.]

5.4. A pressure cooker is designed to cook food under a pressure of 4 atm. To what maximum temperature can water be raised in the cooker? Latent heat of vaporization = 39·7 kJ mol^{-1}.

[*Ans.*: 145°C.]

5.5. Sulphur has a melting point of 119·0°C under 1 atm pressure. Under what pressure will sulphur melt at 123·0°C? Densities are 1·811 g cm^{-3} (l) and 2·050 g cm^{-3} (s). The latent heat of fusion is 55·2 J g^{-1}.

[*Ans.*: 86 atm.]

CHAPTER 6

The Solid State

SOLIDS can exist in two states of aggregation, viz. *crystalline* and *amorphous* states. In the former state the basic units (molecules, atoms or ions) are arranged in space in an ordered fashion and because of this possess a high degree of symmetry. In the latter symmetry is less apparent because the units are arranged in a somewhat less regular pattern.

The manner in which the units arrange themselves in a crystal is largely determined by the nature of the forces acting between them. It is therefore convenient to classify solids in the following way:

Ionic solids—in which ions are attracted to each other by strong coulombic forces.

Covalent solids—in which atoms are strongly bonded to each other by covalent forces.

Molecular solids—in which molecules are bonded to each other by weak van der Waals forces or by dipole–dipole attractions as in, for example, hydrogen bonding.

Metallic solids—in which metallic ions are held together by the strong metallic bond.

These four classes we shall consider in turn.

6.1. Ionic solids

The most significant feature of the ionic bond is its non-directionality. Because of this, ions will tend to pack together as efficiently as possible. Thus, for sodium chloride (an AB salt type in which both ions carry the same charge) the most efficient close packing of the ions is achieved when (i) each sodium ion is in contact with six chloride ions positioned at the corners of a regular octahedron, and (ii) each chloride ion is in contact with six sodium ions similarly positioned at the corners of a regular octahedron (Fig. 6.1).[†] The above configuration of ions is common to many inorganic compounds, e.g. the halides of lithium, sodium, potassium and rubidium. It does not, however, apply to the chloride,

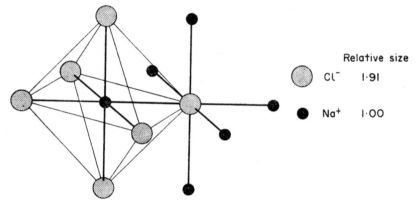

Relative size

Cl⁻   1·91

Na⁺   1·00

FIG. 6.1. The sodium chloride structure.

bromide and iodide of caesium, the fifth member of the alkali metal group of elements. The reason for this is that the ionic radius of the caesium ion is greater than the other alkali metal ions, so much so that it is comparable in size with the negative halide ions. Under these circumstances it becomes possible for each ion to pack around itself eight oppositely charged ions situated at the corners of a cube (Fig. 6.2).

So far we have considered materials in which the charge on both the positive and negative ions has been the same. However, for salts of the AB_2 type this is not the case and the structure is accordingly more complex. A common structure for salts of this type is the fluorite (CaF_2) structure (Fig. 6.3). In this arrangement eight fluoride ions positioned at

---

† In discussing the structure of solids, some difficulty is encountered in describing the structures pictorially. This problem is most easily overcome by "exploding" the structure, i.e. separating the ions by distances which do not represent the true separation. This is done in Fig. 6.1 in the interest of clarity.

Relative size

Cl⁻ 1·07

Cs⁺ 1·00

Fig. 6.2. The caesium chloride structure.

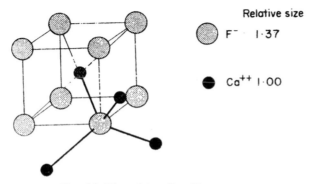

Relative size

F⁻ 1·37

Ca⁺⁺ 1·00

Fig. 6.3. The calcium fluoride structure.

the corners of a cube are packed symmetrically around and in contact with the central calcium ion. The number of calcium ions surrounding any one fluoride ion is then predetermined at four because of the necessary condition of electrical neutrality. These four are situated at the corners of a regular tetrahedron, the calcium ion occupying the central position in the tetrahedron.

6.2. Covalent solids

A covalent bond is formed in a direction which gives maximum overlap of the electron clouds of the combining atoms (Section 1.7). The covalent bond is therefore a directed bond. In the case of carbon, four equivalent bonds are directed towards the corners of a regular tetrahedron. In combination with hydrogen, each sp^3 hybrid orbital of carbon overlaps with a hydrogen s orbital to produce methane. If now we imagine replacing each hydrogen atom in methane by a carbon atom, the resulting molecule C_5 has, unlike methane, unsatisfied valencies. These can become satisfied by the further addition of carbon atoms to produce a vast interlocking structure, a fragment of which is shown in Fig. 6.4. The above arrangement, showing the spatial relationship be-

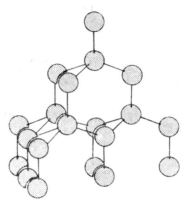

FIG. 6.4. The diamond structure.

tween carbon atoms in *diamond*, is, because of the strength and directional character of the covalent bond, a particularly rigid and hard structure— hence the use of diamonds in cutting and drilling tools. Similar structures are possessed by the elements silicon and germanium, and by the compounds zinc blende (ZnS) and silicon carbide (SiC).

Carbon, like a number of other elements (e.g. Sn, S, P), can have more than one crystalline form, a phenomenon referred to as *allotropy*. Different crystalline modifications of the same material are called *allotropes*; the second naturally occurring allotrope of carbon is *graphite*.

The structure of graphite is very different from that of diamond. It exhibits a layer structure in which each carbon atom is bonded to three others by sp^2 hybrid orbitals to form a planar network of hexagonal rings which can be thought of as being a fused aromatic ring system (Fig. 6.5). Whereas in benzene the non-hybridized p electrons are delocalized and free to move from carbon atom to carbon atom within the *ring*, so in the above fused ring system, the non-hybridized p electrons, being similarly delocalized, have freedom of movement within the *layer*.

FIG. 6.5. Planar network of hexagonal rings in graphite.

This accounts for the fact that graphite is a conductor of electricity, whereas diamond, with all electrons localized about a particular bond, is an insulator (Section 6.4).

The layers in graphite are aligned parallel to each other and are bonded together by van der Waals forces (Fig. 6.6). Because such forces are

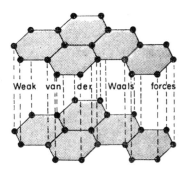

FIG. 6.6. The graphite structure.

weak, little effort is required to displace the layers relative to each other— hence the use of graphite as a high temperature lubricant (Section 29.3).

6.3.   Molecular solids

If a molecular species is of a polar nature then dipole–dipole inter-actions are established between molecules. For example, in ice the positively charged hydrogen of one molecule assumes a position relative to the negatively charged oxygen in another molecule such that a net attraction results and a hydrogen bond forms. This leads to a structure which can best be appreciated by considering the spatial relationships between the oxygen atoms of different water molecules. Thus in Fig. 6.7,

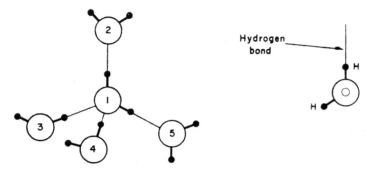

Fig. 6.7. The ice structure.

the central oxygen atom belonging to molecule 1 is surrounded tetra-hedrally by four other oxygen atoms belonging to molecules 2, 3, 4 and 5. Because the hydrogen bond has, like the covalent bond, directional characteristics, close packing of the water molecules is not achieved. Indeed the ice structure is very much an open one and this accounts for the fact that ice contracts rather than expands on melting. At the melting point some of the hydrogen bonds are broken, thus allowing the water molecules to pack more closely together.

For molecules having no polar character whatsoever (e.g. CO_2, Cl_2, O_2, N_2, C_6H_6) van der Waals forces act to hold the molecular units together and again, because these forces are particularly weak, such molecular solids have, in the main, low melting and boiling points.

van der Waals forces, like coulombic forces, are non-directed and the way in which the molecules pack together is largely determined by their shape. This is reflected in the structure assumed by fairly simple molecules as compared with more complex ones in which the molecular shape is unsuited to efficient packing.

6.4. Metallic solids

Properties such as malleability, ductility, and good electrical and thermal conductivity are highly characteristic of metallic solids. These properties are a consequence of (a) the metallic bond and (b) the metallic structure.

(a) *The metallic bond*

None of the bonds so far considered can account for the fact that the atoms in a metallic crystal are very obviously attracted to each other by strong cohesive forces. To account for this it is necessary to introduce a new cohesive force called the *metallic bond*. This bond can be thought of as being similar to a covalent bond, in that there is a sharing of electrons. Whereas, however, in a true covalent bond the valency electrons are shared between two atoms, in the metallic bond the valency electrons can be thought of as being shared by all other atoms in the metal. It is this electron sharing that is responsible for the strong attractive forces in the metallic bond. We can therefore picture the situation in metals as follows: each atom loses control of its valency electrons and in so doing becomes a positive ion. The released electrons are mobile and are free to move throughout the entire structure of the metal. Thus, at any one point in time, one valency electron can be shared by two positive ions, while at some later point, it will be shared by other positive ions elsewhere in the structure.

The mobility of the valency electrons satisfactorily explains both the high thermal and electrical conductivity of metals. In a solid heat is transferred from one vibrating molecule to the next in a chain process. At high temperatures, the molecules, atoms or ions will vibrate more vigorously than at lower temperatures. The ease with which energy is transferred from one vibrating molecule to the next depends on the degree of perfection of the crystal structure and is strongly influenced by any structural imperfections such as the presence of impurity atoms.

Although the metallic structure is a particularly efficient one, it is by no means unique; the crystal structure of argon is similar to that of many metals yet its thermal conductivity is not to be compared with that of metals. Clearly some other factor must operate to make metals unique as far as this property is concerned. We can now appreciate what the nature of this other factor is, for in a metal energy can be transferred from one part of a metal to another through the agency of the mobile electrons in much the same sort of way that gas molecules transport energy, viz. through collision.

Electrical conductivity has much in common with thermal conductivity. The ability of metals to conduct a current is due to the fact that the mobile electrons, moving about quite randomly, will acquire a preference for moving in one direction only if placed under the influence of an applied voltage. (Ionic, covalent and molecular solids do not possess such mobile electrons and are therefore non-conductors, i.e. insulators.) The electrons now moving preferentially in one direction are, in their journey, obstructed by imperfections in the crystal structure and by the thermal vibrations of the metallic ions. Such vibrations become more and more energetic as the temperature rises and it is therefore hardly surprising that the electrical resistance should increase with increase of temperature. Conversely as the temperature is decreased, so also is the resistance. Most metals show electrical resistance even at temperatures approaching absolute zero—some, however, behave quite remarkably in this region of temperature, when their resistance disappears completely. In this condition the metals are said to be *superconductors*.

(b) *The metallic structure*

Most metals crystallize in one of three forms. These are:

(i) the face-centred cubic close-packed (fcp) structure (Fig. 6.8(a)), e.g. Cu, Ag, Au, Pt.
(ii) the hexagonal close-packed (hcp) structure (Fig. 6.8(b)), e.g. Cr, Mo, W, Ti.

(iii) the body-centred cubic (bcc) structure (Fig. 6.8(c)), e.g. Li, Na,
K, Cs.

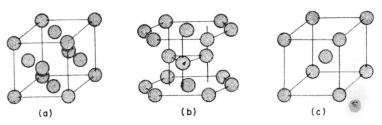

(a)   (b)   (c)

FIG. 6.8. Face-centred cubic, hexagonal and body-centred close-packed structures.

Because the metallic bond, like the ionic bond, is non-directional, the structure assumed by metals is principally determined by the various ways in which the spherical ions can pack together. The fcp and hcp structures are of special interest in this respect, for they represent the two ways in which equally sized spheres can be packed together in the most efficient manner, viz. one which makes the volume occupied by a large number of spheres a minimum. In both these structures the spheres occupy about 74% of the total volume.

Let us now consider how these structures arise. First the spheres are arranged in layers (Fig. 6.9) in such a way that each sphere is in contact with six others. The second layer is then placed on top of the first in such a way that all the second-layer spheres lie in the cavities formed by the first layer of spheres. When a third layer is built up two possibilities exist, depending on which cavities are used in the second layer of spheres to accommodate the spheres of layer 3. Thus, in one case, third-layer spheres can be placed in second-layer cavities in such a way that (i) each sphere of layer three is directly above one in layer one to yield an hcp structure (Fig. 6.9(a)), or (ii) the configuration of third-layer spheres relative to the spheres of layers one and two is as shown in Fig. 6.9(b). This latter arrangement corresponds to an fcc structure.

One important difference between the fcp and hcp metals is that in the former, the metals are both malleable and ductile (i.e. they can be either hammered into thin sheets or drawn out into a wire), while in the

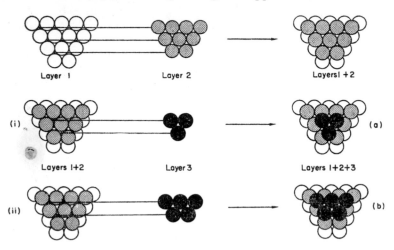

FIG. 6.9. Hexagonal and cubic close-packing.

latter, the metals tend to be both hard and brittle. This is accounted for by the fact that planes of ions can more easily glide over each other in the fcp structure than is the case in the hcp structure.

In the bcc structure, the spheres, occupying about 68% of the total volume, are arranged in a less efficient manner, in that each sphere in a layer has four rather than six nearest neighbours (Fig. 6.10). As before, second-layer spheres fall into the cavities formed by the first-layer spheres. The addition of a third layer presents no problem in this case for in the

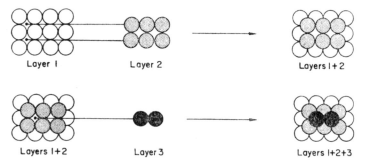

FIG. 6.10. Body-centred close-packing.

third layer, each sphere must occupy a cavity which places it directly above one in layer one.

The alkali metals crystallize in this form and their relative softness is attributed to the more open (i.e. less close-packed) structure.

6.5. Imperfections in metallic crystals

There are basically two types of imperfections in metallic crystals. In one, the imperfection is of a physical character and can be introduced by imposing some strain on the metal. This creates a less-ordered structure while at the same time altering the properties of the metal. Thus when a metal is *cold-worked* (i.e. hammered, rolled or pulled in the cold) it breaks up into small irregularly shaped crystals called *crystallites* which are oriented with respect to each other in a manner reminiscent of a mosaic pattern (Fig. 6.11). Within each crystallite the atoms are ar-

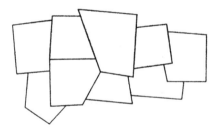

FIG. 6.11. Mosaic pattern of crystallites.

ranged in a regular fashion (viz. close-packed or body-centred) but as a result of the irregular way in which the crystallites are oriented relative to each other, the susceptibility of the metal to flow is reduced, for although planes of atoms can glide over each other within a crystallite when the metal is under some sort of strain, this flow cannot be communicated from crystallite to crystallite. Thus cold working has the effect of strengthening the metal, at the same time making it less ductile.

The reversal of the above process can be accomplished by a heat treatment referred to as *annealing*. Here thermal energy is sufficient to cause the mosaic structure to return to the more regular structure of a large crystal, thus reducing the strength of the metal and making it more ductile.

The second type of imperfection is of a chemical nature. In this case, the structural regularity of the pure metal is disturbed by the presence of "foreign" atoms of either a metallic or a non-metallic element. The resulting substance is called an *alloy*.

There are two ways in which a "foreign" atom can be accommodated into the structure of a metal. In one, the "foreign" atoms enter at random the spaces or interstices between the "host" atoms to yield what are called *interstitial alloys*. In the second case, some of the "host" atoms are replaced by the "foreign" atoms to give *substitutional alloys*.

(a) *Interstitial alloys*

For an alloy of this type to be formed, the "foreign" atoms must be small. The elements hydrogen, boron, nitrogen and carbon are sufficiently small to be accommodated in this way. The inclusion of carbon into the structure of iron is of particular interest for the resulting alloy, *steel*, is the most important of all the alloys. (Metals such as chromium, tungsten, manganese, and molybdenum are also present in some steels depending on the desired properties of the steel.) Thus when carbon is added to molten iron it dissolves and on cooling, the melt crystallizes to an iron–carbon alloy with the carbon atoms occupying some of the interstices

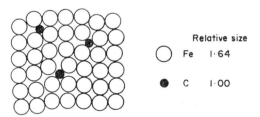

Relative size

◯ Fe   1·64

● C   1·00

FIG. 6.12. Iron–carbon alloy.

in the iron structure. As can be seen from Fig. 6.12, some distortion of the structure is inevitable and this has the effect of reducing the flow tendency. In turn this makes the iron less ductile but very much stronger.

There is obviously a limit on the extent to which a "foreign" atom can be accommodated into the "host" structure—it is in fact reached before all the available spaces become occupied. In this condition the alloy will have a definite composition but should not be thought of as a true compound.

(b) *Substitutional alloys*

This type of alloy is formed when "host" and "foreign" atoms are of comparable size. Thus when a copper–nickel melt is cooled, a substitutional alloy is formed in which the nickel atoms are statistically distributed throughout the structure of the copper (Fig. 6.13). In this particular case

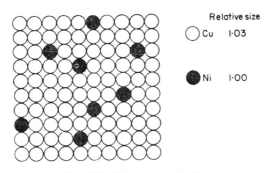

FIG. 6.13. Copper–nickel alloy.

the atomic sizes of the two atoms are so similar that a substitutional alloy of any composition can be produced. This, however, is rather rare—usually there is a limit on the degree to which one atom can substitute for another. This situation arises when the difference in atomic sizes is in excess of about 15%. Thus silver atoms can be substituted into the structure of copper (or vice versa) to a limited extent only, because of the distortion created by the substituent atoms (Fig. 6.14). As before, the distorted structures will have properties quite different from those of the parent structure.

Up till now we have not mentioned the effect of alloying on the thermal and electrical conductivities of the metal. However, it should be clear

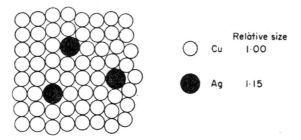

FIG. 6.14. Copper–silver alloy.

that the presence of "foreign" atoms constitutes an imperfection which will therefore cause both conductivities to fall. Thus the thermal and electrical conductivities of copper are reduced considerably when it is alloyed with zinc (as in brass) or tin (as in bronze). Stainless steel, which because of the interstitial carbon has exceptional mechanical strength, exhibits very poor thermal and electrical conductivity for the same reason.

There are two other alloy types which cannot be discussed under the heading of crystal imperfections. In one type, the two metals form what are called *intermetallic compounds*. The formation of these compounds often fails to obey the normal valency rules. A lengthy discussion of such alloys is beyond the scope of this book. The remaining alloy type is one in which the alloy is simply an intimate mixture of the two metals. Each metal still retains its own identity and because of this the properties of the alloy are not appreciably different from those of the parent metal and in the main such alloys are of little industrial importance.

6.6. Semiconductors

Elements like boron, silicon, germanium and selenium have properties which do not allow them to be placed definitely in either the metallic or non-metallic group. They possess the typical metallic lustre of metals but exhibit very poor electrical conductivity. In conductivity terms their behaviour is intermediate between, on the one hand, metals (i.e. conductors) and, on the other hand, non-metals (i.e. insulators) and for this reason they are called *semiconductors*.

One very surprising feature of semiconductors is that an increase in temperature brings about an increase in conductivity. The explanation for this is that the valency electrons are, at normal temperatures, almost exclusively associated with atom pairs (as in diamond) and are therefore immobile and incapable of transporting electricity. However, as the temperature is raised a number of valency electrons break away from their restrictive environment and become mobile—indeed, the greater the thermal energy, the greater is their number. (That this situation does not arise in diamond is due to the firmness with which the valency electrons are shared—in silicon these electrons are less effectively localized.) As in the case of a metal, a temperature increase will cause the atoms of the semiconductor to vibrate with more vigour. This by itself would cause the conductivity to decrease. However, the superimposition of these two effects leads to an overall increase in conductivity with increasing temperature. This behaviour makes semiconductors particularly useful in the role of high precision thermometers where their high temperature coefficient of resistance makes them much more sensitive than normal resistance thermometers.

The valency electrons in a semiconductor can also be released through the absorption of radiant energy of a sufficiently low wavelength (i.e. high energy). When the conductivity is improved in this way the effect is called *photoconductivity*. This phenomenon finds many applications in, for example, light meters, infrared detectors and photocopying.

The conductivity of semiconductors like silicon and germanium can be greatly enhanced by forming a substitutional alloy of these elements with the group V elements phosphorus, arsenic and antimony, the latter being present to the extent of only a few parts per million. Four of the five valency electrons of these impurity elements can be considered to be localized, forming four tetrahedrally directed covalent bonds. The remaining fifth electron cannot be so accommodated and being delocalized is free to act as a conducting electron (Fig. 6.15(a)). This type of alloy is called an *n*-type semiconductor (*n* for negative).

A similar change in the properties of silicon or germanium can be accomplished by introducing trace amounts of the group III elements boron and aluminium to produce what is called a *p*-type semiconductor

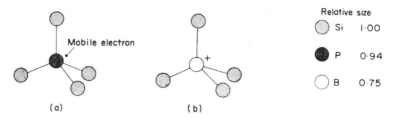

FIG. 6.15. *n*-type and *p*-type semiconductors.

(*p* for positive). In this case, the impurity element has a deficiency of electrons and there is therefore introduced into the structure a "positive hole" which can be considered to be capable of migrating from one site to another under the influence of an applied voltage (Fig. 6.15(b)). It should not be thought that the two alloys considered possess any resultant charge—they are both quite neutral. The presence of negative and positive sites in the crystals follows as a consequence of the fact that the impurity elements are unable to satisfy the normal valency requirements in the "host" structure.

   A most unusual effect is observed when an *n*-type and a *p*-type semiconductor are combined. The combination can be accomplished by "doping" one side of a wafer of very pure silicon or germanium with a group III element, thus creating a *p*-type surface; an *n*-type surface is then built up by "doping" the other side of the wafer with a group V element. Now when this *n–p* combination is placed in an electrical circuit it is found that the resistance of this "cell" is direction dependent, being low when the current is moving in one direction, but very high when the

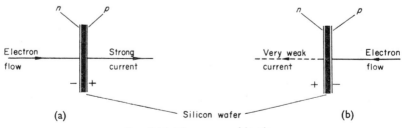

FIG. 6.16. The *n–p* combination.

flow is reversed. In other words, the *n–p* "cell" has the rectifying properties of a diode valve. Thus, if the polarity is as shown in Fig. 6.16(a), electrons will flow towards that part of the system which is electron deficient, i.e. from *n* to *p*. However, if the polarities are reversed (Fig. 6.16(b)), the electrons will tend to move towards that part of the system which is associated with an excess of electrons, i.e. from *p* to *n*. This unnatural tendency is resisted strongly with the result that only a very small current leak can be detected.

Just as an *n–p* combination of semiconductors has the rectifying properties of a diode, so an *n–p–n* or *p–n–p* combination has both the rectifying and amplifying properties of a triode. Such a combination is known as a transistor, the development of which has revolutionized the electronics industry.

Exercises

6.1. How many atoms are there in 1 lb of copper?

$$M \text{ (Copper)} = 63 \cdot 5 \text{ g mol}^{-1}; \quad 1 \text{ lb} = 454 \text{ g}.$$

[*Ans.*: $43 \cdot 1 \times 10^{23}$.]

6.2. In the uranium oxide structure (UO_2), each uranium atom is surrounded by eight oxygen atoms. How many uranium atoms surround each oxygen atom?

[*Ans.*: 4.]

6.3. A white crystalline substance X melts below 100°C. It does not conduct electricity either as a solid or when melted. What type of solid is X?

[*Ans.*: Molecular.]

6.4. The element X has a high melting point and low thermal conductivity. Its electrical conductivity increases with increase in temperature and it crystallizes in the diamond structure. Is the element X titanium, silicon, sulphur or carbon?

[*Ans.*: Silicon.]

8*

6.5. The relative sizes of gold and silver atoms are 1·014 and 1·000 respectively. What type of alloy is most likely to be formed?

[*Ans.*: Substitutional.]

6.6. Using simple geometric principles, show that the identical atoms of a metal occupy 74% of the total volume when the packing is face-centred cubic and 68% when the packing is body-centred cubic.

CHAPTER 7

Cements

ALTHOUGH the term "cement" may be broadly used to describe any adhesive material, it is more generally used to describe the lime-based products used in civil engineering to bind together coarse aggregates, and it is this type of cement we shall discuss here. In particular we shall concern ourselves with *hydraulic cements*, i.e. cements which are capable of setting and hardening under water due to the chemical interactions between water and the constituents of the cement.

7.1. Portland cement

By far the most important cement is *Portland cement*, which is the essential binding material in concrete. This is manufactured by igniting a mixture of limestone (calcium carbonate) and clay (alumino-silicates) in a rotary kiln at temperatures up to 1500°C. At these high temperatures partial fusion of the material occurs, and on cooling *cement clinker* is obtained. This is mixed with gypsum (calcium sulphate) and ground to a fine powder.

The material obtained after the roasting process is fundamentally different from the mixture fed into the kiln. At the high temperature used, water and carbon dioxide are driven off and the solid material remaining reacts to form a variety of chemical compounds, depending on the exact composition of the original mixture. Portland cement contains four main constituents, tricalcium silicate, β-dicalcium silicate, tricalcium aluminate and a ferrite phase which is a solid solution of somewhat variable composition. The formulae of these materials are shown in Table 7.1.

TABLE 7.1. PRINCIPAL CONSTITUENTS OF PORTLAND CEMENT

Constituent	Formula	Abbreviated formula
Tricalcium silicate	$3CaO \cdot SiO_2$	C_3S
β-Dicalcium silicate	$2CaO \cdot SiO_2$	C_2S
Tricalcium aluminate	$3CaO \cdot Al_2O_3$	C_3A
Ferrite phase	$4CaO \cdot Al_2O_3 \cdot Fe_2O_3$	C_4AF

Two points are worthy of note at this stage. Although the formulae of the compounds listed in the above table are written as mixed oxides, it must not be inferred that these oxides exist as such in the actual compound. The first compound listed, tricalcium silicate, could equally well be represented by the formula Ca_3SiO_5. The formula simply gives the information that the compounds in question contain certain elements in fixed proportions. The mixed oxide formulae are normally further abbreviated by the cement chemist as shown in the final column of Table 7.1, where each oxide is represented by a single letter, e.g.

$$CaO \quad SiO_2 \quad Al_2O_3 \quad Fe_2O_3 \quad H_2O$$
$$C \quad\quad S \quad\quad A \quad\quad F \quad\quad H$$

7.2. Hydration reactions

It has already been stated that cementing action arises from the chemical interaction of water with the compounds present in the original cement. The four principal constituents as formed in the roasting process are anhydrous materials which are unstable in the presence of water. If therefore water is added to these materials, the system is not in a state of equilibrium and chemical reaction must occur.

Tricalcium silicate

When finely ground tricalcium silicate is added to water a fairly rapid hydration reaction occurs with the formation of a gelatinous hydrated calcium silicate and crystals of calcium hydroxide. The formation of the hydrated silicate on the surface of the anhydrous particles tends to slow the hydration reaction considerably. A suggested equation is

$$2Ca_3SiO_5 + 6H_2O \rightarrow Ca_3Si_2O_7 \cdot 3H_2O + 3Ca(OH)_2$$

β-Dicalcium silicate

This ingredient reacts very slowly with water, possibly because it has a more closely packed structure than the tricalcium silicate. Again, the products are lime and a hydrated silicate which analyses $Ca_{3.3}Si_2O_{7.3} \cdot 3 \cdot 3H_2O$. Although the two hydrated products are not chemically identical their physical structures are remarkably similar.

Tricalcium aluminate

The reaction of this material with water is rapid and is accompanied by a strong evolution of heat. The reaction differs from the previous ones in that no lime is produced, the main product being the crystalline hydrate $3CaO \cdot Al_2O_3 \cdot 6H_2O$.

Ferrite

As with the previous material reaction with water is rapid, the products including the compounds $3CaO \cdot Al_2O_3 \cdot 6H_2O$ and $3CaO \cdot Fe_2O_3 \cdot 6H_2O$.

The development of compressive strength on hydration for each of the individual compounds is shown in Fig. 7.1, where it can be seen that tricalcium silicate develops a high strength extremely rapidly, dicalcium silicate rather more slowly, and the remaining materials never develop a comparable strength.

If we now consider the addition of water to cement itself, i.e. to a ground mixture of the four main constituents, we find that for a useful cement the rates of the individual hydration reactions become extremely important. We must at this point distinguish between *setting* and *hardening*. Setting is the initial stiffening of the material and normally occurs within a few hours of mixing, while hardening is the slow development of strength which proceeds over a number of years.

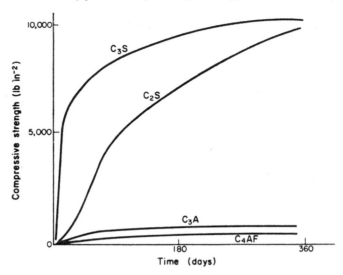

Fig. 7.1. Development of compressive strength of cement compounds. (Reprinted from Bogue and Lerch, *Ind. Eng. Chem.* **26** (1934) 837, by permission of the American Chemical Society.)

When mixed with water, simple cement paste hydrates rapidly with a strong evolution of heat—a process described as *flash set*. Since a cement paste must in practice remain workable over a period of time, gypsum is added to the main ingredients and retards the setting process. The phenomenon of flash set is generally attributed to the fast hydrolysis of tricalcium aluminates to form insoluble complexes of the type $3CaO \cdot Al_2O_3 \cdot 3CaSO_4 \cdot 31H_2O$. The retarding action is probably due to the formation of insoluble coatings on the highly reactive tricalcium aluminate surface, this slowing the hydrolysis reactions. The mechanism is not simple, however, and in the retarding of Portland cement it is desirable to have both gypsum and lime present, since a mixture of these two compounds is far more effective than either acting alone.

Any inorganic calcium salt can in fact act as a retarder in the hydrolysis of Portland cement but some (e.g. $CaCl_2$, $Ca(NO_3)_2$), while retarders at low concentrations, actually accelerate the setting reaction when present in higher concentrations.

7.3. Hardening of Portland cements

From what we have already learned it is obvious that the reactions which can occur when cement powder is mixed with water are both numerous and complex. It is hardly surprising therefore that controversy arose as to the actual mechanism of the cementing action.

One theory, that of crystallization, arose because the setting of plaster of Paris could be attributed to a rapid crystallization process which yielded a cohesive mass of small crystals. Plaster of Paris $\left(CaSO_4 \cdot \frac{1}{2}H_2O\right)$ is much more soluble than gypsum ($CaSO_4 \cdot 2H_2O$), so that when the hemihydrate is mixed with water and becomes hydrated, the solution is supersaturated with respect to the dihydrate. A rapid crystallization results, with the formation of an interlocking mass of very small needle-like crystals to yield a solid of considerable strength. It was believed that a similar situation could arise in the setting of Portland cement, whereby the cement ingredients could dissolve to form a solution supersaturated with respect to the hydration products. Rapid crystallization should then result to give a highly cohesive mass of small crystals.

An alternative view was that the hardening of cement paste was due to the formation of a *gel*, which initially would be grossly swollen with water, but would eventually harden to a solid mass. A gel is a colloidal system of at least two components and mechanically generally behaves as a typical solid. The dispersed component may be either crystalline or

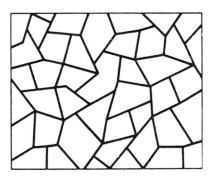

Fig. 7.2. Gel structure.

amorphous, but forms an irregular open framework within which the dispersion medium is contained, as illustrated in Fig. 7.2.

Modern theory amalgamates these two opposing views. The initial reaction does appear to involve solution of the cement, followed by precipitation of hydrated products. Although these complex hydrates may be described as crystalline, the particles are of colloidal size and the structures are normally designated as gels.

Figure 7.3 shows schematically how the anhydrous cement powder may be converted into a hard solid mass. Figure 7.3(a) shows the initial

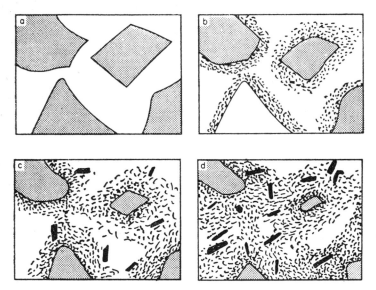

FIG. 7.3. Mechanism of cementing action. (Reprinted from *The Chemistry of Cements*, Vol. 1, H. F. W. Taylor; Academic Press, 1964.)

situation with discrete cement grains dispersed in water. Chemical interaction occurs to give a supersaturated solution from which hydrated products precipitate at the surface of the cement grains as shown in Fig. 7.3(b). This process continues with the formation of a gel, both inwards into the original cement grains and outwards into the voids between the grains, until the products around each grain begin to merge (Fig. 7.3(c)).

This is the setting stage. In Fig. 7.3(d) it can be seen that hardening has occurred, and the spaces between grains have become progressively more closely packed. Some relatively large crystals begin to form at the setting stage, and more develop by recrystallization processes.

7.4. Weathering of cements

Since concrete made with Portland cement is so widely used as a structural material, some knowledge of its resistance to chemical attack is essential. Even pure water is capable of decomposing cement given the correct conditions. The principal mechanism is the leaching of lime from the set cement, leaving eventually a soft spongy structure with no cohesive strength. Fortunately this leaching action is extremely slow unless the water can flow continuously through the cement material, and deterioration is generally negligible.

The action of acidic waters on cement structures is rather more serious, deterioration again being caused by a dissolution of some of the constituents. Acidity in natural waters usually arises from dissolved carbon dioxide or humic acid derived from decayed vegetable matter, the latter not being particularly active. The outer surfaces of hardened cement contain calcium carbonate formed by the action of atmospheric carbon dioxide on the lime in the cement, and this provides some degree of protection against attack by acidic waters. The carbonate can be dissolved, however, by conversion into bicarbonate:

$$CaCO_3 + H_2CO_3 \rightarrow Ca(HCO_3)_2$$

Since the permeability of the set cement is an important factor in water resistance, one obvious method of protection is to render the cement less permeable to water, by either internal or external treatment. Materials added to the cement mix include inert materials which simply act as pore fillers, e.g. clays, ground silica; inert materials which act as water repellants, e.g. waxes, mineral oils and bitumen; and active water repellants which actually react chemically with the cement, e.g. sodium or potassium soaps which react to form insoluble calcium soaps within the cement. It is possible that these internal additives act simply as workability aids, thus giving a more uniform and therefore stronger end product.

Surface treatments include drying oils, bituminous paints and even complete lining by sheet plastics or glass may be used when extremely corrosive materials are present, e.g. in certain industrial applications. These coatings successfully stop deterioration for a time, but since they have only limited resistance to abrasion, periodic renewal of the protective films is necessary.

One of the most serious causes of deterioration of cement structures is the presence of sulphates which can react with both the lime and the hydrated calcium aluminates in the cement, e.g.

$$Ca(OH)_2 + Na_2SO_4 \rightarrow CaSO_4 \cdot 2H_2O + 2NaOH$$

Magnesium sulphate is particularly corrosive in that it attacks the hydrated calcium silicates far more readily than do sodium or potassium sulphates. Apart from the fact that the products of sulphate attack have less binding power than the original materials, they generally occupy a much greater solid volume, thus local expansion and disruption of the mortar occurs.

Deterioration due to sulphate-containing waters is found not only in certain soils, but also where cement is exposed to sea water, which contains considerable amounts of magnesium sulphate as well as a high concentration of chlorides. Again the initial chemical attack is the formation of calcium sulphate and sulphoaluminates, but whereas with simple sulphate waters deterioration is caused by expansion of the mortar, in sea water no expansion is found. Instead, the structure is weakened by a severe leaching process due to the increased solubility of the reaction products in sea water and to the continuous agitation by wave motion.

Other mechanisms operate concurrently with sulphate attack, e.g. the cement structure is disrupted by the crystallization of salts within the pores due to alternate wetting and drying by waters of high salt content. Reinforced concrete is particularly susceptible to attack by sea water, since corrosion of the reinforcing metal causes further cracking of the structure, thus accelerating the deterioration.

One method of increasing the sulphate resistance of Portland cement is to increase the amount of the ferrite phase (C_4AF) at the expense of the calcium aluminate (C_3A). Alternatively a different type of cement such as alumina cement (Section 7.5) may be used where sulphate waters are present. A further method is to cure the cement in steam at temperatures greater than 100°C. This process brings about several changes in the cement constituents, e.g. the free lime forms calcium silicates, while the tricalcium and dicalcium silicates react to form more stable crystalline hydrates, as does tricalcium aluminate. The overall effect is therefore to produce a cement much more resistant to sulphate attack.

7.5. Special purpose cements

Alumina cement

This is manufactured by roasting limestone and bauxite ($Al_2O_3 \cdot xH_2O$) to give a product containing about equal amounts of alumina and lime, together with a substantial amount of iron and a small percentage of silica. The principal constituent after roasting is monocalcium aluminate (CA), the iron occurring as a ferrite solid solution of composition between C_4AF and C_6AF_2. The hydration reactions are similar to those occurring in Portland cement, the main hydrate being the compound CAH_{10}, although other hydrates may be produced if the temperature is greater than 25°C. This material is far more crystalline than Portland cement, but still the crystal particles are extremely small.

Although there is little difference in the setting time of alumina cement as compared to that of Portland cement, the former thereafter develops high strength very rapidly, nearing its maximum strength even after only 24 hours. As already described alumina cement has a high resistance to sulphate attack.

Another feature of alumina cement is its use in refractory concrete. The concrete is mixed in the normal way, but the aggregate itself must have heat-resisting properties, e.g. crushed firebrick. When the concrete is heated it gradually loses strength until a minimum is reached at about 900°C. At temperatures greater than this strength again develops due to the formation of new bonds between the cement products and the aggregate. Concretes based on alumina cements together with pure alumina aggregate can be used up to 1800°C. These high-temperature cements are widely used in the manufacture of kilns and furnaces.

White and coloured cements

Ordinary Portland cement is grey because of its iron content and for a white cement this must be kept as low as possible. The raw materials must therefore be a suitable limestone, e.g. chalk, and china clay, both of very low iron content. The ultimate strength of such a cement is generally lower than that of normal Portland cement.

Although Portland cement can be coloured by pigments, a wide range of colours is not practicable because the original grey colour tends to persist. Instead, white cement with an appropriate pigment is generally used. The achievement of lasting colour is difficult, not because of any defect in the pigment itself, but because of the development of a white calcium carbonate film over the surfaces. This can be removed by treatment with dilute mineral acids, but the treatment must be carefully controlled due to the possible attack of the acid on the cement itself.

Expansive cements

Normal cements undergo a contraction in volume on drying, and this causes obvious manipulative difficulties. Although no cement has yet been discovered which dries without any volume change, some progress in this direction has been made, mainly by using additives to Portland cement. The aim has been to produce a cement which expands initially, the expansion counteracting the final shrinkage on drying, so that no net volume change occurs.

By igniting a mixture of bauxite, limestone and gypsum, a calcium sulphoaluminate is obtained which when ground with Portland cement clinker gives the desired properties. The material is not particularly easy to use and careful control of the curing process is essential.

Low heat cements

The hydration of the anhydrous compounds of normal Portland cement is an exothermic process and liberates a considerable quantity of heat. This can be of particular importance in constructions where large masses of concrete are involved, leading to large temperature gradients within the concrete mass. Temperature rises of up to 50°C have been observed. The principal danger is that the eventual cooling of the structure will be accompanied by crack formation, which could lead to early failure of the material.

The main sources of heat are the hydration reactions of tricalcium aluminate and silicate. The amount of the former can fairly readily be decreased without any detriment to strength by increasing the Fe/Al ratio, thus converting the C_3A material to C_4AF. The effect of the silicate can be reduced by increasing the proportions of β-C_2S at the expense of C_3S, although this does reduce the early strength of the set cement (see Fig. 7.1).

Pozzolanic cements

In Portland cement anhydrous calcium silicates are prepared by roasting, and the subsequent hydration reactions give a material with binding properties. Pozzolanas are materials which, although not in themselves cements, will react with lime and water at normal temperature to produce a cement-like material. These materials when occurring naturally are usually of volcanic origin, but can be made artificially by roasting such materials as clays, shales and pulverized fuel ash.

Pozzolanic cements are made by grinding together pozzolanas with Portland cement to a fine powder, which is then treated as a normal cement. The Portland cement constituents appear to hydrate normally, but the lime produced in the hydration reactions further reacts with the pozzolanas in the presence of water.

These cements possess several desirable characteristics, e.g. the heat of hydration is low, their resistance to chemical attack is very good, and their ultimate strength can be greater than that of conventional Portland cement. They do, however, suffer from the disadvantage that their rate of hardening is rather slow. At low temperatures this characteristic is so accentuated that in certain climates their use is not practicable.

Exercises

7.1. Define the following terms as used in cement chemistry:

(i) hydraulic cement;
(ii) Portland cement;
(iii) cement clinker;
(iv) flash set.

7.2. Write brief notes on each of the following types of cement:

(i) alumina cement;
(ii) low-heat cements;
(iii) expansive cements.

7.3. State the four main constituents of Portland cement and describe the role of these components in the hardening process.

CHAPTER 8

Electrochemistry and Metallic Corrosion

8.1. Oxidation and reduction

Although at a very elementary level it is convenient to define oxidation as simply the gain of oxygen or loss of hydrogen, such a definition has only limited application. If, for example, we wish to describe the conversion of ferrous chloride to ferric chloride

$$FeCl_2 + \tfrac{1}{2}Cl_2 \rightarrow FeCl_3$$

as an oxidation reaction we must obviously redefine the term oxidation. The most convenient definition is that *oxidation involves the loss of electrons, reduction involves the gain of electrons.*

If we re-examine the simple reaction above we find we can consider the overall process to be the sum of two parts. The ferrous iron is oxidized to ferric iron, involving the loss of one electron,

$$Fe^{2+} \rightarrow Fe^{3+} + 1e$$

while the oxidizing agent chlorine gains an electron and is reduced to the negatively charged chloride ion

$$\tfrac{1}{2}Cl_2 + 1e \rightarrow Cl^-$$

Obviously the two simple reactions, called *half-reactions*, are complementary and must occur in equivalent amounts, i.e. the electrons generated in the oxidation reaction must be consumed in the reduction.

If we now imagine the two half-reactions to occur in separate vessels, but joined by a suitable conductor, we can see that if the overall reaction still takes place, the transfer of electrons will occur through the conductor, i.e. an electric current will have been generated.

Electrochemistry is the study of systems such as this where chemical reactions are used to generate an electric current, i.e. *electrochemical cells*, together with the converse process where an applied current is used to bring about chemical change, i.e. *electrolysis*.

8.2. Standard electrode potentials

If a rod of metal, e.g. zinc, is immersed in water or a solution containing zinc ions, a potential difference is set up between the rod and the solution. Generally the tendency is for ions from the metal rod to go into solution, leaving the metal negatively charged:

$$Zn \rightleftharpoons Zn^{2+} + 2e$$

The size of the *electrode potential*, i.e. the potential between the rod and solution, is determined by the nature of the metal and the concentration of metal ions in the solution.

If we now consider a different metal, e.g. copper, we find that this rod is positively charged with respect to the solution. Here the reverse reaction has occurred, metal ions leaving the solution and being deposited on the metal surface where they acquire electrons

$$Cu^{2+} + 2e \rightleftharpoons Cu$$

FIG. 8.1. The zinc–copper cell.

If these two *half-cells* are left in isolation, equilibrium is rapidly attained and reaction ceases. They can, however, be coupled together as is shown in Fig. 8.1. The two half-cells are separated by a porous partition, and the two metal rods joined by a wire. The high concentration of electrons generated in the reaction

$$Zn \rightleftharpoons Zn^{2+} + 2e$$

no longer builds up on the zinc rod, because the electrons are now free to travel along the wire to the copper electrode where they are used in the reaction

$$Cu^{2+} + 2e \rightleftharpoons Cu$$

The overall reaction occurring in the cell is

$$Zn + Cu^{2+} \rightleftharpoons Cu + Zn^{2+}$$

i.e. the zinc is being oxidized and the cupric ion reduced.

The tendency for any given element to be oxidized or reduced in a galvanic cell can be measured by coupling together two half-cells, one of whose potentials is already known. By convention, the reference half-cell is the standard hydrogen electrode which is assigned a potential of zero. The hydrogen cell, illustrated in Fig. 8.2, consists of a platinum

FIG. 8.2. The hydrogen electrode.

electrode coated with platinum black, partially immersed in a $1 \cdot 0 \, \text{mol} \, l^{-1}$ hydrogen ion solution, while hydrogen gas at a pressure of one atmosphere is bubbled over the electrode surface. The standard reference temperature is 25°C.

We can now see that in order to determine the standard electrode potential of, say, zinc, we would connect a standard hydrogen cell to a standard zinc cell made by immersing a zinc rod in a solution $1 \cdot 0$ molar in zinc ions at 25°C, and measure the potential of the cell. This cell potential is found to be $0 \cdot 76$ V, and the sign convention we shall use (though not universally accepted) is that the standard electrode potential of zinc is negative, i.e. $-0 \cdot 76$ V.

On this basis standard electrode potentials of the majority of elements yielding ions in solution have been obtained. These elements can then be listed in order, forming the *electrochemical series*, as is shown in Table 8.1. This is an extremely useful table since it lists elements according to their ease of oxidation and can be used as a rough guide as to how certain elements and ions will react. If we consider reactions involving positive ions, any element will displace any other below it in the series. As we have already seen, zinc displaces copper ions:

$$Zn + Cu^{2+} \rightarrow Zn^{2+} + Cu$$

TABLE 8.1. STANDARD ELECTRODE POTENTIALS

Reaction	$E°$ (V)	Reaction	$E°$ (V)
$Li^+ + e \rightleftharpoons Li$	$-3 \cdot 05$	$Pb^{2+} + 2e \rightleftharpoons Pb$	$-0 \cdot 13$
$K^+ + e \rightleftharpoons K$	$-2 \cdot 93$	$2H^+ + 2e \rightleftharpoons H_2$	$0 \cdot 00$
$Ca^{2+} + 2e \rightleftharpoons Ca$	$-2 \cdot 87$	$Cu^{2+} + 2e \rightleftharpoons Cu$	$+0 \cdot 34$
$Na^+ + e \rightleftharpoons Na$	$-2 \cdot 71$	$I_2 + 2e \rightleftharpoons 2I^-$	$+0 \cdot 54$
$Mg^{2+} + 2e \rightleftharpoons Mg$	$-2 \cdot 37$	$Ag^+ + e \rightleftharpoons Ag$	$+0 \cdot 80$
$Al^{3+} + 3e \rightleftharpoons Al$	$-1 \cdot 66$	$Br_2 + 2e \rightleftharpoons 2Br^-$	$+1 \cdot 07$
$Zn^{2+} + 2e \rightleftharpoons Zn$	$-0 \cdot 76$	$Cl_2 + 2e \rightleftharpoons 2Cl^-$	$+1 \cdot 36$
$Fe^{2+} + 2e \rightleftharpoons Fe$	$-0 \cdot 44$	$Au^{3+} + 3e \rightleftharpoons Au$	$+1 \cdot 50$
$Ni^{2+} + 2e \rightleftharpoons Ni$	$-0 \cdot 25$	$F_2 + 2e \rightleftharpoons 2F^-$	$+2 \cdot 65$
$Sn^{2+} + 2e \rightleftharpoons Sn$	$-0 \cdot 14$		

Similarly, for reactions involving negative ions, any element will displace one above it in the series, e.g. chlorine will displace iodine:

$$\tfrac{1}{2}Cl_2 + I^- \rightarrow \tfrac{1}{2}I_2 + Cl^-$$

8.3. Non-standard conditions

In the previous section all measured electrode potentials referred to a standard state of 25°C and molar solutions of metal ions. Since we cannot expect that in practice metals will always be in contact with solutions which are 1·0 molar with respect to the appropriate ions, we must consider the effect of deviating from these idealized conditions.

For a metal electrode where the half-cell reaction is written

$$M^{n+} + ne \rightarrow M$$

the electrode potential for any concentration is given by the equation

$$E = E^\circ + \frac{RT}{nF} \ln [M^{n+}]$$

where E° is the standard electrode potential,
  R is the gas constant,
  T is the temperature (K),
  F is the Faraday (96,500 coulombs),
  $[M^{n+}]$ is the concentration of M^{n+} (mol l^{-1}).

At 25°C this equation simplifies to

$$E = E^\circ + \frac{0·059}{n} \log [M^{n+}]$$

We can now consider the combination of two Cu/Cu^{2+} half-cells, as in Fig. 8.3 where the electrode is in each case a copper rod, but the solution of cupric ions is not the standard 1·0 molar in each half-cell. We discover that a galvanic cell has been produced, just as happened when a copper and a zinc rod were connected.

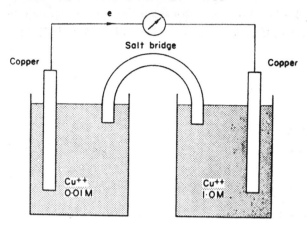

FIG. 8.3. A Cu/Cu^{2+} concentration cell.

The potential in one half-cell is

$$E_1 = E^\circ + \frac{0 \cdot 059}{2} \log 0 \cdot 01$$

while that in the other half-cell is

$$E_2 = E^\circ + \frac{0 \cdot 059}{2} \log 1 \cdot 00$$

The overall potential of the cell is the difference between the two half-cell potentials

$$E_2 - E_1 = \frac{0 \cdot 059}{2} \log \frac{1 \cdot 0}{0 \cdot 01}$$

$$= 0 \cdot 059 \ \text{V}$$

Thus a galvanic cell has been established such that copper will dissolve at electrode 1 and cupric ions will deposit at electrode 2 until the concentration of cupric ions is equalized in the two half-cells.

We have now learned what is meant by galvanic action, and how to predict the relative ease of oxidation of a given element. It must be emphasized, however, that although a study of electrode potentials may be a useful guide as to the feasibility of a reaction taking place, such a study tells us nothing about the rate at which the reaction proceeds. In certain cases the rate of a perfectly feasible chemical reaction may be so slow as to appear to us not to occur.

8.4.   Electrolysis

Just as we have described an oxidation-reduction reaction in terms of two half-reactions of a galvanic cell, we can consider oxidation-reduction reactions to occur in electrolysis. Here, when direct electric current is passed through an electrolyte, cations migrate to the *cathode* (negative electrode) where they accept electrons and are reduced, while anions migrate to the *anode* (positive electrode) where they lose electrons and are therefore oxidized.

If we consider the electrolysis of molten sodium chloride the following processes occur:

(i) migration of ions to the electrodes
(ii) at the cathode

$$Na^+ + 1e \rightarrow Na$$

(iii) at the anode

$$Cl^- \rightarrow \tfrac{1}{2}Cl_2 + 1e$$

Thus the overall reaction may be written

$$Na^+ + Cl^- \rightarrow Na + \tfrac{1}{2}Cl_2$$

In the above example the only ions present are those of sodium (Na^+) and chloride (Cl^-), and these inevitably must be the ions discharged. When more than one ion of given sign is present, one particular ion is always more readily discharged than the others. The ease of discharge of an ion is largely determined by the value of its *discharge potential*, which is the minimum potential which must be applied between the solution and the electrode before the ion is discharged. The numerical value of the discharge potential is usually approximately equal to the standard electrode potential.

A complication arises when discharge reactions involving the formation of gases are considered. Here the discharge potential may be much greater than the electrode potential, the difference in value being called the *over-voltage*. Its magnitude depends on several factors, the most important being the nature of the electrode.

An example of the significance of this is found in the common lead storage battery. This cell is composed of two electrodes, lead and lead dioxide, in contact with sulphuric acid. When supplying current, the lead is oxidized to lead ions (Pb^{2+}), while the lead dioxide is reduced to lead ions (Pb^{2+}). The half-reactions may be formally represented:

$$Pb \rightarrow Pb^{2+} + 2e$$
$$\underline{PbO_2 + 4H^+ + 2e \rightarrow Pb^{2+} + 2H_2O}$$
$$PbO_2 + 4H^+ + Pb \rightarrow 2Pb^{2+} + 2H_2O$$

i.e. $$Pb + PbO_2 + 2H_2SO_4 \rightarrow 2PbSO_4 + 2H_2O$$

In the discharge reaction, sulphuric acid is used up and lead sulphate is produced at each electrode.

If we now consider the reverse process of charging the battery, a potential is applied and electrolysis occurs. At the cathode two reactions are possible:

$$Pb^{2+} + 2e \rightarrow Pb$$
$$H^+ + 1e \rightarrow \tfrac{1}{2}H_2$$

The electrochemical series would suggest that the discharge of hydrogen would be the favoured reaction. However, the over-voltage of hydrogen on lead raises the discharge potential of hydrogen to a value greater than that of lead, and lead is in fact deposited at the electrode. Similarly at the anode, a possible reaction would be the evolution of oxygen:

$$2H_2O \rightarrow O_2 + 4H^+ + 4e$$

but again this does not occur because of the over-voltage effect and the reaction which is preferred is the oxidation of lead ions (Pb^{2+}) to lead dioxide.

The process of electrolysis has many technical applications, e.g. in metal refining which will be discussed in Chapter 9, in the preparation of certain inorganic and organic materials, and in electroplating where a base metal is coated with a thin layer of another suitable metal, e.g. silver or chromium, by electrolytic deposition.

8.5.   Fuel cells

A different type of cell which is potentially of great importance is the *fuel cell*, which is a means of converting conventional fuels directly into electric current. The fuel is supplied to one electrode and an oxidant to the other.

Perhaps the simplest cell of this type is the hydrogen–oxygen cell, which was first successfully demonstrated more than 100 years ago.

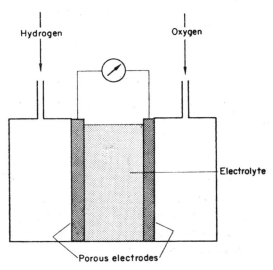

FIG. 8.4. A H_2/O_2 fuel cell.

Here the "fuel" is hydrogen gas which is supplied to the anode, while oxygen is supplied to the cathode as shown in Fig. 8.4. At the anode hydrogen is adsorbed and gives up electrons to the electrode, forming hydrogen ions. At the cathode, adsorbed oxygen accepts electrons to form hydroxyl ions:

Anode	Cathode
$2H_2 \rightarrow 4H$ (adsorbed)	$O_2 \rightarrow O_2$ (adsorbed)
$4H$ (adsorbed) $\rightarrow 4H^+ + 4e$	O_2 (adsorbed) $+ 2H_2O + 4e \rightarrow 4OH$

$$H^+ + OH^- \rightarrow H_2O$$

The reaction product is water, just as in the normal combustion of hydrogen.

Gaseous fuels require porous electrodes so that the reaction takes place at a gas–liquid–solid interface. Gaseous fuels other than hydrogen which have been successfully used include the readily available hydrocarbons methane, ethane and propane. In these cases the reaction products include carbon dioxide as well as water, e.g. at the anode:

$$C_3H_8 + 6H_2O \rightarrow 3CO_2 + 20H^+ + 20e$$
Propane

A gaseous fuel of a rather different type is ammonia, which is oxidized to nitrogen:

$$2NH_3 + 6OH^- \rightarrow N_2 + 6H_2O + 6e$$

Liquid fuels offer some advantages, e.g. the electrode system is usually simpler and storage is less of a problem. Apart from higher molecular weight hydrocarbons, methyl alcohol has been used:

$$CH_3OH + H_2O \rightarrow CO_2 + 6H^+ + 6e$$
Methyl alcohol

It can be seen that a variety of materials can be used as fuels for these cells, but although they have some attractive features, e.g. high efficiency and no need of recharging, a number of technical difficulties must be overcome before their use is widespread.

8.6. Corrosion

A knowledge of the causes of corrosion is of great importance to anyone involved in the engineering sciences. Corrosion is defined here as the transformation of a metal from the elemental to the combined state, although in its broadest sense the word could also be applied to the deterioration of any material. Under the appropriate conditions all metals will undergo chemical attack leading ultimately to severe deterioration of strength or of appearance.

The study of metallic corrosion is complicated by two principal factors. Firstly, corrosion is rarely a simple process and the interplay of a variety of possible mechanisms must be considered. Secondly, corrosion is, in general, a fairly slow process and accelerated tests have to be devised. This leads to the difficulty of deciding to what extent conclusions drawn from such tests are valid for the slow naturally occurring reaction.

Although it is probably true that all types of corrosion could be described as electrochemical reactions, it is convenient to classify corrosion into two main types, *direct* and *indirect*. In the former case we consider a localized reaction to occur, whereas in the latter, galvanic cells are established such that the complementary oxidation and reduction reactions can take place in different parts of the system.

8.7. Direct corrosion

Direct chemical attack of a metal is usually a fairly straightforward reaction and relatively easy to control. In this type of attack all chemical reaction must occur at the surface of the metal, and the corrosion product is deposited there. Since the reaction is a surface one, corrosion can continue only if the attacking species can successfully diffuse through the layer of reaction product. Because of this many direct corrosive reactions are self-inhibiting.

The most important type of direct attack is *atmospheric corrosion*. The main constituents in the atmosphere which can cause corrosion are oxygen, water and carbon dioxide, although in industrial atmospheres the presence of soot and sulphur compounds can cause severe attack. Direct oxidation of most metals occurs only to a slight extent at normal temperatures, but as the temperature is increased they will oxidize readily. In this respect we can classify metals into two main categories, those which form oxide films with a volume less than that of the underlying metal, and those which form an oxide film of greater volume.

The metals of the first two groups of the periodic table are of the first type, where the corrosion product has an open porous structure and is therefore not a protective film. These metals then are very susceptible to aerial oxidation. Metals of the second type may oxidize readily at first, but since the corrosion product hinders the diffusion of oxygen inwards (or the diffusion of metal ions outwards) the rate of corrosion often decreases markedly.

A complicating factor must now be considered. The protective oxide layer is effective because it occupies a greater volume than the underlying metal. It is quite likely therefore that this protective layer is under considerable lateral stress, and cracking of the oxide film may occur, particularly when the film is thick. If this happens the metal is no longer protected and corrosion may then proceed once again. These types of corrosion are illustrated in Fig. 8.5, where curve *A* shows how corrosion

FIG. 8.5. Rate of atmospheric corrosion.

proceeds when non-protective films are produced, while curve *B* is typical of protective film formation followed by rupture of the protective coating.

As mentioned already, humidity is an important factor in determining the extent of corrosion. In general, the rate of corrosion increases with increasing humidity. Indeed, for many metals a *critical humidity* exists, below which corrosion occurs relatively slowly, but above which a marked acceleration of corrosion rate occurs.

We must also remember that in industrial atmospheres the presence of soot and sulphur compounds is important. Sulphur compounds are partially oxidized by the air to form highly corrosive sulphuric acid. The corrosive effect is even more pronounced if soot particles are present, because moisture and sulphur dioxide are readily absorbed by carbon, thus forming locally very high concentrations of corrosive materials.

8.8.   Indirect corrosion—simple galvanic cells

The process of corrosion is said to be indirect if it can be divided into two separate reactions, an oxidation and a reduction, both involving the transfer of electrons. We learned earlier in this chapter that this is just what occurs in a simple cell. Obviously such a situation may often arise, i.e. different metals or alloys will be in contact and if immersed in a suitable electrolyte will function as a galvanic cell. The more active metal will act as the anode and will be oxidized, while the more noble metal will act as a cathode. Thus at the anode we expect reactions of the type

$$M \rightarrow M^{2+} + 2e$$

whilst at the cathode hydrogen gas may be evolved, particularly from acidic solutions

$$2H^+ + 2e \rightarrow H_2$$

While the current continues to flow corrosion will persist, but the hydrogen gas may accumulate at the cathode and cause *polarization*, thus preventing further corrosion. In such cases corrosion can only continue if the cell is depolarized by chemical removal of the hydrogen, e.g. by the action of dissolved oxygen or some other oxidizing agent. The rate of reaction is then determined by the rate at which oxygen diffuses to the cathode, and hence corrosion is in these circumstances fairly slow.

This type of corrosion is extremely common where metals are riveted or screwed together. Since the severity of corrosion depends on the current density, a large cathode area coupled to a small anode area will lead to severe corrosion. In a conducting medium a steel rivet in a copper sheet will be quickly corroded because of this effect. Corrosion of this type can usually be prevented by suitably insulating the junction of the two metals and preventing the current from flowing.

A different example of the same phenomenon is found in the effect of impurities in metal. Very pure zinc dissolves extremely slowly in non-oxidizing acids because the over-voltage effect prevents the easy evolution of hydrogen gas. If iron is present as an impurity, however, a series of galvanic cells is produced such that the hydrogen gas is readily evolved from the cathodic impurities and the zinc, acting as the anode, then readily dissolves.

In aqueous solution there is always the possibility that the anode reaction may involve the discharge of hydroxyl ions rather than the dissolution of metal:

$$M \rightarrow M^{2+} + 2e$$
$$OH^- \rightarrow OH + e$$

If this occurs, a protective layer of oxide or hydroxide may be deposited on the metal surface, preventing further corrosion. The metal is said to be rendered *passive*.

We must remember therefore that oxygen can act in two possible ways in corrosion. It may accelerate corrosion by depolarizing the cathode, or it may inhibit the anodic reaction by rendering the metal passive.

8.9.   Concentration cells

We learned in Section 8.3 that when metals are in contact with electrolytic solutions of differing concentrations galvanic action will result if the metals are connected by a suitable conductor. The metal in contact with the more dilute solution will dissolve, the cell current being such that it decreases the cell potential. Metal ion concentration cells of this type are quite common since a concentration gradient can arise in a variety of ways. In particular, recessed stagnant areas are very often places where the metal ion concentration can accumulate.

Probably the most commonly encountered concentration cell is the oxygen concentration cell, usually termed the *differential aeration cell*. We know that oxygen can act as a depolarizing agent for the cathode of a galvanic cell, and can also produce protective oxide coatings. Consequently a metal surrounded by a high oxygen concentration tends to become more "noble", or cathodic, and conversely areas with low oxygen concentration become anodic.

This type of concentration cell can be readily demonstrated by assembling two identical metal electrodes as illustrated in Fig. 8.6, such that

FIG. 8.6. Differential aeration cell.

one electrode compartment can be aerated. When air is bubbled over one electrode surface a galvanic cell is established and a current flows, the aerated rod being cathodic, the unaerated anodic. The cell can be reversed by bubbling air around the other electrode. At the cathode hydroxyl ions are produced, while at the anode the metal dissolves.

$$\text{Cathode} \qquad\qquad \text{Anode}$$
$$\tfrac{1}{2}O_2 + H_2O + 2e \rightarrow 2OH^- \qquad M \rightarrow M^{2+} + 2e$$

This type of concentration cell can lead to corrosion in a variety of situations. If, for example, we imagine a sheet of iron which has developed a small crack, then the bulk of the metal surface is more efficiently aerated than the small crevice and consequently tends to be cathodic. Galvanic corrosion will occur, and as this proceeds iron hydroxide

$(Fe(OH)_2)$ will be formed as a product of the ferrous ions diffusing from the anode and the hydroxyl ions formed in the cathode reaction. This solid corrosion product tends to be deposited in such a way that it enhances the differential aeration effect, and corrosion becomes more severe. It is this tendency for the corrosion product to be formed at a distance from the point of active corrosion that can make galvanic corrosion so much more serious than direct chemical attack, because there is little possibility of forming a protective layer.

Another example of this type of corrosion is waterline attack where severe corrosion can occur at a liquid–air interface. The oxygen concentration in the bulk of the liquid is much less than that of the surface layer and again a corrosion cell is established.

8.10.   Stress corrosion

We can consider stresses to be of two main types, a static stress or a repeated stress. When the former type is coupled with corrosion we find the phenomenon of *stress corrosion*, while the latter type leads to *corrosion fatigue*.

While it is true that stressed metal is rendered less noble than unstressed and thus galvanic action is possible, the effect is only slight and serious corrosion does not normally occur. If, however, some degree of cracking occurs the coexistence of corrosive action and stress can lead to rapid deterioration of the metal. Sometimes the failure is mainly mechanical, the function of corrosion being simply to initiate crack formation or to destroy the sound material which limits the amount of cracking. On other occasions a steady electrochemical action occurs, the stress preventing the formation of any protective film and concentrating the point of attack at the very tip of the crack.

8.11.   Protection against corrosion

The corrosion of metal structures can be prevented, or at least slowed down, by a number of techniques, one of which is the use of *inhibitors*. It has already been pointed out that one of the characteristics of galvanic corrosion is that the anodic and cathodic reactions can occur at different parts of the system and that therefore no protective film of corrosion product is formed. The oxidation and reduction are, however, mutually dependent and if one of these reactions can be prevented then the overall corrosion reaction must cease. Corrosion inhibitors therefore are chemicals which act either at the anode or the cathode to hinder the corrosive reaction.

Chemicals which function at the anode are termed *anodic inhibitors*, and can be either organic or inorganic reagents. A large number of inorganic compounds have been found to act as inhibitors, but the most common are sodium carbonate, sodium hydroxide, sodium phosphate, sodium silicate and potassium chromate. The mechanism whereby these inhibitors render the anode passive is not altogether certain, but all seem to be capable of forming insoluble films on the metal surface. The effective film may not be an insoluble phosphate or carbonate, but may be a hydroxide layer since most of the common inorganic inhibitors yield alkaline solutions.

Organic materials can also act as anodic inhibitors, e.g. thiophenol ($C_6H_5 \cdot SH$) attaches itself through the electronegative sulphur atom, thus rendering that particular metallic region more negative and so unlikely to corrode. The degree of protection can be enhanced by introducing suitable substituents in the benzene ring.

The use of anodic inhibitors can be a dangerous technique. The addition of inhibitor simply restricts the area at which corrosion occurs, and it is therefore possible if the correct amount of inhibitor is not added that very intense attack could occur at the small anodic areas which were not protected. Apart from this danger, many of the inorganic inhibitors are oxidizing agents and can therefore act as depolarizers and consequently increase the rate of corrosion.

Some inorganic salts act as *cathodic inhibitors*, e.g. zinc, magnesium and calcium salts. This effect is generally due to the formation of relatively thick insoluble films which hinder the diffusion of ions at the cathode and thus slow the corrosive reaction down.

Organic inhibitors can also be used to form protective layers at the cathode by adsorption. Organic bases, e.g. pyridine or other amines, form positive ions which attach themselves through the amino nitrogen atom to the cathode areas, forming protective films. The efficiency of this type of inhibitor depends on its "covering power", e.g. high molecular weight amines are more effective than those of low molecular weight, and highly branched tertiary amines are better inhibitors than straight chain primary ones.

A second way of protecting against corrosion is by the use of protective coatings, which may be *metallic, inorganic* or *organic*.

Metallic coatings are commonly used and can be applied in many different ways. At first sight the most obvious choice for the coating material would be a noble metal, because of their known chemical resistance. While these are used for some special purposes, e.g. silverplating, they suffer two main disadvantages, viz. their high cost and their tendency to set up extremely corrosive galvanic cells at any discontinuity in the protective coating.

Commonly used metal coatings are chromium, tin and zinc. Chromium plating—a thin coating of chromium electrolytically deposited over a layer of nickel—is widely used because of its hardness, attractive lustre and tarnish resistance. Chromium is itself highly resistant to corrosion, readily becoming passive due to protective layer formation, but because of this it permits the formation of galvanic cells with underlying less noble metals, e.g. steel. It is important therefore that a strongly adhering film free of cracks be deposited on the metal to be protected.

Zinc and tin are rather similar in that both can be deposited either electrolytically or by a dipping process. Both metal coatings are resistant to general corrosion and are anodic to iron.

Examples of inorganic and organic protective coatings are found in *enamels* and *paints*. Here the obvious mechanism is simply to protect the underlying metal by a coating of some stable material which is impermeable to moisture. In fact the process is not so simple as this, and it is essential to pretreat the metal with an inhibitor before applying the outer protective coat, e.g. "priming" with inhibitive pigments such as red lead or zinc chromate.

The final method of protection to be discussed is termed *cathodic protection*, where inhibition is achieved by using the current produced by galvanic action to protect vital parts of the system. This is illustrated in Fig. 8.7. Initially let us imagine that the cell consists simply of two electrodes A and C, the latter being the more noble. If galvanic action were to occur, metal ions would pass into solution at the anode, and electrons would flow from the anode to the cathode as shown.

If we now add a third electrode B to the system, where this electrode B is a metal even more reactive than A, a corrosion cell still functions

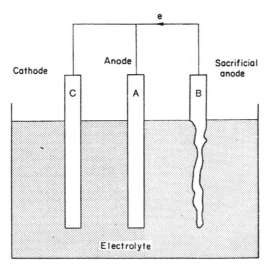

FIG. 8.7. Sacrificial anode.

but the electrons now flow from the anode B. All corrosion is confined to the electrode B which is described as a *sacrificial anode*. This is designed to be easily renewed when necessary, e.g. magnesium bars bolted to the sides of ships act as sacrificial anodes and thus protect the hulls from corrosion.

It can be seen from Fig. 8.7 that the function of the anode B is to maintain a supply of electrons to the cathode. This protective current could be supplied in another way, viz. from a d.c. generator supplying current to a non-sacrificial anode. A current supplied in this way is termed an *impressed current*.

10*

Both of these methods are widely applied to control corrosion of large metal structures such as ships, pipelines, etc., and are normally used to supplement the normal types of protective coatings described previously.

Exercises

8.1. Explain the meaning of the following terms:

    (i) oxidation;
    (ii) half-cells;
    (iii) standard electrode potential;
    (iv) discharge potential;
    (v) over-voltage;
    (vi) sacrificial anode;
    (vii) polarization;
    (viii) impressed current.

8.2. Describe the main types of metallic corrosion and explain the principal methods used to protect metallic structures against corrosion.

8.3. Using the data in Table 8.1, calculate the value of E^0 for the following reactions:

    (i) $Mg(s) + 2H^+ \rightarrow Mg^{2+} + H_2(g)$;
    (ii) $Zn(s) + Cl_2(g) \rightarrow Zn^{2+} + 2Cl^-$.

[*Ans.*: (i) 2·37 V; (ii) 2·12 V.]

8.4. Using the data in Table 8.1, calculate the value of E for the reactions:

    (i) $Pb^{2+} + 2e \rightarrow Pb$, where the concentration of Pb^{2+} is 0·02 mol l^{-1};
    (ii) $Pb^{2+} + 2e \rightarrow Pb$, where the concentration of Pb^{2+} is 0·2 mol l^{-1}.

[*Ans.*: (i) 0·18 V; (ii) −0·12 V.]

8.5. Sketch the hydrogen electrode and explain its particular importance in the field of electrochemistry.

CHAPTER 9

The Metallic Elements

9.1. Extraction of metals

In a text of this size it is not possible to consider in any detail the chemistry of all the elements in the periodic table. Instead, in this chapter, we shall study a selection of elements, and in particular the more important metals. This being the case, we shall start by considering the occurrence of metals, and briefly discuss how they can be obtained in a pure state.

We have already learned (Chapter 8) of the significance of the electrochemical series and therefore can appreciate that only a very few metals can be expected to exist naturally in a pure state. Those that do, viz. metals below hydrogen in the series, such as copper, silver, gold and platinum, are termed *noble metals*.

All other metals are found chemically combined with other elements, the natural materials from which the metals can be economically extracted being termed *ores*. The process of obtaining a pure metal from an ore generally involves three distinct processes:

(i) *preliminary treatment;*
(ii) *reduction;*
(iii) *purification.*

The preliminary treatment of an ore does not generally involve any chemical properties of the material, but rather depends on the physical characteristics of the ore. Typically, the initial treatment would involve pulverizing the ore followed by removal of waste material. If the ore particles are fairly dense, the concentration of the ore may be achieved simply by washing, thus removing the less dense waste.

The converse process of *flotation* is sometimes used, particularly with sulphide ores. In froth flotation a froth is formed by blowing air into a pulp of pulverized ore and water to which a small amount of oil has been added as a frothing agent. Minerals that have an affinity for the air bubbles rise to the surface in the froth, while the waste material (e.g. sand, clay) is more readily wetted by the water and sinks to the bottom. The process has been considerably refined by the development of reagents specific to one mineral or class of minerals which form water-repellent films on the mineral surface. These reagents are usually polar organic molecules (e.g. amines, phenols, xanthates) which attach themselves with their polar ends to the mineral surface, thus producing a non-polar surface film around the ore particle, which is then not readily wetted by water. In this way it is even possible to separate one ore from another by the correct choice of suitable reagents.

Before the second stage, some ores undergo a *roasting process* to convert them to a compound more suitable for the reduction step, e.g. carbonates and sulphides are converted to the corresponding oxides by heating.

$$CaCO_3 \rightarrow CaO + CO_2$$
$$2ZnS + 3O_2 \rightarrow 2ZnO + 2SO_2$$

The reduction of the ore to the crude metal can be carried out in a variety of ways, depending on the chemical properties of the metal and the general economics of the process. Generally, oxide ores of metals which are not too firmly bound are reduced with carbon, or in some cases with reducing agents such as aluminium or hydrogen:

$$ZnO + C \rightarrow Zn + CO$$
$$Cr_2O_3 + 2Al \rightarrow 2Cr + Al_2O_3$$

The ores of the active metals which are high in the electrochemical series, e.g. the metals of groups I and II in the periodic table, are not so easily reduced by chemical means. In these cases an electrolytic technique is usually adopted.

Purification, or *refining*, can also be achieved in a variety of ways. For example, if the metal has a low boiling point, purification may involve only a simple distillation, although the most common method is probably electro-refining where the impure metal is used as the anode of a suitable electrolytic cell. Metals of high purity can be obtained by *zone refining*. Here a narrow molten zone of metal is made to move slowly along the length of the material as shown in Fig. 9.1, the impurities being swept to one end of the rod. A number of sweeps may be necessary to obtain the desired purity. This method is based on the fact that when a liquid mixture is cooled, the solid separating out usually has a different

FIG. 9.1. Zone refining.

composition from the liquid. Since the impurities are generally preferentially dissolved in the liquid zone, the recrystallized solid will contain less of the impurity. Other classical methods of separation such as *ion exchange* or *solvent extraction* may be used. Occasionally the crude metal is reacted to form a compound which is more easily purified than the metal itself. Nickel, for example, is readily converted to a volatile carbonyl compound:

$$Ni + 4CO \rightarrow Ni(CO)_4$$

The nickel carbonyl, which boils at 43°C, can be distilled off and decomposed at higher temperatures (150–200°C) to yield a high purity metal.

9.2. Group IA elements

The elements of group IA in the periodic table are lithium, sodium, potassium, rubidium, caesium and francium, and are commonly referred to as the *alkali metals*. Their electronic configurations and some of their physical properties are shown in Table 9.1.

TABLE 9.1. THE ALKALI METALS

Element	Electronic configuration	Atomic radius (Å)	M.P. (°C)	B.P. (°C)
Li	2.1	1·55	186	1336
Na	2.8.1	1·90	97·6	880
K	2.8.8.1	2·35	62·3	760
Rb	2.8.18.8.1	2·48	38·5	700
Cs	2.8.18.18.8.1	2·67	28·5	670

The alkali metals are the most electropositive elements known and very readily lose one electron to form a monopositive ion:

$$M \rightarrow M^+ + 1e$$

This is so for two reasons. From Table 9.1 it can be seen that the electronic configurations of the alkali metals are such that each metal has one electron in excess of an inert gas configuration, therefore by losing a single electron a very stable configuration is attained. The size of the atoms is also an important factor. The alkali metals have a very large atomic radius and consequently the outermost electrons are not firmly held by the nucleus and are fairly easily lost. Because of this ease of loss of the valency electron the alkali metals are extremely reactive and will react with even the weakest oxidizing agents. This reactivity is reflected in the high position these metals occupy in the electrochemical series.

The reactivity of these metals causes problems of various kinds; for example, there is no possibility of finding these metals in the uncombined state in nature. On the contrary, they are generally found as complex silicates, e.g. $LiAlSi_2O_6$, $KAlSi_3O_8$, or at least as very stable salts which are difficult to reduce. Because of their high chemical reactivity the general method of obtaining the metal is by electrolysis of a suitable salt. Sodium, for example, is usually manufactured by the *Downs process*, where a mixture of fused sodium chloride and calcium chloride is electrolysed using a carbon anode and a copper or iron cathode. Since sodium chloride has rather a high melting point (801 °C) the calcium chloride is added to reduce the melting point of the electrolyte to around 600°C. Sodium is produced at the cathode:

$$Na^+ + 1e \rightarrow Na$$

while chlorine gas is evolved at the anode:

$$2Cl^- \rightarrow Cl_2 + 2e$$

Again at this stage of the extraction process problems arise because of the extreme reactivity of these metals. The electrolysis cell must be so designed that the metal produced does not immediately react to form a compound either with the byproduct of the electrolytic reaction or with air. Once obtained, the metals are usually stored under kerosine.

The alkali metals crystallize with a body-centred cubic lattice (see Chapter 6) and are all extremely soft, malleable and ductile. They are good conductors of both heat and electricity. The most important members of the group are sodium and potassium, but from what has been said above it will be realized that they cannot readily be used as conventional metals because of their high chemical activity. The pure metals tarnish very quickly in air due to the formation of oxide films, and react violently with water with the evolution of hydrogen gas:

$$2M + 2H_2O \rightarrow 2M^+ + 2OH^- + H_2$$

The metals are used technically for some rather specialized purposes, e.g. liquid sodium and potassium alloys have been used as heat transfer agents in nuclear reactors. Sodium in particular is widely used in metallurgy and forms alloys with many metals. Sodium dissolves in mercury to form an *amalgam* which is a commonly used reducing agent in chemical reactions. Its main advantage is that the presence of the mercury makes the sodium far less reactive so that the reactions are more easily controlled.

Of the remaining elements in the group, lithium is the most common. Again its main use is in metallurgy where it is used in metal refining because of its ability to combine with oxygen and nitrogen, and in the formation of alloys with metals such as magnesium and aluminium.

Caesium and rubidium are not very abundant in nature and have a chemical reactivity even greater than that of sodium or potassium. They have, however, a very useful property in that they can be induced to ionize, i.e. to eject an electron, simply by stimulation with light. This is termed the *photoelectric effect*, and these metals are used in the manufacture of *photocells*, which are devices which utilize this property to convert a light signal into an electric pulse.

The last member of the series, francium, occurs to only a minute extent as a short-lived radioactive isotope.

Technically the most important alkali metal compounds are the hydroxides and carbonates, in particular sodium hydroxide (*caustic soda*) and sodium carbonate, commonly referred to as *soda ash*, or simply *soda*. Both these chemicals are extensively used in the manufacture of other chemicals, textiles, soap, paper and in petroleum refining. Their manufacture will be fully discussed in Chapter 15.

9.3.   Group IIA elements

The group IIA elements, beryllium, magnesium, calcium, strontium, barium and radium, which are often referred to as the *alkaline earth* elements, differ from the group IA metals in that they all possess two valency electrons, as can be seen from Table 9.2.

TABLE 9.2. THE ALKALINE EARTH METALS

Element	Electronic configuration	Atomic radius (Å)	M.P. (°C)	B.P. (°C)
Be	2.2	0·89	1280	2970
Mg	2.8.2	1·36	650	1107
Ca	2.8.8.2	1·74	842	1240
Sr	2.8.18.8.2	1·91	800	1150
Ba	2.8.18.18.8.2	1·98	850	1140
Ra	2.8.18.32.18.8.2	2·00	700	1140

We have already learned that atoms which are fairly large and which possess a small number of electrons in excess of a stable electronic configuration generally lose their electrons readily to form ionic compounds. This is the case with the alkaline earth metals which are similar in behaviour to the group IA elements, except that they are not quite so reactive. The early members of the group, beryllium and to some extent magnesium, are much smaller than the later members and behave rather differently in that they tend to form covalent bonds.

From their position in the electrochemical series it can be seen that the alkaline earth metals will be found only in the combined state in nature, in compounds which will be fairly difficult to reduce by chemical means. Like the alkali metals, these metals are found as complex silicates (e.g. $Mg_3Si_4O_{10}(OH)_2$), but also as insoluble carbonates and sulphates. As with the group IA metals, reduction of the ore is generally achieved by an electrolytic process. In the extraction of magnesium from sea water, for example, the magnesium is first precipitated as the hydroxide, converted to the chloride by treatment with hydrochloric acid and the dry chloride, fused with other salts to lower the melting point, electrolysed at about 700°C to yield high purity metal.

Beryllium and magnesium

Of these two metals magnesium is the more important technically. Although beryllium is used in some specialized alloys, its compounds are highly toxic and thus not suitable for general use.

Magnesium is the lightest of all structural metals, but since it is such a reactive metal its corrosion resistance is not very high. It has, however, a good resistance to atmospheric corrosion because of the formation of protective films. Since it is a very light metal and is easily cast and welded it is widely used as a constructional material, particularly in the form of alloys, those with aluminium and zinc being the most important.

Chemically, a useful feature of magnesium is its great affinity for oxygen. Because of this magnesium can be used to reduce metal oxides to the corresponding metal.

Calcium, strontium and barium

These metals are more reactive than beryllium or magnesium, and are readily attacked by both oxygen and water, and therefore find no application as structural materials. Of the three, calcium is the most widely used, mainly in metallurgy where it is used as an alloying agent. Because of its high chemical reactivity, e.g. the ease of formation of oxides, nitrides, etc., calcium is also used to remove impurities from metals.

The most important compounds of the group IIA metals are the oxides, hydroxides and sulphates. Calcium oxide (quicklime) is obtained by thermal decomposition of the carbonate. The oxide formed is readily converted to the hydroxide (slaked lime) by treatment with water.

$$CaCO_3 \rightarrow CaO + CO_2$$
$$CaO + H_2O \rightarrow Ca(OH)_2$$

Since it is considerably cheaper than sodium hydroxide, lime is widely used as an industrial base and is an important constituent of cements (Chapter 7).

Calcium sulphate occurs in nature as *anhydrite* ($CaSO_4$) and as *gypsum* ($CaSO_4 \cdot 2H_2O$). Gypsum is widely used, for example, in the manufacture of cements, paper, paints and pharmaceuticals. On heating, gypsum loses part of its water of crystallization to form *plaster of Paris*:

$$CaSO_4 \cdot 2H_2O \rightleftharpoons CaSO_4 \cdot \tfrac{1}{2}H_2O + \tfrac{3}{2}H_2O$$

The use of plaster of Paris for making casts and moulds arises from the reversibility of this reaction. Since the reverse reaction is accompanied by an increase in volume, very good reproductions of the mould are obtained.

9.4. Group IIIA elements

This group of the periodic table is composed of the elements boron, aluminium, gallium, indium and thallium, whose physical properties are listed in Table 9.3.

Since it becomes progressively more difficult for an element to transfer electrons, these elements differ from groups IA and IIA in that their

TABLE 9.3. THE GROUP IIIA METALS

Element	Electronic configuration	Atomic radius (Å)	M.P. (°C)	B.P. (°C)
B	2.3	0·80	2300	2550
Al	2.8.3	1·25	660	2300
Ga	2.8.18.3	1·25	29·8	2000
In	2.8.18.18.3	1·50	155	1450
Tl	2.8.18.32.18.3	1·55	304	1460

compounds tend to be covalent rather than ionic in character, and their general properties differ in many respects from those of the preceding groups. Boron in particular differs from the remainder of the group in that it is a non-metal. This change in character as we descend the group is illustrated by the behaviour of the hydroxides of the elements. The hydroxide of boron ($B(OH)_3$) is acidic, the hydroxides of indium and thallium are basic, while the hydroxides of aluminium and gallium are *amphoteric*, i.e. they can act either as a base or as an acid, depending on the reaction conditions.

Aluminium, which is by far the most important member of the group, occurs widely in the form of complex silicates (e.g. $KAlSi_3O_8$) from which it would be difficult to extract the metal, and as *bauxite* ($Al_2O_3 \cdot xH_2O$) which is the important ore. In the extraction process use is made of the amphoteric nature of aluminium. The concentration of the ore is achieved by treatment with caustic soda solution where the aluminium dissolves with the formation of the *aluminate* ion,

$$Al_2O_3 \cdot xH_2O + OH^- \rightarrow [Al(OH)_4]^-$$

which is subsequently precipitated as the hydroxide ($Al(OH)_3$). This is in turn converted to the oxide from which the aluminium may be obtained by electrolysis, using an electrolyte of aluminium oxide dissolved in molten fluorides.

Aluminium metal is a very good conductor of both heat and electricity. Although itself a fairly reactive element, its corrosion resistance is high because of the formation of protective oxide coatings. This property, together with its lightness, makes the metal useful as a structural material. The pure metal is fairly soft, but it very readily forms alloys with, for example, copper, manganese or magnesium to give materials of good mechanical properties. The metal is also a very efficient reflector of heat and light and is therefore widely used as an insulating material.

Like magnesium in the previous group, aluminium has a very high affinity for oxygen, and this property can be utilized to obtain certain metals from their oxides when common reducing agents such as carbon or hydrogen cannot be used, e.g. the reduction of the oxides of manganese, chromium and tungsten. An extension of this is the *thermite reaction* where a mixture of powdered aluminium and ferric oxide is ignited:

$$2Al + Fe_2O_3 \rightarrow 2Fe + Al_2O_3$$

The large amount of heat evolved in this reaction is sufficient to melt the mixture, so that this process can be used for welding operations.

The remaining metals in the group are fairly rare, and are generally obtained as by-products from the manufacture of aluminium or zinc. Perhaps the most important common property of these metals is that they form compounds with arsenic and antimony to give semiconducting materials, of great importance to the electronics industry.

9.5.   Group IVA elements

The group IVA elements, carbon, silicon, germanium, tin and lead, show an even more pronounced change from non-metallic to metallic properties than do the group IIIA elements (Table 9.4). Carbon and silicon are predominantly non-metallic, tin and lead are typical metals and germanium is somewhat intermediate in character.

The first three members of the group tend to form typical covalent bonds, whereas tin and lead form normal cations. The pronounced change in character as we descend the group is reflected in the melting points. Carbon, silicon and germanium form strong interlocked covalent structures, whereas tin and lead are typical low melting metals. From their position in the electrochemical series we can see that compounds of tin

TABLE 9.4. THE GROUP IVA ELEMENTS

Element	Electronic configuration	Atomic radius (Å)	M.P. (°C)	B.P. (°C)
C	2.4	0·77	3500	4200
Si	2.8.4	1·17	1420	2355
Ge	2.8.18.4	1·22	959	2700
Sn	2.8.18.18.4	1·40	232	2270
Pb	2.8.18.32.18.4	1·75	327	1620

and lead should be easily reduced by chemical means. This, together with the fact that they are low melting, has meant that these metals were amongst the first to have been used by man.

Tin

The principal ore is *cassiterite* (SnO_2), which is first roasted to remove arsenic and sulphur, and finally reduced by carbon in a blast furnace:

$$SnO_2 + 2C \rightarrow Sn + 2CO$$

Refining of the metal is generally achieved by electrolysis.

Tin exhibits the phenomenon of *allotropy*, i.e. it can exist in more than one form:

$$\text{Grey tin} \underset{}{\overset{13\cdot2°C}{\rightleftharpoons}} \text{White tin} \underset{}{\overset{161°C}{\rightleftharpoons}} \text{Brittle tin}$$

The generally used form of tin is white tin, but we see that this converts to grey tin, which is a powdery form of the element, at temperatures below 13·2°C. In practice this change occurs very slowly, but articles made of tin can be expected to disintegrate if exposed to low temperatures for prolonged periods. Nevertheless, tin is a very versatile metal with some useful properties. Although a fairly weak metal it is easily worked, has a high resistance to corrosion, is non-toxic and has a pleasing appearance. For these reasons the metal is widely used as a surface coating, as was described in Chapter 8.

The other main application for tin is its use in making alloys with a high resistance to corrosion. Three main types exist:

(i) *bronzes* containing copper and tin;
(ii) *solders* containing lead and tin;
(iii) *white metals* containing lead, antimony, copper and tin. These are used for bearings because of their strength and good antifriction properties.

The nature of the material depends on the composition of the alloy. So-called *tinman's solder* (35% Pb) is solid at 183°C, but liquid at 184°C, and is used, for example, in soldering electrical equipment where a high fluidity is desirable. *Plumber's solder*, on the other hand, does not melt sharply at any given temperature, but rather becomes plastic over a temperature range (183–250°C) and is therefore easily shaped and manipulated.

Lead

The principal lead ore is *galena* (PbS), which is first roasted to form the oxide and then reduced by carbon. The metal is soft, dense, low melting and very easily worked. Like tin, its resistance to corrosion is very high due to the formation of protective carbonate coatings. For these reasons lead has been widely used as a cable sheathing, for roofing, piping, battery grids and as a corrosion lining in chemical plant. Because of its high density, it is also used as a shielding material against radiation. Since lead and its salts are highly toxic, it is no longer used as a piping material for drinking water. Soft waters generally attack lead sufficiently to make its use for this purpose dangerous.

Like tin, lead is widely used in alloys such as solders, bearing metals and antimony alloys, where the incorporation of a small amount of antimony increases the hardness and strength of the lead. Both tin and lead show variable valency, forming both two- and four-valent compounds. Thus lead dioxide (PbO_2), which is the main constituent of the anode of the lead storage battery, is reduced during the discharge reaction to plumbous ions (Pb^{2+}), as explained in Section 8.4.

Another oxide of lead which is of technical importance is *red lead* (Pb_3O_4), which is widely used as an undercoat for painting steel structures. Since red lead is an oxidizing agent it renders the iron passive and so helps to prevent corrosion. Other lead compounds which are used as pigments are *white lead* ($Pb_3(OH)_2CO_3$) and *lead chromate* ($PbCrO_4$). Since lead salts are highly toxic, however, their use is being superseded by other pigments.

The last member of the group which can be classed as a metal is germanium. Chemically this is a very unreactive element and is used mainly by the electronics industry in the manufacture of transistors.

9.6. Early transition elements—groups IIIB–VIIB

We have already learned that the transition metals are those which possess incomplete inner electron shells, and that the first transition series starts at the element scandium and finishes with copper. The element zinc, however, has much in common with the transition metals and will therefore be considered under this general heading. Since the metals of the first transition series are by far the most abundant we shall consider their properties in some detail and refer only briefly to a few of the heavier transition metals.

As can be seen from Table 9.5 the elements we are considering here are scandium, titanium, vanadium, chromium and manganese. It will be noted that with the exception of chromium each metal contains two outer electrons, and we would expect therefore that the chemical behaviour of these elements would be rather similar. In the case of chromium greater

TABLE 9.5. EARLY TRANSITION METALS

Element	Electronic configuration	Atomic radius (Å)	M.P. (°C)	B.P. (°C)
Sc	2.8.8.1.2	1·44	1200	2400
Ti	2.8.8.2.2	1·32	1800	3000
V	2.8.8.3.2	1·22	1710	3000
Cr	2.8.8.5.1	1·17	1890	2480
Mn	2.8.8.5.2	1·17	1260	1900

stability is attained by accommodating one of the $4s^2$ electrons in a $3d$ orbital so that each of the $3d$ orbitals contains one electron.

The first transition metal of real technical importance is titanium, which is a fairly abundant element found in the ores *ilmenite* ($FeTiO_3$) and *rutile* (TiO_2). The metal itself is difficult to obtain in a state of high purity because of its high melting point and because it reacts readily with non-metals such as oxygen, nitrogen, hydrogen and carbon. It can, however, be obtained by the reduction of titanium halides with active metals such as sodium or magnesium. As a metal it is strong, light and has excellent corrosion resistance, and is consequently widely used in making alloys suitable for the construction of jet engines and aircraft. The most important compound is the dioxide (TiO_2) which is an excellent white pigment and is now very widely used in the manufacture of paints.

The other members of the titanium subgroup are zirconium and hafnium, which like titanium have a high resistance to corrosion. Both these metals are used in nuclear engineering, zirconium because of its low neutron capture cross-section being used as a sheathing material for fuel elements, and hafnium because of its capacity for absorbing neutrons, as control rods in a nuclear reactor.

The vanadium subgroup, which includes the elements niobium and tantalum, are of minor importance, their main use being in the formation of alloys with iron to yield steels of special properties, and need not be discussed further here.

Chromium occurs mainly as the ore *chromite* ($FeCr_2O_4$) from which it is normally extracted as an alloy with iron (ferrochromium) simply by reduction with carbon. The pure metal can be obtained by reduction of the oxide with aluminium powder.

$$FeCr_2O_4 + 4C \rightarrow Fe + 2Cr + 4CO$$
$$Cr_2O_3 + 2Al \rightarrow Al_2O_3 + 2Cr$$

The main uses of chromium are in the formation of electroplated coatings or, as with so many of the early transition metals, in the formation of high strength corrosion resistant alloys. Chromium itself is a very hard metal with a high metallic lustre which owes its resistance to corrosion to the formation of a stable protective oxide film. The metal alloys readily with iron and nickel to form stainless steels, and with nickel to form an alloy with excellent high temperature characteristics (nichrome), which is used in electrical heating devices.

The other members of the chromium subgroup are molybdenum and tungsten, the former being an important component of steels which retain their strength at high temperatures. Tungsten is one of the hardest and heaviest of metals, is the highest melting metal, and chemically is highly inert. As an alloy with iron it is used in the production of steels with good temperature characteristics, and as the pure metal is used as filaments in electric lamps.

Manganese, the final member of this series, occurs in the ore *pyrolusite* (MnO_2) from which the pure metal may be obtained by reduction with aluminium.

$$3MnO_2 + 4Al \rightarrow 3Mn + 2Al_2O_3$$

Again, this metal is widely used in the manufacture of steels where it is used as a "scavenger" by combining with oxygen and sulphur, and also to form an alloy with iron which is both hard and tough.

Perhaps the most important compound is manganese dioxide which is used as a decolorizing agent in the glass industry, as a depolarizer in dry cells, and for the production of other manganese salts.

9.7. Group VIIIB elements

The so-called group VIIIB metals are, strictly, three separate groups which are considered together because of their similarity. Thus the metals iron, cobalt and nickel, whose physical properties are listed in Table 9.6, are often described as the *first transition triad*.

TABLE 9.6. GROUP VIIIB METALS

Element	Electronic configuration	Atomic radius (Å)	M.P. (°C)	B.P. (°C)
Fe	2.8.8.6.2	1·16	1535	3000
Co	2.8.8.7.2	1·16	1495	2900
Ni	2.8.8.8.2	1·15	1455	2900

Apart from similarities of chemical and physical properties, these elements are unique in that they are *ferromagnetic*, i.e. they are strongly attracted to magnetic fields and can themselves be permanently magnetized.

Iron is the most important of all the metals and is by far the most widely used. It is the second most abundant metal in the earth's crust and can be obtained from its ore by a simple reduction treatment with carbon. The metal itself is very versatile with good mechanical properties, which can be further improved by the addition of small amounts of "impurities" to form alloys. Unfortunately the metal is fairly reactive chemically and hence is particularly susceptible to corrosion.

The most important source minerals are *hematite* (Fe_2O_3) and *magnetite* (Fe_3O_4), which are reduced in a blast furnace (Fig. 9.2). The charge of ore, carbon coke and flux (limestone) is introduced at the top of the

FIG. 9.2. Blast furnace.

furnace, where the temperature is around 200°C. Hot air (in modern practice this is enriched with oxygen and fuel oil) is forced into the base of the furnace where the coke reacts to form carbon monoxide, at the same time liberating a large quantity of heat.

$$2C + O_2 \rightarrow 2CO$$

This region of the furnace has the highest temperature (2000°C) and as the gases rise up the furnace they eventually cool to around 200°C.

The overall reduction of the ore to metal occurs in a series of reactions, depending on the temperature range.

$$\sim 250°C \quad 3Fe_2O_3 + CO \rightarrow 2Fe_3O_4 + CO_2$$
$$\sim 500°C \quad Fe_3O_4 + CO \rightleftharpoons 3FeO + CO_2$$
$$\sim 700°C \quad FeO + CO \rightleftharpoons Fe + CO_2$$

In addition to these reactions the limestone is thermally decomposed and the carbon dioxide reacts with coke to form carbon monoxide.

$$CaCO_3 \rightarrow CaO + CO_2$$
$$CO_2 + C \rightarrow 2CO$$

The former reaction produces lime which reacts with the silica impurities of the ore to form a molten slag, which floats at the surface of the molten iron and can be tapped off at the bottom of the furnace.

$$CaO + SiO_2 \rightarrow CaSiO_3$$

This slag may be used in the manufacture of cement (Chapter 7). The latter reaction helps to maintain the high concentration of carbon monoxide which is necessary for the reduction reactions in the upper zones of the furnace.

The metal obtained direct from the blast furnace is called *pig-iron* and still contains a fair percentage of impurities, particularly carbon together with smaller amounts of manganese, silicon and phosphorus. This crude material is then further refined to form any one of a large number of possible products.

Cast-iron is produced by remelting and cooling pig-iron, the nature of the product being determined mainly by the rate of cooling and by the chemical composition. The most important factor here is the way the carbon present is incorporated in the final material. The carbon may exist as flakes of graphite distributed throughout the metal, giving a relatively soft material called *grey cast-iron*, or in the form of *cementite* (Fe_3C) producing a hard brittle metal described as *white cast-iron.*

If molten pig-iron is cooled quickly the cementite has no time to decompose and the resulting product is white cast-iron. Slow cooling, on the other hand, assists the formation of graphite, thus producing grey cast-iron. Apart from carbon, cast-iron contains relatively small amounts of silicon and sulphur, both of which affect the type of iron produced. The presence of silicon aids the formation of graphite, so producing a grey iron, while sulphur has the effect of stabilizing the cementite, thus producing a white iron.

Steel is manufactured by further refining the pig-iron produced by the blast furnace. The aim is to remove the impurities either as gases or as slags and then to add carefully controlled amounts of, for example, carbon, chromium or manganese to give a steel with the required properties.

Cobalt and nickel tend to be used as alloys rather than in the pure state, although nickel itself has many properties similar to mild steel. Alloyed with aluminium and iron, cobalt and nickel form the material *alnico* which is widely used in the manufacture of permanent magnets. Nickel alloys can be made with the same coefficient of expansion as glass and can be used as "seal-in" wires in, for example, electric light bulbs; while cobalt alloyed with iron is used in the manufacture of high speed cutting tools and surgical instruments.

9.8.   Group IB elements

The elements of group IB, copper, silver and gold, are generally referred to as the *coinage metals,* and have long been used by man because of their pleasing appearance and resistance to corrosion. Table 9.7 shows that while we might expect some similarity to group IA metals in that they

TABLE 9.7. THE COINAGE METALS

Element	Electronic configuration	Atomic radius (Å)	M.P. (°C)	B.P. (°C)
Cu	2.8.18.1	1·17	1083	2336
Ag	2.8.18.18.1	1·34	961	1950
Au	2.8.18.32.18.1	1·34	1063	2600

all possess a single outer valence electron, in fact the two groups are very dissimilar. The group IA elements were light, low melting metals, and were extremely reactive. The coinage metals, on the other hand, are dense, high melting materials and chemically unreactive—so much so that they are found in the native state or as easily reducible ores.

The principal ores of copper are *chalcocite* (Cu_2S), *copper pyrites* ($CuFeS_2$) and *malachite* ($CuCO_3 \cdot Cu(OH)_2$), from which the metal is readily obtained by a simple smelting process, followed by electrolytic refining. Technically, copper probably comes second only to iron in importance. The pure metal is an excellent conductor of both heat and electricity and is consequently widely used in the electrical industry. It is also a very easily worked metal, with a very high resistance to corrosion. These properties make it a valuable constructional material, e.g. for chemical plant, boilers, roofing. Apart from the many uses for the pure metal, copper forms a large number of useful alloys, e.g. *brass* (Cu–Zn), *bronze* (Cu–Zn–Sn), *German silver* (Cu–Zn–Ni).

Because of its low electrode potential, copper does not dissolve in acids with the liberation of hydrogen gas, but will dissolve in oxidizing acids or in acids where it can form complex ions:

$$3Cu + 8HNO_3 \rightarrow 3Cu(NO_3)_2 + 4H_2O + 2NO$$

The most important salt is the sulphate which is used extensively as a fungicide and germicide, e.g. *Bordeaux mixture* is a mixture of copper sulphate and lime.

Silver occurs mainly as the native metal or as the ore *argentite* (Ag_2S). One method of extracting silver from its ores depends on the formation of soluble silver complexes with cyanide. Thus either crude silver or the sulphide ore can be made to dissolve in a sodium cyanide solution through which air is blown:

$$4Ag + 8NaCN + O_2 + 2H_2O \rightarrow 4Na[Ag(CN)_2] + 4NaOH$$
$$2Ag_2S + 8NaCN + O_2 + 2H_2O \rightarrow 4Na[Ag(CN)_2] + 4NaOH + 2S$$

The solution containing the complex silver ions may then be reduced with zinc or aluminium to yield metallic silver:

$$Zn + 2Na[Ag(CN)_2] \rightarrow Zn(CN)_2 + 2NaCN + 2Ag$$

The metal has a very high lustre, is easily worked, and is the best known conductor of heat and electricity. Although the latter properties are not generally utilized because the metal is expensive, silver is widely used as a decorative metal because of its attractive appearance. It does not corrode readily except in the presence of sulphur-containing compounds where it tarnishes fairly readily due to the formation of black silver sulphide:

$$4Ag + 2H_2S + O_2 \rightarrow 2Ag_2S + 2H_2O$$

Silver is a very soft metal and normally is used as an alloy (e.g. with copper) or as a thin coating plated on a base metal. This can readily be achieved by electroplating, whereby the article to be coated is made the cathode of an electrolytic cell, with pure silver as the anode, and a solution of the silver cyanide complex ($NaAg(CN)_2$) as electrolyte.

One interesting application of silver salts is in photography, where use is made of the light sensitivity of certain silver salts, and of their ability to form complexes. The photographic film is a plastic or glass plate coated with an emulsion of silver bromide in gelatin. When the film is exposed the silver bromide granules are affected by the light and a small amount of reduction takes place, with the formation of metallic silver. Development of the film is simply a process of reduction, where a chemical reducing agent (e.g. pyrogallol) is used to accentuate the reduction process which occurred on exposure. The areas of the plate subjected to intense light are rapidly reduced, whereas areas not subjected to the same extent reduce much more slowly. The fixing process is one which removes the silver bromide remaining unchanged on the plate. This is dissolved off by treatment with sodium thiosulphate ("hypo"), giving a plate which may then be safely exposed to light.

$$3AgBr + 3Na_2S_2O_3 \rightarrow 2NaBr + Na_4[Ag_2(S_2O_3)_3]$$

9.9. Group IIB elements

Although the metals of this group, zinc, cadmium and mercury, each possess two outer valence electrons, they resemble the transition metals rather more closely than they do the group IIA metals, which have similar electronic configurations (Table 9.8).

TABLE 9.8. THE GROUP IIB METALS

Element	Electronic configuration	Atomic radius (Å)	M.P. (°C)	B.P. (°C)
Zn	2.8.18.2	1·25	419	907
Cd	2.8.18.18.2	1·41	321	767
Hg	2.8.18.32.18.2	1·44	− 38·9	357

The principal zinc ore is *zinc blende* (ZnS) from which the zinc may readily be obtained. After a roasting process to convert the sulphide to oxide, the material is reduced with coke to form the metal which is then simply distilled out of the furnace.

Although zinc is chemically a fairly reactive metal it is widely used as a protective coating for iron because of its resistance to atmospheric corrosion due to protective film formation, and because it is anodic to iron. Zinc is also widely used in the formation of alloys, e.g. with copper to give *brass,* or with small amounts of aluminium to give useful die-casting alloys. Both zinc oxide and sulphide are used as white pigments, while the sulphide, which acts as a phosphor, is used for making fluorescent screens.

Cadmium is chemically very similar to zinc and the two elements are usually found together. Like zinc, cadmium may be used as a protective coating and is used in the formation of bearing alloys with a low coefficient of friction.

Mercury, the only metal which is a liquid at room temperature, is occasionally found in the native state but more generally as the ore *cinnabar* (HgS). The metal may be obtained simply by roasting, since the oxide initially formed is thermally unstable and decomposes:

$$HgS + O_2 \rightarrow Hg + SO_2$$

The metal is chemically inert, has a high density, but is for a metal a poor conductor of electricity. It is nevertheless used in the manufacture of scientific equipment, e.g. thermometers, barometers, vacuum pumps, and in general for making electrical contacts where its mobility outweighs its poor conductivity. A very large number of metals dissolve in mercury to form *amalgams*, which may be either liquid or solid.

Mercury compounds, particularly organometallics, are widely used as fungicides and herbicides, e.g. the oxide is used as an antifouling paint. Mercury and its soluble compounds are all highly toxic and great care must be taken in their use.

Exercises

9.1. Define the terms:

    (i) flotation process;
    (ii) zone refining;
    (iii) alkali metals;
    (iv) thermite reaction;
    (v) allotropy.

9.2. Summarize the main steps in the preparation of a pure metal from its ore, and outline the general chemical principles involved in each of these steps.

9.3. Compare and contrast the physical and chemical properties of the elements of group IA with those of group IB.

9.4. Suggest reasons for each of the following observations:

   (i) Beryllium tends to form covalent bonds whereas calcium, strontium and barium typically form ionic bonds.

  (ii) The metals iron, cobalt and nickel are chemically very similar whereas sodium, magnesium and aluminium differ widely in their chemistry.

 (iii) The transition metals show variable valency whereas the alkaline earth elements do not.

CHAPTER 10

Solutions

A SOLUTION is a homogeneous dispersion of two or more substances. The substance dispersed is called the *solute* whilst the dispersion medium is the *solvent*. In true solutions the solute is dispersed as individual molecules or ions. Common examples of solute–solvent pairs are:

gas–liquid	e.g. air in water,
liquid–liquid	e.g. ethanol in water,
solid–liquid	e.g. salt in water.

Of the above three types, the most important are liquid–liquid and solid–liquid solutions, and in this chapter some of the more important properties of such solutions will be discussed.

10.1. Solid–liquid solutions—vapour pressure

When a non-volatile solute is dissolved in a suitable solvent the vapour pressure of the solvent is lowered. This is the result we would expect for in a solution the solvent molecules experience not only a mutual attraction for each other but also an attraction for the solute molecules. (If no such attraction existed then the solute would not have dissolved in the solvent in the first place.) As a result, the "escaping tendency" of the surface solvent molecules is lowered. On this basis, we could reasonably predict that the magnitude of the effect would depend on the solute concentration. This dependence is described by *Raoult's law* which states that *the relative lowering of the vapour pressure is equal to the mole fraction of solute in the solution.* Thus, if $P°$ and P are the vapour pressures of pure solvent and solution (both solvent and solution being at the same temperature), then

$$\frac{P^\circ - P}{P^\circ} = \frac{\Delta P}{P^\circ} = x_2$$

where $x_2 =$ mole fraction of solute,
$\Delta P =$ actual lowering of vapour pressure,
$\Delta P/P^\circ =$ relative lowering of vapour pressure.

A solution which accurately obeys Raoult's law at all concentrations is said to be an ideal solution. Such ideal behaviour will manifest itself provided the attractive interactions between the two different molecular species are of the same magnitude as those between the similar species. In reality, deviations from ideal behaviour can arise, but if the solution is dilute, such deviations are usually (but not always) sufficiently small to be ignored.

For a particular solvent at constant temperature, P° is a constant. Hence the magnitude of the vapour pressure lowering, ΔP, will depend exclusively on the concentration of the solute (i.e. the number of molecules or ions in a given amount of solvent) but not at all on the nature of the solute.

Now, although the value of ΔP is not influenced by the nature of the solute, its value is dependent on the nature of the solvent. Thus, for a given solute concentration the value of ΔP will depend on the vapour pressure of the pure solvent, i.e. $\Delta P \propto P^\circ$ when temperature and solute concentration x_2 are constant.

Properties of solutions which exhibit the above features are termed *colligative properties*.

10.2. Solid–liquid solutions—boiling point

A direct consequence of Raoult's law is that the boiling point of a pure liquid is elevated by the addition to it of a non-volatile solute. In Fig. 10.1 the vapour-pressure curve of solvent is compared with the corresponding curve for the solution. At temperature T_b, the vapour

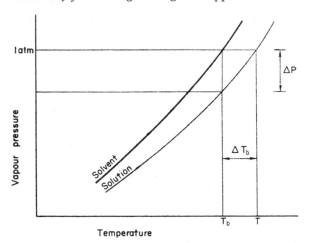

FIG. 10.1. Variation of vapour pressure with temperature for both
solvent and solution.

pressure of the pure solvent is 1 atm. Consequently, if the liquid is heated
under a constant pressure of 1 atm, it will boil at this temperature.
However, the solution at temperature T_b exerts a vapour pressure less
than 1 atm and if it is to be brought to the boil its temperature must be
raised to a value at which there is once more equality between vapour
pressure and external pressure. This condition is satisfied at temperature T.

Experimentally it is found that the elevation of the boiling point,
ΔT_b, is directly proportional to the lowering of the vapour pressure,
ΔP, which has already been shown to be proportional to the solute
concentration expressed by x_2.

Thus $\Delta T_b \propto x_2 = kx_2$.

It is customary to relate the boiling-point elevation to the solute
concentration in terms of *molality* (*m*) rather than mole fraction, where
the molality is defined as the number of moles of solute dissolved in
1000 g of solvent, i.e. *m* is expressed in mol kg^{-1}. This substitution of
molality for mole fraction is only justifiable in the case of a dilute solution.
To illustrate this, let us consider w_2g of solute dissolved in w_1g of solvent.

It follows, by simple proportion, that the weight of solute contained in 1000 g of solvent is $1000w_2/w_1$.

∴ Moles of solute in 1000 g of solvent $= \dfrac{1000w_2}{w_1M_2}$

where M_2 is the molar mass of solute.

This quantity is, by definition, the molality (m) of the solution, i.e.

$$m = \frac{1000w_2}{w_1M_2}$$

Multiplication of both sides of the above equation by M_1, the molar mass of the solvent, yields

$$mM_1 = \frac{1000w_2M_1}{w_1M_2} = \frac{1000n_2}{n_1}$$

where n_1 and n_2 are the moles of solvent and solute respectively.

If the solution is dilute, then

$$\frac{n_2}{n_1} \approx x_2, \quad \text{where} \quad x_2 = \frac{n_2}{n_1+n_2} \quad \text{and} \quad n_1 \gg n_2$$

∴ For a *dilute* solution $x_2 = \dfrac{mM_1}{1000}$

Substitution of this value in the equation

$$\Delta T_b = kx_2$$

yields

$$\Delta T_b = \frac{kM_1m}{1000}$$

As k and M_1 are constants characteristic of the solvent, they can be combined to give

$$\Delta T_b = K_b m, \quad \text{where} \quad K_b = \frac{kM_1}{1000}$$

This new constant K_b, referred to as the *molal elevation constant*, is equivalent to the boiling-point elevation of a one molal solution.

EXAMPLE 10.1. *Benzene has a normal boiling point of 80·2°C. When 3·0 g of a non-volatile substance of molecular weight 82 is dissolved in 100 g of benzene the solution is allowed to boil until a temperature of 81·7°C is reached. What weight of benzene is removed as vapour? K_b for benzene* = 2·53 K.

When the above solution boils, only benzene vapour is lost; therefore, as boiling continues the solution becomes progressively more concentrated and boils at a progressively increasing temperature, so that when $\Delta T_b = 1.5$ K,

$$m = \frac{\Delta T_b}{K_b} = \frac{1.5}{2.53} = 0.593 \text{ mol kg}^{-1}$$

∴   Solute concentration at 81·7°C

$$= 0.593 \text{ mole per 1000 g of benzene}$$
$$\equiv 0.593 \times 82 \text{ g per 1000 g of benzene}$$
$$= 3 \text{ g per 61·7 g of benzene}$$

∴   Wt. of solvent lost $= 100 - 61.7 = 38.3$ g

10.3.   Solid–liquid solutions—freezing point

Reference to Fig. 10.2 shows that the freezing point of a solution is depressed below that of the pure solvent. Such is always the case provided only the solvent crystallizes out from solution.

The depression ΔT_f is, like ΔT_b, found to be proportional to the lowering of vapour pressure.

Thus $\Delta T_f \propto \Delta P \propto x_2$ and $\Delta T_f = k'x_2$

where k' has, as before, a value specific to the solvent.

∴ $\Delta T_f = \dfrac{k'M_1}{1000} m = K_f m,$ where $K_f = \dfrac{k'M_1}{1000}$

The constant K_f is called the *molal depression constant*, and is equivalent to the freezing-point depression for a one molal solution.

Fɪɢ. 10.2. Influence of pressure on the freezing point of both solvent and solution.

Both boiling-point elevation and freezing-point depression are, like vapour-pressure lowering, colligative properties, in that the magnitude of the property is determined not by the nature of the solute but by its concentration, the latter being expressed in terms of the molality of the solution. Also, for a given solute concentration both ΔT_b and ΔT_f will depend on the value of K_b and K_f respectively, i.e. on the nature of the solvent.

EXAMPLE 10.2. *0·2 g of calcium chloride is dissolved in 100 g of water. At what temperature will the resulting solution freeze? K_f for water =* 1·86 K.

$$\text{Molality of calcium chloride} = \frac{\text{wt.}}{\text{mol. wt.}} \times \frac{1000}{100} = \frac{0·2}{111} \times \frac{1000}{100}$$

$$= 0·018 \text{ mol kg}^{-1}$$

At first sight we would expect the depression in the freezing point to be $1·86 \times 0·018 = 0·03348$ K. In fact, such a solution as the one above would give a depression substantially greater than $0·03348$ K. To understand why this should be so we must bear in mind that calcium chloride

12

is an ionic substance and the process of solution is one in which the water molecules separate ions already present in the solid state. Thus, for every 1·0 mole of $CaCl_2$ dissolved, there are present in solution 1·0 mole of Ca^{++} and 2·0 moles of Cl^-.

$$CaCl_2 \rightarrow Ca^{++} + 2Cl^-$$

Hence the *actual* molality of the solution

$$= \text{molality of } Ca^{++} + \text{molality of } Cl^-$$
$$= 0·018 + 0·036$$
$$= 0·054 \text{ mol kg}^{-1}$$

$\therefore \Delta T_f = 1·86 \times 0·054 = 0·1004 \text{ K } (= 3 \times 0·03348)$.
$\therefore$ The solution will freeze at $-0·1004°C$.

Even this answer is only approximate; in reality the observed depression would be slightly less than 0·1004 K on account of interionic attraction between the positive and negative ions. Such attraction prevents the ions from acting completely independently of each other. However, this effect is influenced by the distance separating the ions. Thus if a solution is of high concentration the ions will, on average, be near each other and interionic attraction will be strong. In a dilute solution, however, the ions will be separated by much larger distances; interionic effects will then be weak, and as a result, the calculated depression will be very nearly equal to the observed value.

10.4. Liquid–liquid solutions—vapour pressure

Up till now, the solute has been considered to be non-volatile, and our only concern has been the influence of solute on the vapour pressure of solvent; this was expressed by the equation

$$\frac{\Delta P}{P^\circ} = x_2$$

For our present purpose, it is convenient to alter the form of this relationship. Thus

$$\frac{\Delta P}{P^\circ} = \frac{P^\circ - P}{P^\circ} = 1 - \frac{P}{P^\circ} = x_2$$

But $x_1 + x_2 = 1$, where x_1 = mole fraction of solvent

$\therefore$ $$P = x_1 P^\circ$$

At constant temperature P° is a constant

$\therefore$ $$P \propto x_1$$

i.e. *the vapour pressure of solvent in solution is directly proportional to the mole fraction of solvent in the solution.* This statement is often quoted as an alternative form of Raoult's law.

If a solution is made up of two volatile components,[†] e.g. two liquids A and B, a complication is introduced, for we must now consider not only the influence that B has on the vapour pressure of A but also the effect that A has on the vapour pressure of B. This added complication is most easily dealt with if we imagine for the moment that liquid B is non-volatile. In this event the vapour pressure of liquid A in solution is related in the ideal case to the mole fraction of A by the equation:

$$P_A = x_A P_A^\circ$$

where P_A° = vapour pressure of pure A,
P_A = partial pressure of A in solution,
x_A = mole fraction of A in solution.

We must now establish the contribution made by liquid B to the total pressure. This is done by imagining liquid A to be the non-volatile component. Thus if

P_B° = vapour pressure of B,
P_B = partial pressure of B in solution,
x_B = mole fraction of B in solution,

then $P_B = x_B P_B^\circ$ in the ideal case.

Applying Dalton's law of partial pressures, it follows that

$$P_{total} = P_A + P_B = x_A P_A^\circ + x_B P_B^\circ$$

[†] For such a solution, the terms solute and solvent cease to have any very obvious meaning. In this event, the material present in excess is normally called the solvent and the other the solute.

12*

This equation reveals the way in which the total pressure varies with the composition of the solution. The relationship is made clearer when $1 - x_A$ is substituted for x_B.

Thus
$$P_{total} = x_A\, P_A + (1 - x_A)P_B^\circ$$
$$= (P_A^\circ - P_B^\circ)x_A + P_B^\circ$$

At constant temperature, P_A° and P_B° are themselves constant and so the equation in this form is the equation of a straight line when P_{total} is plotted against x_A (Fig. 10.3) with the gradient $= P_A^\circ - P_B^\circ$ and the intercept $= P_B^\circ$. This line is called the *liquid composition line* (L). Also shown in Fig. 10.3 are the individual contributions made to the total pressure by A and B separately. Thus, *for ideal liquid–liquid solutions,*

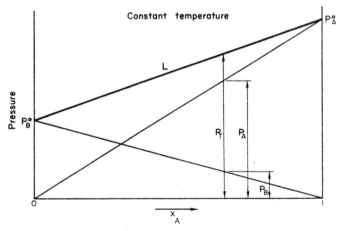

FIG. 10.3. Variation of vapour pressure with liquid composition.

the vapour pressure is always intermediate between the vapour pressures of the pure components.

If we continue our analysis of this type of system, it is possible to show that the vapour produced from and in equilibrium with an ideal liquid mixture is not of the same composition as the original liquid mixture but in fact is always richer in the more volatile component. This fact can be demonstrated in the following manner:

$$P_{total} = x_A P_A^\circ + x_B P_B^\circ$$

From Dalton's law $P_A = x'_A P_{total}$ where x'_A is the mole fraction of A in the *vapour*,

i.e.
$$P_{total} = \frac{P_A}{x'_A}$$

$\therefore$
$$\frac{P_A}{x'_A} = x_A P^\circ_A + x_B P^\circ_B$$

$\therefore$
$$\frac{x_A P^\circ_A}{x'_A} = x_A P^\circ_A + x_B P^\circ_B \qquad P_A = x_A P^\circ_A$$

$\therefore$
$$\frac{x_A \text{ (mole fraction of A in the liquid)}}{x'_A \text{ (mole fraction of A in the vapour)}} = x_A + x_B P^\circ_B / P^\circ_A$$

Now $x_A + x_B = 1$. In the event of P°_B being equal to P°_A (at one particular temperature), x_A will equal x'_A. Such a condition is not impossible but it is distinctly improbable. If, on the other hand, P°_A is greater than P°_B then x_A must be less than x'_A, i.e. *the vapour contains a higher proportion of A, the more volatile component, than does the original liquid mixture.*

EXAMPLE 10.3. *A solution of 0·6 mole fraction heptane and 0·4 mole fraction octane boils at 40°C at a pressure of 67·3 mmHg. At this temperature, the vapour pressure of pure heptane is 91·5 mm. What is the composition of the vapour initially produced from the boiling liquid?*

The liquid mixture boils at 40°C under a pressure of 67·3 mmHg. It therefore follows that the total pressure exerted by the solution at 40°C is also equal to 67·3 mmHg.

Ideally
$$P_{total} = x_H P^\circ_H + x_O P^\circ_O$$

where x_H and x_O are the mole fractions of heptane and octane in the solution, and P°_H and P°_O are the vapour pressures of pure heptane and octane respectively.

$$67·3 = 0·6 \times 91·5 + 0·4 P^\circ_O$$
$$P^\circ_O = 31 \text{ mm}$$

We now have all the information required to compute the initial vapour composition using the equation

$$\frac{x_H \text{ (mole fraction of heptane in the liquid)}}{x_H' \text{ (mole fraction of heptane in the vapour)}} = x_H + x_O \frac{P_O^o}{P_H^o}$$

$$\therefore \qquad x_H' = 0.8158$$

If values for P_A^o and P_B^o, x_A and x_B are known and substituted in the above equation, corresponding x_A' values can be obtained. These can then be plotted against the total pressure (Fig. 10.4) to yield a curve called the *vapour composition curve (V)*.

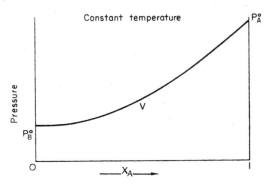

Fig. 10.4. Variation of vapour pressure with vapour composition.

The results expressed graphically in Figs. 10.3 and 10.4 are combined in Fig. 10.5.

10.5.   Liquid–liquid solutions—boiling point

Figure 10.5 represents the liquid–vapour equilibrium by means of a pressure–composition diagram. An alternative description can be used in which the boiling point is plotted against the composition (Fig. 10.6). Thus if A is the more volatile component, $P_A^o > P_B^o$ and $T_A < T_B$, where T_A and T_B are the boiling points of pure A and B. The lower curve (L) describes the liquid composition, while the upper curve (V) describes the vapour composition. Thus, a liquid mixture of composition

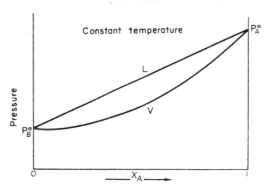

FIG. 10.5. Liquid and vapour composition curves.

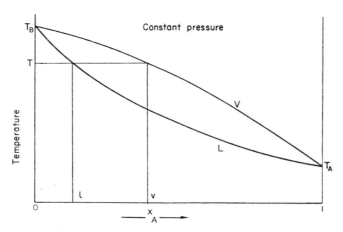

FIG. 10.6. Boiling point composition diagram.

l will boil at temperature T to produce an initial vapour of composition *v*. This vapour is seen to be richer in component A, the more volatile component.

The type of solution so far discussed is an ideal one, i.e. one in which each component is considered to obey Raoult's law at all concentrations. As with solid–liquid solutions, few liquid–liquid solutions behave completely ideally and very marked deviations from ideality are not at all uncommon. A lengthy discussion of such systems is beyond the scope of this book, although reference is made to non-ideal liquid–liquid systems in Chapter 17.

10.6. Liquid–liquid solutions—freezing point

In Section 10.3 we saw that the freezing-point depression could be thought of as being a consequence of vapour pressure lowering. While such a treatment is perfectly valid for solid–liquid solutions, it is in-applicable to liquid–liquid systems, which accordingly must be treated differently. However, it is found that, provided only one component crystallizes out from solution, the same relationship holds good, viz.

$$\Delta T_f = K_f m$$

10.7. Antifreezes

In the internal combustion engine, only a portion of the energy released in the combustion of gasoline is converted into power; the remainder is dissipated as heat. Part of this heat is removed in the exhaust gases whilst the remainder is transferred to the cooling system, in which (in most cars) water is caused to circulate through the engine block. Water, how-ever, freezing at 0°C, can easily "ice-up" during the winter months and cause the block to crack. To overcome this problem a variety of sub-stances called *antifreezes* can be added to the water to reduce its freezing point to a level below that of temperatures likely to be encountered during the winter months.

The two most popular antifreeze compounds in use at present are methyl alcohol and ethylene glycol, although ethyl alcohol, isopropyl alcohol, glycerol and propylene glycol have some properties which make them moderately suitable as antifreeze agents. On a weight to weight basis the efficiency of these materials, in their ability to lower the freezing point, is related to the molecular weight of the antifreeze.

Thus if $\qquad\qquad\qquad \Delta T_f = K_f m$

then $\qquad\qquad\qquad \Delta T_f = \dfrac{1000 K_f w_2}{M_2 w_1}$

Now if the ratio w_2/w_1 is made the same for each antifreeze in turn, then inspection of the above equation shows that the depression is inversely proportional to the molecular weight of antifreeze,

i.e. $$\Delta T_f \propto 1/M_2$$

As the relationship has been developed from equations which strictly apply only to ideal, dilute solutions, it should be applied with care. However, if two materials have markedly different molecular weights, it is possible to assess their relative efficiencies in functioning as anti-freezes. Thus, on this basis alone, methyl alcohol (mol. wt. = 32) is clearly a more efficient antifreeze than ethylene glycol (mol. wt. = 62) which suffers from the further disadvantage of being dearer than the very cheaply available methyl alcohol. However, it too has its disadvantages, viz. its greater tendency to give corrosive products, the toxicity of its fumes and its high volatility. Ethylene glycol in these respects is the more suitable antifreeze and overall is favoured more than methyl alcohol. Its low volatility is a particularly strong argument in favour of its use, for when a solution of ethylene glycol (b.p. 197°C) in water boils, it does so at a temperature in excess of 100°C, whereas a solution of methyl alcohol (b.p. 64°C) in water boils at a temperature intermediate between 64°C and 100°C. Now when a car engine is turned off, the engine transfers some of its heat to the antifreeze in the cooling system, thereby raising its temperature. If methyl alcohol is used, this increase in temperature may cause some of the antifreeze to "boil off". Loss of radiator fluid does not occur if a relatively non-volatile material such as ethylene glycol is used.

Irrespective of what antifreeze is used, extensive corrosion will occur in the cooling system. To counteract this, *inhibitors* (Section 8.11) are added to reduce the extent of corrosion to a tolerable level.

EXAMPLE 10.4. *Calculate the freezing point of a methyl alcohol–water solution containing 15% by weight of methyl alcohol. What weight per cent of ethylene glycol would be required to give the same depression in the freezing point? K_f for water = 1·86 K.*

Every 100 g of this solution contains 15 g of methyl alcohol and 85 g of water.

$$\therefore \qquad \text{Molality of solution} = \frac{15 \times 1000}{32 \times 85} = 5 \cdot 516 \text{ mol kg}^{-1}$$

$$\therefore \qquad \Delta T_f = 1 \cdot 86 \times 5 \cdot 516 = 10 \cdot 26 \text{ K}$$

$$\therefore \qquad \text{Freezing point of solution} = -10 \cdot 26°\text{C}$$

If ethylene glycol is to afford the same protection, the molality of the ethylene glycol–water solution must be as before, viz. $5 \cdot 516$ mol kg^{-1},

i.e.   Ethylene glycol concentration

$$= 5 \cdot 516 \text{ moles per 1000 g of water}$$
$$\equiv 5 \cdot 516 \times 62 \text{ g per 1000 g of water}$$
$$= 342 \cdot 1 \text{ g per 1000 g of water}$$

$$\text{Per cent by weight of ethylene glycol} = \frac{342 \cdot 1}{1342 \cdot 1} \times 100$$

$$= 25 \cdot 49\%$$

EXAMPLE 10.5. *What weight of ice would crystallize out if 100 g of the ethylene glycol solution of Example 10.4 was cooled to* $-15°C$?

When the above solution is cooled it begins to freeze at $-10 \cdot 26°$C, i.e. ice crystals appear for the first time at this temperature. As the temperature is further reduced more ice separates out, the solution becomes more concentrated and the freezing point is progressively depressed. It is thus seen that an aqueous solution, unlike pure water, does not freeze solid when its temperature is reduced below its freezing point; instead it becomes progressively more concentrated as more and more ice separates out. Now as the solid phase separating out is pure water it follows that, as the temperature drops from $-10 \cdot 26°$C to $-15°$C, the solution must throughout contain the initial amount of ethylene glycol. By comparing the initial glycol concentration with the glycol concentration of the solution at $-15°$C we can deduce the weight of ice separating out. Thus:

Initial glycol concentration $= 25 \cdot 49$ g per $74 \cdot 51$ g of water

When $\Delta T = 15$ K, $m = \dfrac{15}{1 \cdot 86} = 8 \cdot 065$ mol kg^{-1}

$\therefore$ Glycol concentration at $-15°C$

$$= 8 \cdot 065 \text{ moles per } 1000 \text{ g of water}$$
$$\equiv 500 \text{ g per } 1000 \text{ g of water}$$
$$= 25 \cdot 49 \text{ g per } 51 \cdot 98 \text{ g of water}$$

$\therefore$ Wt. of ice separating out $= 74 \cdot 51 - 51 \cdot 98$
$$= 22 \cdot 53 \text{ g}$$

Ultimately a stage is reached when the system is entirely solid but this occurs at a temperature well below that at which ice crystals first appear. It is thus seen that the addition of an antifreeze to the water of the cooling system provides a high degree of protection against "icing-up".

Exercises

10.1. The vapour pressure of pure water at 25°C is $23 \cdot 8$ mmHg. Calculate the vapour pressure of an aqueous solution which contains $20 \cdot 0$ g sucrose ($C_{12}H_{22}O_{11}$) in $50 \cdot 0$ g water.

[*Ans.*: $23 \cdot 3$ mmHg.]

10.2. Calculate the boiling point and freezing point of the above sucrose solution.

$$K_b(H_2O) = 0 \cdot 513 \text{ K}; \qquad K_f(H_2O) = 1 \cdot 86 \text{ K}.$$

[*Ans.*: $100 \cdot 60°C$; $-2 \cdot 18°C$.]

10.3. What weight of ethylene glycol ($C_2H_4(OH)_2$) must be added to $10 \cdot 0$ kg of water to prevent the formation of ice crystals down to $-10 \cdot 0°C$?

$$K_f(H_2O) = 1 \cdot 86 \text{ K}.$$

[*Ans.*: $3 \cdot 33$ kg.]

10.4. Camphor has a very high molal depression constant ($K_f =$ 40·0 K). Consequently it is a particularly good solvent for the determination of molecular weight by freezing-point depression. By way of illustration 0·442 g of an unknown organic compound dissolved in 0·9120 g of camphor depressed the freezing point of camphor by 12·6°C. Determine the molar mass and hence the molecular weight of the compound.

[*Ans.*: $M = 154$ g mol; $M_r = 154.$]

10.5. Liquid oxygen and liquid nitrogen form ideal solutions. Calculate the total vapour pressure at $-200°C$ of a solution whose concentration is 30·0 mole % oxygen. Proceed to determine the composition of vapour in equilibrium with this solution.

The vapour pressures of pure oxygen and nitrogen at $-200°C$ are 11·33 kPa and 59·98 kPa respectively.

($1 \text{ N m}^{-2} = 1$ Pa (Pascal).)

[*Ans.*: 45·4 kPa; 7·5 mole % O_2.]

CHAPTER 11

Colloids

WE HAVE now discussed the solid, liquid and gaseous states of matter and have an understanding of the properties of these states and how these properties are related to the nature of the atoms and molecules involved. We also know that a homogeneous dispersion at molecular level of one material in another is called a *solution*. This contrasts with a *suspension* (such as fine mud in water) where relatively large particles are dispersed throughout a second material. The two materials can in this case be relatively easily separated by gravity (i.e. by allowing the large particles to settle out) or by simple filtration.

Between these two extreme types of dispersion are instances where the dispersed particles are significantly larger than typical molecular size but yet are too small to be readily visible. Such systems are called colloids, and their properties are quite different from the states of matter we have considered so far.

11.1. Colloidal dimensions

The term colloid was first used by Graham in the mid-nineteenth century to describe certain substances such as glue or albumin which in solution did not diffuse through a parchment membrane nearly so readily as did other dissolved materials such as salts or sugar. In fact we now know that it is wrong to classify substances as "colloidal" or "non-colloidal" since most materials can if treated correctly be made to exhibit colloidal properties. The colloidal state is a function of the size of the particles in question rather than of their chemical formulae.

We know that typically molecules have diameters of around 1×10^{-10} m. Similarly, in an ordinary suspension the particle size will be somewhat greater than 1×10^{-6} m. However, in the range of particle sizes lying between these two extremes we find the colloidal state. It should perhaps be pointed out that within that size range we find particles which are aggregates of many molecules but also some which are single molecules, i.e. polymers.

The essential feature then of the colloidal state is the size of the particles involved, and the most important characteristic of any material dispersed in this way is that the surface area is extremely large. If, for example, we consider a cube of a solid material of side 1 cm, then the surface area is 6 cm². If, however, that cube is subdivided into much smaller cubes each of side 1×10^{-7} cm, then we will have produced 10^{21} cubes each with a surface area of 6×10^{-14} cm², i.e. a total surface area of 6×10^{7} cm². In fact the important properties of the colloidal state result from this very large surface area.

11.2. Optical properties

One of the obvious questions we must consider is how to observe colloidal particles. Individual particles in a normal suspension may be seen with the naked eye, or at the very least by means of a microscope. Particles of colloidal size are too small to be observed in this way. They may, however, be observed by means of the *ultramicroscope*, which is an instrument that allows the object to be observed at right angles to the illuminating beam. Although the particles are too small to be observed directly they may be large enough to scatter some of the incident light, and this is what the observer actually sees. The same effect can be observed when very small dust particles, normally invisible, are readily seen in a beam of sunlight.

FIG. 11.1. The Tyndall effect.

An extension of this forms the basis of one of the simplest tests which will distinguish between a true solution and a colloid. This is the *Tyndall effect*, which is illustrated in Fig. 11.1.

When a beam of light is passed through a true homogeneous solution the path of the beam is not visible since the dissolved particles are too small to scatter the incident light. However, when the same beam is passed through a colloidal solution the light is scattered by the colloidal particles and the beam is clearly visible when viewed from the side.

Techniques of the type described give useful information on the size of colloidal particles and further allow the movement of single particles to be followed. When observed in this way colloidal particles are seen to exhibit *Brownian motion* in a fashion analogous to that described in Chapter 4 on gases (Section 4.1).

11.3. Classification of colloids

Before considering in more detail the characteristics of colloidal systems it will be helpful to define a number of terms commonly used and to classify colloids into a number of common types. Colloidal systems contain particles of one material (the *disperse phase*) dispersed throughout another (the *dispersion medium* or *continuous phase*). Table 11.1 indicates how colloids may be classified in terms of the disperse phase and the dispersion medium.

A further classification of colloids in which the dispersion medium is a liquid is on the basis of the extent of the attraction between the disperse

TABLE 11.1. CLASSIFICATION OF COLLOIDS

Disperse phase	Dispersion medium	Name	Example
Solid	Gas	Aerosol	Smoke
Solid	Liquid	Sol	Metal in water
Liquid	Gas	Aerosol	Mist
Liquid	Liquid	Emulsion	Milk
Gas	Liquid	Foam	Whipped cream

phase and the dispersion medium. On this basis we can define colloids as *lyophilic*, where there is a high degree of compatibility between the two phases, and as *lyophobic*, where there is little or no attraction between the two.

11.4. Preparation of colloids

There are two general methods for the preparation of colloids—by *dispersal* and by *condensation*. In dispersal methods the starting material is in the bulk state and is progressively broken down into smaller and smaller particles until a colloidal size is attained. All such methods involve significant expenditure of energy. The simplest mechanical method is grinding the disperse phase and dispersion medium together in an agate mortar, but this is suitable only for small-scale laboratory preparations. Commercially a *colloid mill* is used, where the ingredients are subjected to shearing rather than grinding forces between a rotating cone and a closely fitting tapered annulus.

Another method, known as the *Bredig arc process*, involves the dispersal of metals in liquids by striking a high frequency arc between two metal rods immersed in the liquid-dispersion medium. This method is widely used in the preparation of platinum, silver and gold sols.

A somewhat different process is that of *peptization*, which refers to the dispersal of one material by the addition of an electrolyte. In general the presence of a common ion encourages peptization, provided that the total concentration of electrolyte is not too high. For example, alkali hydroxides peptize insoluble metal hydroxides, and a small excess of chloride ions can peptize freshly precipitated silver chloride.

In condensation methods the material from which the sol is to be prepared is present initially as a true solution, and as a result of changing the physical or chemical environment insoluble particles of colloidal size are produced. The size of the particle produced in a process of this kind depends on the relative rates of nucleation and of crystal growth. It is important to obtain a high rate of nucleation if very small particles are to be produced, and the experimental conditions must be carefully controlled.

An example of this process is the formation of colloidal As_2S_3 in very dilute solution by the process:

$$As_2O_3 + 3H_2S \rightarrow As_2S_3 + 3H_2O$$

Another example is the hydrolysis of ferric chloride to form a reddish-brown colloidal solution of ferric hydroxide:

$$FeCl_3 + 3H_2O \rightarrow Fe(OH)_3 + 3HCl$$

An interesting feature of this example is that the chemical reaction which produces the colloid at the same time generates an impurity (in this case HCl) which renders the colloidal system unstable. This impurity can be removed by the process of *dialysis*, where the impure colloid is placed in a container and separated from pure solvent by a suitable membrane made of cellophane or parchment. The very small ions forming the impurity can diffuse through the membrane into the solvent, whereas the larger colloidal particles cannot. A more sophisticated version of this technique is *electrodialysis* where the purification process is greatly accelerated by applying a potential difference between two compartments containing solvent placed on either side of the parchment vessel containing the colloid, as illustrated in Fig. 11.2. When the circuit

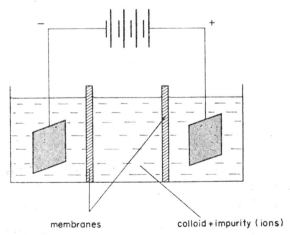

membranes   colloid + impurity (ions)

Fig. 11.2. Electrodialysis cell.

is closed, positively charged ions in the central compartment diffuse through the membrane towards the negative electrode, while negatively charged ions diffuse through the membrane towards the positive electrode.

11.5.   Electrical properties of colloids

Before we consider in detail the stability of colloidal systems it will be useful to consider briefly one of the electrical characteristics of colloids, and this is that particles of the disperse phase are found to be charged with respect to the dispersion medium although the system as a whole is electrically neutral. The point has already been made that a characteristic of colloidal systems is that the surface area of the disperse phase is extremely high. We also know (see Chapters 5 and 6) that forces of attraction exist at any surface, and so it will be readily accepted that a colloid will have a strong tendency to adsorb[†] other material—normally ions—from the immediate surroundings.

Thus Fig. 11.3 illustrates a colloidal particle on whose surface has been adsorbed a fixed layer of positive ions. In turn this positively charged

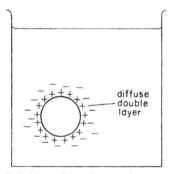

diffuse double layer

FIG. 11.3. Colloid particle with duplicate double layer.

species will attract a secondary, more diffuse, layer of negatively charged ions, but the particle will still possess a charge relative to the bulk of the solution surrounding it. This charge resulting from the *diffuse double layer*

---

† Adsorption is the term used to describe the adhesion of molecules on the *surface* of a solid. Any increase in surface area will lead to an increase in adsorption.

is responsible for the stability of many colloids, since the particles are prevented from coalescing to form conventional suspensions by the repulsive forces between the charged particles. The net charge may be either positive or negative, and there is a tendency for colloids to preferentially adsorb from solution ions which are common to them. For example, AgI in NaI solution will adsorb I^- ions, giving a negative particle, whereas AgI in $AgNO_3$ solution will adsorb Ag^+ ions, giving a positive particle.

11.6. Stability of colloidal systems

Lyophobic sols

These may be prepared from most inorganic materials and their stability depends on the existence of adsorbed charged particles as described in the previous section. The ions involved in the diffuse double layer usually come from dissolved electrolytes, and this gives rise to two methods of bringing about the coagulation of lyophobic sols—either by removing electrolyte or by adding electrolyte.

We have already learned of the process of dialysis whereby excess electrolyte can be removed from a colloidal system to effect purification. If, however, this process is overdone and too much of the electrolyte is removed, then the adsorbed electrical layer begins to break down and coagulation results.

The same effect can be achieved by the addition of electrolyte. Table 11.2 shows the concentration of various electrolytes necessary to bring about coagulation of different colloids.

TABLE 11.2. COAGULATION OF COLLOIDS

As_2S_3 sol (negatively charged)		Fe(OH)$_3$ sol (positively charged)	
NaCl	55 mmole dm^{-3}	KCl	60 mmole dm^{-3}
MgCl$_2$	0·7 mmole dm^{-3}	K$_2$SO$_4$	0·12 mmole dm^{-3}
Al Cl$_3$	0·09 mmole dm^{-3}	K$_3$Fe(CN)$_6$	0·05 mmole dm^{-3}

It is immediately obvious that the effective species is the ion whose charge is of opposite sign to that of the colloidal particles. Further, divalent ions are more effective than monovalent, and trivalent more effective than divalent. It appears, therefore, that a small excess of electrolyte will neutralize the ionic atmosphere around each colloid particle, thus allowing these particles to coalesce on collision.

A further example of this effect is seen in mutual precipitation, where the addition of a positive to a negative colloid results in the precipitation of both.

Lyophilic sols

Commonly these are naturally occurring materials, e.g. starch and proteins, of high molecular weight, and have inherently a much higher stability than have lyophobic sols. As their name implies, they form strong interactions with the solvent, and it is this solvation effect which stabilizes the colloidal particles. The extent of this interaction with the solvent can be demonstrated by a study of the relative viscosity of colloidal systems. Typically, the viscosity of a lyophobic colloidal solution is not markedly different from that of the pure solvent, but lyophilic colloidal solutions have abnormally high viscosities.

Lyophilic sols are much less susceptible to precipitation by electrolytes, although coagulation can be brought about in this way. The mechanism is not simply the breakdown of the electrical double layer, but involves removing the solvation layer from the particles. Thus we find that the effectiveness of ions is in the order:

$$Mg^{2+} > Ca^{2+} > Ba^{2+} > Li^+ > Na^+ > K^+$$

and it is significant that the Mg^{2+} ion is very strongly hydrated; that is, has a high affinity for water molecules.

Emulsions

The term *emulsion* may strictly be applied to any colloidal system composed of one liquid dispersed in another, but in practice it usually refers to oil–water systems. As we found with lyophobic sols, the presence of other components is necessary to achieve stability. In the case of emulsions these are called *emulsifying agents* and are present in relatively large concentrations (normally (1–5%).

Emulsions can be further classified into two types: oil-in-water (O/W) or water-in-oil (W/O), depending on whether the aqueous or the oil component forms the continuous phase. Which of these is actually formed depends on a number of factors, e.g. the relative amounts of each component, the nature of the emulsifying agent and even in some cases on the order of mixing. The conversion of one type to the other is called *inversion*. Thus cream is an aqueous emulsion containing about 30% fat, stabilized by the protein casein. Vigorous agitation of cream causes the emulsion to break down and the fat particles coalesce to form butter where the fat now forms the continuous phase and the system holds about 15% of water dispersed throughout the butter fat. That is, inversion has occurred.

In discussing the stability of emulsions we must distinguish between "creaming" and coagulation. Creaming is the accumulation of droplets at the top of the emulsion due to density differences. The most common example is, of course, creaming of fresh milk. Since the tendency to cream is a function of particle size (large-particle emulsions being more susceptible), stability can be enhanced by breaking down the particles still further. Thus milk is frequently homogenized to make the emulsion structure more stable.

As has already been explained, the general stability of any emulsion system depends on the action of emulsifying agents. For W/O emulsions typical emulsifying agents are the heavy metal salts of fatty acids (e.g. zinc stearate) or long-chain alcohols (e.g. cetyl alcohol). For O/W emulsions the common stabilizing agents are proteins, gums, and natural and synthetic soaps. The mechanism of this type of emulsification is fully discussed under the heading "Detergent action" in Chapter 29.

11.7. Miscellaneous applications

All paints are examples of colloidal systems. The opaque films are achieved by the incorporation of either light-scattering pigments such as TiO_2 or light-absorbing pigments such as carbon black and coloured pigments. Broadly paints are of two basic types—aqueous-based or oil-based.

Aqueous-based paints (commonly called "emulsion" paints) are stabilized dispersions of polymer particles and pigments. After application, the water forming the dispersion medium evaporates off and the polymer particles coagulate to form the desired continuous film containing dispersal pigment.

Oil-based paints consist of pigments dispersed in a polymer solution containing a drying oil alkyd resin (see Chapter 28) dissolved in an aliphatic hydrocarbon solvent. The alkyd molecule, which possesses both polar and non-polar sites, at the same time acts as an absorber for the pigment and as a stabilizer, because of its affinity for the solvent.

Other areas where there are many examples of the use of colloids are the pharmaceutical and cosmetic industries. Modification of the physical properties by using preparations in colloidal dispersions can have many advantages. For example, in order that the active ingredient may be quickly absorbed by the body it is administered in as finely a dispersed state as possible, either in solution or as a sol. When the active ingredient is a non-polar organic material, including it as a component of a colloid is a useful way of diluting the material and does, for example, help to mask the unpleasant taste of the oily material. Obviously in these cases care must be taken to use stabilizers which are non-toxic.

Another interesting application of the principles of colloid chemistry is found in the use of materials in the form of *pastes*, which may be regarded as highly concentrated suspensions. Many examples readily come to mind, e.g. putty, clays, dough, grinding and cleaning pastes. These materials are all characterized by the property of plasticity or mouldability, and their behaviour is determined by the concentration of the material, by the size range of the solid particles and by the general stability of the colloid system.

Some of these pastes possess a property known as *thixotropy*, where on being physically disturbed the semi-solid system breaks down and becomes fluid, but returns to its original condition once the applied stress is removed. A well-known example of this is quicksand. Again there are many ways of using this property to advantage in the field of technology.

For example, in thioxotropic paints the paint flows easily under the action of a brush but "sets" quickly thereafter; in "drilling muds" the mud is used to prevent the debris produced by a drill from falling to the bottom and jamming the drill itself when the boring action has ceased.

Exercises

11.1. Define the following terms:

    (i) Tyndall effect;
    (ii) aerosol;
    (iii) dispersion medium;
    (iv) Brownian motion;
    (v) lyophilic sol;
    (vi) dialysis;
    (vii) thixotropic gel;
    (viii) inversion.

11.2. Compare the properties of lyophilic and lyophobic sols and account for any notable differences.

11.3. In many rivers the water contains a high concentration of suspended material much of which is of colloidal dimensions. State how this may in part explain the formation of river deltas where the rivers flow into the sea.

11.4. Summarize the methods available for the preparation of colloids.

CHAPTER 12

Thermochemistry

12.1. Energy of a chemical system

The energy content of any system is partly kinetic, partly potential. A gas, for example, will possess kinetic energy by virtue of the chaotic motion of the molecules (the amount of kinetic energy being dependent on the temperature); and it will possess potential energy as a result of a particular arrangement of protons, neutrons and electrons within the molecules (this arrangement being specific to a particular substance). In short, its energy content will depend on the state of the system, i.e. on the temperature, pressure, volume, amount and nature of the gas in question. The same is true for liquids and solids.

A measure of this energy content is a quantity referred to as the *enthalpy* (H) of the system.[†] This quantity, whose value is a function of the state of the system, is sensibly called a *state function*.

There are two important consequences of the fact that enthalpy is a state function:

(i) its value will not depend on the way in which the state has been reached;

(ii) when the system experiences some change in its state (e.g. as a result of reaction occurring when reactant material is replaced by product material), the change in enthalpy (ΔH) will depend only on the initial and final states of the system,

i.e. $$\Delta H = H_{\text{final}} - H_{\text{initial}}$$

[†] An absolute value for H cannot be obtained. Fortunately, however, this is no cause for concern, for the chemist is interested primarily in a change in energy content, i.e. a change in enthalpy.

A system which provides a useful analogy to explain the nature of a state function is illustrated in Fig. 12.1, where a ball is pictured as lowering its position by x cm, in the first case by rolling down an incline, in the second case by dropping vertically. In each case, the change in potential energy is the same, i.e. the potential energy of the ball depends only on its position, or *state*, and not on the route by which it has reached that position. The potential energy of the ball is a state function, and any

FIG. 12.1. The energy of a ball as a function of its position.

change in this quantity will be given simply by the difference between the initial and final values of the function,

i.e.
$$\Delta PE = PE_{final} - PE_{initial}$$

Both the distance travelled and the time taken for the ball to change its position are route dependent—these are not state functions.

12.2. Exothermic and endothermic reactions

Consider the general reaction

$$aA + bB \rightarrow cC + dD$$

to take place under conditions of constant pressure and temperature. Then we can say that

$$\text{enthalpy of reactants} = aH_A + bH_B = H_{reactants} = H_R$$
and
$$\text{enthalpy of products} = cH_C + aH_D = H_{products} = H_P$$

where H_A, H_B, H_C, H_D are the enthalpies per mole of materials A, B, C and D.

$\therefore$ Change in enthalpy $= \Delta H = cH_C + dH_D - (aH_A + bH_B)$
$$= H_P - H_R$$

Suppose now that H_R is greater than H_P by an amount x J. The principle of energy conservation demands that the *total* energy of the system remain constant. An energy balance would therefore require that

$$H_R = H_P + x$$

The quantity x J is energy liberated as heat.

Now $H_P - H_R = \Delta H = -x$ J

Hence a change in enthalpy is seen to be identical in magnitude to the quantity of heat liberated under the conditions of constant pressure and temperature. It is, however, of opposite algebraic sign. For this reason, the heat change accompanying reaction can be expressed in one of two ways:

(i) $aA + bB \rightarrow cC + dD + x$ J

or

(ii) $aA + bB \rightarrow cC + dD$ $\Delta H = -x$ J

The convention in chemistry (which will be used in this text) is to view heat changes from the point of view of the reacting system. A system which loses heat is therefore given a negative value of ΔH, and vice versa. Reactions characterized by a negative value of ΔH are described as being *exothermic*. Conversely, reactions for which ΔH is positive ($H_P > H_R$) are accompanied by an absorption of heat and are referred to as being *endothermic*. For example, if the above reaction could be reversed, x J would be absorbed and ΔH would equal $+x$ J

$$cC + dD \rightarrow aA + bB - x \text{ J} \qquad \Delta H = x \text{ J}$$

We thus see that any chemical reaction which is exothermic in a forward direction will be endothermic to precisely the same extent if reversed.

12.3. Heat of formation

It is of considerable interest to the chemist to know the degree to which a reaction is exothermic or endothermic. This requires a knowledge of ΔH. For a number of reasons it is not always possible to obtain experimentally a value of ΔH; indeed it would be extremely laborious to perform an experiment each time a ΔH value was required. One method of calculating ΔH which might occur to the reader would be to use the equation

$$\Delta H = H_P - H_R$$

If the enthalpies of all the materials involved in the reaction were in fact known, it would be no more than a simple arithmetical exercise to calculate ΔH. However, as absolute values of enthalpies are not known, an alternative approach which makes use of *heats of formation* is used.

The heat of formation of a compound is the change in enthalpy accompanying the formation of one mole of the compound from its constituent elements in their standard states in which they usually occur under normal conditions of pressure and temperature. It is given the symbol ΔH_f, and is measured at 25°C (298 K) and 1 atm.

To avoid ambiguity it is wise in thermochemical equations to use the letters s, l, or g to denote the physical state of each substance. Thus the heat of formation of liquid water is indicated by

$$H_2(g) + \tfrac{1}{2}O_2(g) \rightarrow H_2O(l) \qquad \Delta H_f = -285 \cdot 8 \text{ kJ}$$

while the heat of formation of water vapour is indicated by

$$H_2(g) + \tfrac{1}{2}O_2(g) \rightarrow H_2O(g) \qquad \Delta H_f = -241 \cdot 8 \text{ kJ}$$

As a consequence of the way in which heat of formation is defined, it follows that the heat of formation of elements in their standard states is zero. For example, the heat of formation of gaseous oxygen is, by definition, the enthalpy change accompanying the formation of one mole of gaseous oxygen from the state in which it normally occurs, viz. the gaseous state,

i.e. $$\Delta H_{f(O_2,\ g)} = 0$$

Generalizing, $\Delta H_{f(\text{elements in their standard states})} = 0$

Table 12.1 lists the heats of formation of a variety of compounds at 298 K.

TABLE 12.1. HEATS OF FORMATION, ΔH_f, kJ mol^{-1}
AT 298 K

$H_2O(l)$	$-285\cdot8$	$C_2H_4(g)$	$52\cdot3$
$H_2O(g)$	$-241\cdot8$	$CH_3OH(l)$	$-238\cdot7$
$CO(g)$	$-110\cdot5$	$C_2H_5OH(l)$	$-277\cdot6$
$CO_2(g)$	$-393\cdot5$	$SO_2(g)$	$-296\cdot8$
$C(graphite)$	0	$SO_3(g)$	$-395\cdot2$
$C(diamond)$	$1\cdot9$	$HCl(g)$	$-92\cdot3$
$CH_4(g)$	$-74\cdot8$	$HBr(g)$	$-36\cdot4$
$C_2H_6(g)$	$-84\cdot7$	$HI(g)$	$26\cdot5$

12.4.   Heat of reaction

In what way are the heats of formation of compounds involved in a chemical reaction related to the overall heat of reaction?

Consider the following reaction in which carbon disulphide is burned in oxygen to yield carbon dioxide and sulphur dioxide. Let the enthalpy change be ΔH_1.

$$CS_2(l) + 3O_2(g) \rightarrow CO_2(g) + 2SO_2(g) \qquad \Delta H_1 = -1076\cdot8 \text{ kJ}$$

Imagine now the reaction to proceed by an alternative route. Let the first step be the conversion of reactants to their elements in their standard states.

$$CS_2(l) \rightarrow C(s) + 2S(s) \qquad \Delta H_2 = -89\cdot7 \text{ kJ}$$

where

$$\Delta H_2 = H_{(C)} + 2H_{(S)} - H_{(CS_2)} = -\Delta H_{f(CS_2)}$$

Let the second step be recombination of the elements to form products:

$$C(s) + O_2(g) \rightarrow CO_2(g) \qquad \Delta H_3 = -393\cdot5 \text{ kJ}$$

where

$$\Delta H_3 = \Delta H_{f(CO_2)}$$

$$2S(s) + 2O_2(g) \rightarrow 2SO_2(g) \qquad \Delta H_4 = -593\cdot6 \text{ kJ}$$

where

$$\Delta H_4 = 2 \Delta H_{f(SO_2)}$$

The overall enthalpy change $= \Delta H_2 + \Delta H_3 + \Delta H_4 = -1076\cdot8 \text{ kJ}$

Thus the overall enthalpy change for the above sequence of reactions is seen to be equivalent to the enthalpy change for the direct reaction between carbon disulphide and oxygen. Such an equality follows as a consequence of enthalpy being a state function.

Thus

$$\Delta H_1 = \Delta H_2 + \Delta H_3 + \Delta H_4$$
$$= -\Delta H_{f(CS_2)} + \Delta H_{f(CO_2)} + 2\,\Delta H_{f(SO_2)}$$
$$= \Delta H_{f(CO_2)} + 2\,\Delta H_{f(SO_2)} - \Delta H_{f(CS_2)}$$
$$= \Delta H_{f(products)} - \Delta H_{f(reactants)} \qquad (\Delta H_{f(O_2,\,g)} = 0)$$

The reaction considered is one of combustion; the enthalpy change in a combustion reaction is referred to as the *heat of combustion*, and is defined as *the enthalpy change accompanying the complete oxidation of one mole of a compound.*

For *any* reaction the overall enthalpy change ΔH is related to the heats of formation of the various compounds involved by the simple equation

$$\text{heat of reaction,} \quad \Delta H = \Delta H_{f(products)} - \Delta H_{f(reactants)}$$

This important relationship may be used in one of two ways.

(i) Knowing the heats of formation, the heat of reaction can be rapidly calculated, thus avoiding needless experimentation.

EXAMPLE 12.1. *The hydrocarbon 1,3-butadiene, used in the manufacture of synthetic rubber (Section 28.9), is made by the dehydrogenation of 1-butene:*

$$CH_2:CH \cdot CH_2 \cdot CH_3(g) \rightarrow CH_2:CH \cdot CH:CH_2(g) + H_2(g)$$

Determine the enthalpy change (at 25°C) in this reaction. The heats of combustion of 1-butene and 1,3-butadiene are, respectively, $-2716 \cdot 2$ and $-2541 \cdot 0\ kJ\ mol^{-1}$ at 25°C. The heat of formation of $H_2O(l)$ is $-285 \cdot 8\ kJ\ mol^{-1}$ at 25°C.

For the dehydrogenation

$$\Delta H = \Delta H_{f(products)} - \Delta H_{f(reactants)}$$
$$= \Delta H_{f(H_2)} + \Delta H_{f(1,3\text{-butadiene})} - \Delta H_{f(1\text{-butene})} \qquad (1)$$

Values for the respective heats of formation have not been given—this is not uncommon. They can, however, be deduced, in this case from information regarding heats of combustion, as follows.

The combustion of 1-butene (a) and 1,3-butadiene (b) yields as products carbon dioxide and water.

(a) $CH_2 : CH \cdot CH_2 \cdot CH_3 + 6O_2 \rightarrow 4CO_2 + 4H_2O(l)$

$$\Delta H_1 = -2716 \cdot 2 \text{ kJ}$$

Now $\Delta H_1 = 4\,\Delta H_{f(CO_2)} + 4\,\Delta H_{f(H_2O, l)} - \Delta H_{f(1\text{-butene})} - 6\,\Delta H_{f(O_2, g)}$

$\therefore \quad \Delta H_{f(1\text{-butene})} = 4\,\Delta H_{f(CO_2)} + 4\,\Delta H_{f(H_2O, l)} - \Delta H_1 \quad (\Delta H_{f(O_2, g)} = 0)$

(b) $CH_2 : CH \cdot CH : CH_2 + \frac{11}{2}O_2 \rightarrow 4CO_2 + 3H_2O(l)$

$$\Delta H_2 = -2541 \cdot 0 \text{ kJ}$$

Now $\Delta H_2 = 4\,\Delta H_{f(CO_2)} + 3\,\Delta H_{f(H_2O, l)} - \Delta H_{f(1,3\text{-butadiene})}$

$\therefore \quad \Delta H_{f(1,3\text{-butadiene})} = 4\,\Delta H_{f(CO_2)} + 3\,\Delta H_{f(H_2O, l)} - \Delta H_2$

Substitution of the two ΔH_f values in eqn. (1) gives

$$\Delta H = \Delta H_1 - \Delta H_2 - \Delta H_{f(H_2O, l)} \quad (\Delta H_{f(H_2, g)} = 0)$$
$$= -2716 \cdot 2 + 2541 \cdot 0 + 285 \cdot 8$$
$$= 110 \cdot 6 \text{ kJ}$$

(ii) Knowing the heat of reaction, the heat of formation of one of the compounds can be determined. This is the way in which the heats of formation of most organic compounds are obtained; the values simply cannot be obtained by a direct route in which the constituent elements combine to form the compound.

EXAMPLE 12.2. *For acetone, the heat of combustion is* $-1779 \cdot 4$ *kJ mol*$^{-1}$ *at 25°C. Calculate the heat of formation of acetone at 25°C.*

$$CH_3 \cdot CO \cdot CH_3(l) + 4O_2(g) \rightarrow 3CO_2(g) + 3H_2O(l)$$
$$\Delta H = -1779 \cdot 4 \text{ kJ mol}^{-1}$$

$\therefore \quad \Delta H = 3\,\Delta H_{f(CO_2)} + 3\,\Delta H_{f(H_2O, l)} - \Delta H_{f(CH_3 \cdot CO \cdot CH_3)} \quad (\Delta H_{f(O_2, g)} = 0)$

$\therefore \qquad -1779 \cdot 4 = -1180 \cdot 5 - 857 \cdot 4 - \Delta H_{f(CH_3 \cdot CO \cdot CH_3)}$

$\therefore \qquad \Delta H_{f(CH_3 \cdot CO \cdot CH_3)} = -258 \cdot 5 \text{ kJ mol}^{-1}$

12.5. Fuels: water gas, producer gas, semi-water gas

When steam is blown through a bed of hot coke, *water gas*, essentially a mixture of carbon monoxide and hydrogen, is formed.

$$C(s) + H_2O(g) \rightarrow CO(g) + H_2(g) \qquad \Delta H = 131 \text{ kJ}$$

This mixture is of medium calorific value[†] and is a valuable fuel. Because of the endothermic character of the reaction, the continual passage of steam causes the hot coke to cool gradually. If reaction is to continue, an undue drop in the temperature must be prevented. This is achieved by blowing air through the coke, during which time the supply of steam is cut off. The oxygen in the air is first considered to oxidize the carbon to carbon dioxide.

$$C(s) + O_2(g) \rightarrow CO_2(g) \qquad \Delta H = -394 \text{ kJ}$$

The carbon dioxide, passing up through the coke, is in turn reduced to the monoxide.

$$C(s) + CO_2(g) \rightarrow 2CO(g) \qquad \Delta H = 173 \text{ kJ}$$

The overall reaction is therefore

$$2C(s) + O_2(g) \rightarrow 2CO(g) \qquad \Delta H = -221 \text{ kJ}$$

Thus during the air blow a mixture of gases consisting mainly of carbon monoxide and nitrogen (from the air) is formed. This mixture is called *producer gas*. Its calorific value is poor because of the diluting effect of the non-combustible nitrogen. Its efficiency as a fuel, however, is improved if it is burned in the hot raw state so that the heat released in its formation ($110 \cdot 5$ kJ mol^{-1}) is added to that released during its combustion.

$$CO(g) + \tfrac{1}{2}O_2(g) \rightarrow CO_2(g) \qquad \Delta H = -283 \text{ kJ}$$

† The efficiency of a fuel is expressed in terms of its *calorific value*, which is the amount of heat liberated when a given amount of fuel is burned completely. Whilst the chemist is inclined to quote a calorific value in terms of kJ g^{-1} (for solid and liquid fuels) and kJ cm^{-3} (for gaseous fuels), the engineer expresses such a quantity in terms of Btu lb^{-1} or Btu ft^{-3}.

The heat which is released during the formation of producer gas causes a rise in bed temperature which, if uncontrolled, can lead to distortion of grate bars and melting of ash. To overcome these difficulties the air supply is cut off at about 1400°C and steam once more blown through. An arrangement of automatically operated valves ensures the separation of the water gas and producer gas formed alternately in the cyclic process.

Even more convenient in many cases than the cyclic process described above is an arrangement whereby a mixture of steam and air is blown continuously through the hot coke bed in proportions necessary to maintain a constant predetermined temperature. The resulting mixture of gases is referred to as *semi-water gas*. The following example demonstrates how the proper proportion of steam to air may be calculated.

EXAMPLE 12.3. *Determine the proportion by volume of steam to air which would be required to maintain the temperature of the coke at* 1000°C. *Oxygen content in the air* = 20% *by volume.*

$$C(s) + H_2O(g) \rightarrow CO(g) + H_2(g) \qquad \Delta H = 131 \text{ kJ}$$

i.e. for every mole of steam decomposed 131 kJ of heat are absorbed.

$$2C(s) + O_2(g) \rightarrow 2CO(g) \qquad \Delta H = -221 \text{ kJ}$$

i.e. for every mole of oxygen decomposed 221 kJ of heat are liberated.

∴ $\frac{131}{221}$ mole of oxygen decomposing will liberate 131 kJ

This is the quantity of oxygen necessary to counteract the cooling which results from reaction between steam and coke.

∴   Moles of air required = $5 \times \frac{131}{221}$ = 2·96

∴   Molar proportion of steam to air = corresponding proportion by volume

$$= 1 : 2 \cdot 96$$

12.6. Solar heating

Heat from the sun can be used to heat houses. Such dwellings, referred to as solar houses, require the sun's energy to be concentrated and stored so that the energy absorbed during the day may be utilized to heat the house during the night. One such heat-storage unit consists of sealed vessels containing sodium sulphate decahydrate, $Na_2SO_4 \cdot 10H_2O$. At temperatures in excess of $32 \cdot 4°C$ this salt undergoes decomposition according to the equation

$$Na_2SO_4 \cdot 10H_2O \rightarrow Na_2SO_4 + 10H_2O \qquad \Delta H = 81 \cdot 1 \text{ kJ} \qquad (2)$$

The reaction is attended by an enthalpy change of $81 \cdot 1$ kJ. At temperatures below $32 \cdot 4°C$ the reaction proceeds completely in the reverse direction, the consequent enthalpy change being $-81 \cdot 1$ kJ.

$$Na_2SO_4 + 10H_2O \rightarrow Na_2SO_4 \cdot 10H_2O \qquad \Delta H = -81 \cdot 1 \text{ kJ} \qquad (3)$$

Thus during the day the temperature of the sealed containers is maintained at a value in excess of $32 \cdot 4°C$, heat is absorbed from the sun and reaction (2) occurs. At night, the temperature drops below $32 \cdot 4°C$ and the reverse reaction takes place, releasing $81 \cdot 1$ kJ for every one mole of $Na_2SO_4 \cdot 10H_2O$ produced.

This system suffers from a number of disadvantages, the most obvious being the dependency on sun exposure. Such a system could hardly be made efficient in the United Kingdom. Even given reasonable exposure to the sun there still remains one serious disadvantage, viz. the problem of heat storage. The reason for this is made clear in the following example.

EXAMPLE 12.4. *Determine the "calorific value" of* $Na_2SO_4 \cdot 10H_2O$ *when the latter is used as a fuel in the context of house heating.*

$$(1 \text{ lb} = 453 \cdot 5 \text{ g, 1 Btu} = 1 \cdot 054 \text{ kJ.})$$

One mole of Na_2SO_4 produced by the dehydration of one mole of $Na_2SO_4 \cdot 10H_2O$ will release during the night $81 \cdot 2$ kJ or $77 \cdot 0$ Btu. One pound of $Na_2SO_4 \cdot 10H_2O$ will therefore be required to generate $108 \cdot 5$ Btu, i.e. the "calorific value" of $Na_2SO_4 \cdot 10H_2O$ is $108 \cdot 5$ Btu lb^{-1}.

14

Now this figure is particularly low relative to the calorific values of conventional solid fuels where the values range from 10,000 to 14,000 Btu lb^{-1}. As a result, an extremely large heat storage unit would be necessary to satisfy the heating requirements of a house of moderate size.

12.7. Rocket propulsion

Thermochemical considerations are important in rocket-propulsion technology. A rocket develops its thrust as a result of gas molecules under high pressure being forced to stream through a properly constructed nozzle. This thrust will in turn be dependent on the speed and number of gas molecules produced as a result of reaction between fuel and oxidizer, i.e. oxygen or a source of oxygen. (The combination of rocket fuel and oxidizer is normally referred to as the propellant.) Consequently, an efficient propellant system should have the following characteristics:

   (i) The reaction between fuel and oxidizer should be strongly exo-thermic.
  (ii) The reaction should be attended by an appreciable increase in the number of gaseous molecules.
 (iii) The reaction must be capable of occurring rapidly, much more so than in, for example, a jet engine.
 (iv) Both fuel and oxidizer should have a high density in order that storage space be minimized.

The first three features ensure a generation of high velocity gas molecules under high pressure.

The efficiency of a propellant system is normally expressed in terms of its *specific impulse,* this being the thrust in pounds developed per pound of propellant consumed per second. An approximate criterion of efficiency more rapidly calculated is the enthalpy change per cubic centimetre of propellant.

EXAMPLE 12.5. *On the basis of the above criterion for the efficiency of propellants, show which of the two following systems is the more suitable rocket fuel.*

liquid fuel   oxidizer

$$N_2H_5OH(l) + 2H_2O_2(l) \rightarrow N_2(g) + 5H_2O(g) \tag{1}$$

Hydroxylamine Hydrogen
            peroxide

solid fuel   oxidizer

$$2LiBH_4(s) + KClO_4(s) \rightarrow Li_2O(s) + B_2O_3(s) + KCl(s) + 4H_2(g) \tag{2}$$

Lithium   Potassium
borohydride perchlorate

$$\Delta H_{f(N_2H_5OH,\, l)} = -242 \text{ kJ mol}^{-1}, \quad \text{density} = 1\cdot03 \text{ g cm}^{-3}$$

$$\Delta H_{f(H_2O_2,\, l)} = -192 \text{ kJ mol}^{-1}, \quad \text{density} = 1\cdot46 \text{ g cm}^{-3}$$

$$\Delta H_{f(H_2O,\, g)} = -242 \text{ kJ mol}^{-1}$$

$$\Delta H_{f(LiBH_4,\, s)} = -188 \text{ kJ mol}^{-1}, \quad \text{density} = 0\cdot66 \text{ g cm}^{-3}$$

$$\Delta H_{f(KClO_4,\, s)} = -435 \text{ kJ mol}^{-1}, \quad \text{density} = 2\cdot52 \text{ g cm}^{-3}$$

$$\Delta H_{f(Li_2O,\, s)} = -594 \text{ kJ mol}^{-1}$$

$$\Delta H_{f(B_2O_3,\, s)} = -1262 \text{ kJ mol}^{-1}$$

$$\Delta H_{f(KCl,\, s)} = -435 \text{ kJ mol}^{-1}$$

(1) The enthalpy change ΔH_1 for reaction (1)

$$= \Delta H_{f(\text{products})} - \Delta H_{f(\text{reactants})}$$

$$\therefore \quad \Delta H_1 = 5\,\Delta H_{f(H_2O,\, g)} - \Delta H_{f(N_2H_5OH)} - 2\,\Delta H_{f(H_2O_2)} = -584 \text{ kJ}$$

This is the quantity of heat liberated when one mole (50 g) of N_2H_5OH reacts with two moles (68 g) of H_2O_2, i.e. 584 kJ are liberated from $(50+68)$ g of propellant, i.e. 584 kJ are liberated from $(48\cdot54 + 46\cdot58)$ cm³ of propellant,

$\therefore$ 6·14 kJ are liberated from 1 cm³ of propellant

$\therefore$ Enthalpy change per cm³ of propellant $= -6\cdot14$ kJ

(2) The enthalpy change ΔH_2 for reaction (2) $= \Delta H_{f(Li_2O)} + \Delta H_{(B_2O_3)} + \Delta H_{f(KCl)} - 2\,\Delta H_{f(LiBH_4)} - \Delta H_{f(KClO_4)} = 1480$ kJ, i.e. 1480 kJ are liberated from $(43\cdot5 + 138\cdot6)$ g of propellant, i.e. 1480 kJ are liberated from $(65\cdot9 + 55\cdot00)$ cm³ of propellant,

$\therefore$ 12·24 kJ are liberated from 1 cm³ of propellant

$\therefore$ Enthalpy change per cm³ of propellant $= -12\cdot24$ kJ

Thus on this basis alone, the second propellant system incorporating a solid fuel is the more efficient. However, combustion of solid fuels is less easy to control than the combustion of a liquid fuel, and this together with other factors must be given due consideration.

12.8. Influence of temperature on equilibrium

In reversible reactions (Chapter 14), the position of equilibrium can be strongly influenced by the temperature. In this connection, a knowledge of both the sign and magnitude of ΔH are of considerable interest. This particular aspect of thermochemistry (important in particular in the field of gaseous fuels) will be discussed fully in Chapter 14.

Exercises

12.1. Assuming the combustion of propane gas to be 90% efficient, how much heat is released by burning 1 m³ of propane at STP? (Molar volume at STP $= 22\cdot4$ dm³ mol^{-1}.)

Substance	ΔH_f/kJ mol^{-1}
$C_3H_8(g)$	$-103\cdot8$
$CO_2(g)$	$-393\cdot5$
$H_2O(l)$	$-285\cdot8$

[*Ans.*: $8\cdot92\times10^4$ kJ.]

12.2. Use the following data to determine the amount of energy released in the combustion of 1 gallon (Imperial) of n-octane, a constituent of petrol.

Substance	ΔH_f/kJ mol^{-1}
$C_8H_{18}(l)$	$-208\cdot2$
$CO_2(g)$	$-393\cdot5$
$H_2O(l)$	$-285\cdot8$

M(n-octane) = 114 g mol^{-1};   density of n-octane = 0·70 g cm^{-3}

1 gallon (Imperial) = 4·546 dm^3

[*Ans.*: 1·539×10^5 kJ.]

12.3. Formaldehyde (H·CHO) is made by the catalytic oxidation of methanol by air according to the equation

$$CH_3OH(l) + \tfrac{1}{2}O_2(g) \rightarrow H \cdot CHO(g) + H_2O(l)$$

Use the following data to determine the enthalpy change for this reaction:

$$H \cdot CHO(g) + H_2(g) \rightarrow CH_3OH(l) \qquad \Delta H = -121 \cdot 5 \text{ kJ}$$
$$H_2(g) + \tfrac{1}{2}O_2(g) \rightarrow H_2O(l) \qquad \Delta H = -285 \cdot 8 \text{ kJ}$$

[*Ans.*: −164·3 kJ.]

12.4. Hydrazine (N$_2$H$_4$) has been tested as a rocket fuel, its reaction with oxygen producing N$_2$(g) and H$_2$O(g). Calculate the thermal energy available per gram of propellant.

$$\Delta H_f(N_2H_4, l) = 138 \cdot 4 \text{ kJ}; \qquad \Delta H_f(H_2O, g) = -241 \cdot 8 \text{ kJ}$$

[*Ans.*: 9·72 kJ g^{-1}.]

12.5. A mixture of methane and acetylene was completely burnt in excess oxygen to yield equal volumes of carbon dioxide and steam. Calculate the composition of the original mixture (in mole per cent) and determine the enthalpy change per mole of mixture.

Substance	ΔH_f/kJ mol^{-1}
CH_4(g)	− 74·8
C_2H_2(g)	226·7
CO_2(g)	− 393·5
H_2O(g)	− 241·8

[*Ans.*: 50 mol % CH$_4$; − 1029 kJ.]

CHAPTER 13

Chemical Kinetics

CHEMICAL kinetics concerns itself with the speed at which reactions occur. This varies enormously from one reaction to another. For example, the reaction between a strong acid and a strong base is over in an instant, whereas the reaction between hydrogen and oxygen at room temperature proceeds so slowly that one could easily be misled into thinking that no reaction occurs at all. Furthermore, the rate of any one reaction is observed to be influenced by (i) reactant concentration, (ii) temperature, and (iii) the presence of a catalyst.

These observations are explained in terms of the *collision theory*, the basis of which is that *the collision of two or more particles is a prerequisite to chemical reaction*. Using this fundamental assumption, we will consider the sequence of events occurring in the course of a collision between molecules and go on to show (at least qualitatively) why reaction rate should be influenced by the above factors.

13.1. Frequency of collision

The basis of the collision theory is that collision is a prerequisite to reaction. If, however, this were the only requirement then the number of molecules colliding $cm^{-3} sec^{-1}$ would equal the number of molecules reacting $cm^{-3} sec^{-1}$, i.e. the rate of collision would equal the rate of reaction. A comparison of these two quantities, both of which can be calculated, reveals that for most reactions the number of molecules colliding is very much greater than the number of molecules reacting.

Clearly, not all colliding molecules do in fact react upon collision and those which do must differ in some respect from those which do not.

13.2. Energy of collision

We can explain the above inequality by considering in some detail the sequence of events when two molecules collide. Imagine a molecule of hydrogen and one of iodine approaching each other on a collision course. At some point during approach repulsive forces will begin to act due to interaction of the similarly charged electron clouds. During this interaction the potential energy of the system (i.e. its enthalpy) will increase at the expense of its kinetic energy, because work is being done in overcoming the repulsive forces. This increase in potential energy is characterized by, and indeed due to, the gradual weakening of H—H and I—I bonds and to the gradual formation of H—I bonds (Fig. 13.1).

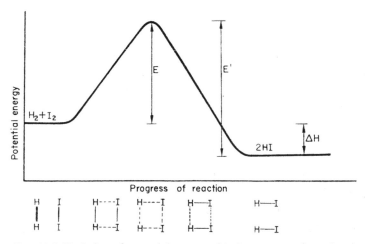

FIG. 13.1. Variation of potential energy with the progress of reaction.

A stage is reached when no distinction exists between any of the four bonds. This arrangement is described as an *activated complex* and E, the energy required to produce this state, is defined as the *activation energy.*

The process of bond breaking and bond forming continues as indicated in the above diagram with consequent decrease in potential energy and increase in kinetic energy until the energy level of the products is reached.

This reaction is reversible (Chapter 14), and if we imagine starting with hydrogen iodide, a similar sequence of events can be traced, producing the same activated complex but a different activation energy E', where $E - E' = \Delta H$.

Hydrogen iodide molecules then can only be formed from hydrogen and iodine molecules via the formation of the activated complex which in turn can only be produced if the energy of the collision is at least equal to the activation energy E. If this is not so then the activated complex will not be formed and the colliding molecules will, having dissipated their kinetic energy, find themselves in the position of possessing potential energy corresponding to the structure

This state is analogous to that of a stretched spring and will, like the spring, tend to change to one of minimum potential energy, i.e. the two molecules will move apart, the kinetic energy increasing at the expense of the potential energy until the latter assumes a value appropriate to that of the reactants. Thus *the energy of the collision determines whether or not reaction will occur*.

Progress is being made but our troubles are not yet over. Why should some collisions be highly energetic while others are not so? You will recall that in a sample of gas the molecules are distributed over a wide range of velocity and hence of kinetic energy (Fig. 4.4). A similar distribution is obtained for collision energies (Fig. 13.2). If E corresponds to the activation energy then the shaded area in Fig. 13.2 represents the fraction of collisions which lead to reaction. For a reaction characterized by a low E value a large proportion of the colliding molecules will suffer decomposition and the rate of the reaction will be high; e.g. the process of neutralization of a strong acid by a strong base requires only the collision between hydrogen and hydroxyl ions, the activation energy being zero and as a result the reaction is so fast as to be almost instantaneous. If, on the other hand, the value of E for a particular reaction is very high only a few collisions will be energetic enough to yield product material

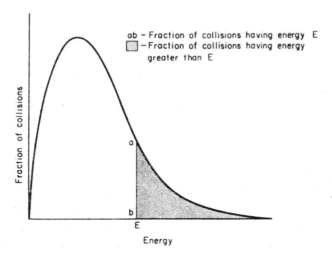

Fig. 13.2. Distribution of collision energies.

and the reaction may then proceed so slowly that no change with time is detectable.

13.3. Geometry of collision

Still one more factor must be considered, viz. an *orientation factor*. Some collisions, though sufficiently energetic, will not result in the formation of an activated complex because the molecules assume during the collision an unsuitable orientation with respect of each other. As would be expected, this effect becomes more significant for a collision occurring between complex molecules.

We may summarize the above by saying that reaction rate is dependent upon

  (i) the frequency of collision;
  (ii) the energy of collision;
  (iii) the geometry of collision,

and any change imposed on a reaction which alters any of these factors will produce a corresponding change in the reaction rate.

13.4. Influence of concentration

Increasing the concentration of reactants has the effect of increasing reaction rate. This is so because a change in concentration produces a proportional change in the frequency of collision. Whilst the *fraction* of effective collisions remains the same, their *number* changes proportionally.

13.5. Influence of temperature

The kinetic theory reveals that heat is to be identified with molecular motion; hence an increase in temperature will result in an increase in collision frequency. For a temperature rise of the order of 10 degrees, this increase is calculated to be small and is totally insufficient in itself to account for the very much larger increases in rate which have been observed (sometimes as much as 300%) for such a rise in temperature.

The influence of temperature on the distribution of collision energies, illustrated in Fig. 13.3, follows as a consequence of the effect of temperature on the spread of molecular energies.

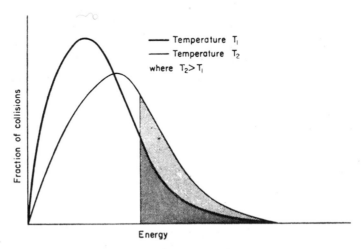

FIG. 13.3. Influence of temperature on the distribution of collision energies.

From Fig. 13.3 we see that temperature has a pronounced effect on increasing the fraction of collisions capable of generating energy in excess of that required to produce the activated state. Thus, although the frequency of collision is not greatly changed by an increase in temperature, a much greater proportion of the collisions are now sufficiently energetic to lead to product formation.

13.6. Influence of catalyst

A *catalyst* is the name given to a substance which increases a reaction rate without being itself consumed. If reactants and catalyst are in the same phase the catalysis is said to be *homogeneous*.

Consider the lead chamber process for the manufacture of sulphuric acid in which sulphur dioxide is oxidized to sulphur trioxide.

$$2SO_2 + O_2 \rightarrow 2SO_3$$

This reaction, occurring in the gas phase, is extremely slow even at moderate temperatures; it is, however, effectively accelerated by the addition of oxides of nitrogen, principally nitric oxide. The manner in which the nitric oxide involves itself in the reaction is formally indicated in the following equations:

$$2NO + O_2 \rightarrow 2NO_2 \tag{1}$$
$$\underline{2SO_2 + 2NO_2 \rightarrow 2SO_3 + 2NO} \tag{2}$$
$$2SO_2 + O_2 \rightarrow 2SO_3 \tag{(1)+(2)}$$

The nitric oxide is consumed in stage 1 and regenerated in stage 2. Hence the amount of nitric oxide is the same at the end as it is at the beginning of the reaction; it is therefore functioning in the role of a catalyst.

In this and similar reactions it is important to note that (i) the catalyst does not initiate reaction, it merely accelerates it, and (ii) the presence of the catalyst causes the reaction to proceed by an alternative route having a quite different rate. These are two important characteristics of all catalysed reactions.

That a reaction can occur in the presence of a catalyst by an alternative route is most significant. In our example, the rate is now determined not by the frequency and effectiveness of collisions between sulphur dioxide and oxygen molecules but by the frequency and effectiveness of collision between (i) nitric oxide and oxygen molecules and (ii) nitrogen dioxide and sulphur dioxide molecules. The concentration of nitric oxide will determine the frequency of collision between reactant molecules (NO and O_2) in stage 1 and reactant molecules (SO_2 and NO_2) in stage 2. Thus the frequency of collision and hence the rate of reaction will be proportional to the concentration of catalyst. This reveals a further characteristic common to all catalysts functioning in a homogeneous capacity.

That the rate of the reaction is appreciably increased where a catalyst is present can only be accounted for by supposing that the alternative

E_1 – Activation energy for uncatalysed reaction
E_2 – Activation energy for catalysed reaction

Fig. 13.4. Influence of catalyst on activation energy.

route provided is one of lower activation energy (Fig. 13.4). As a result the proportion of effective collisions involving oxygen molecules in stage 1 and of sulphur dioxide molecules in stage 2 will be higher than the corresponding proportion of effective collisions between sulphur dioxide and oxygen molecules in absence of catalyst.

The reaction of sulphur dioxide and oxygen is reversible and it should be noted that the activation energy is lowered by an amount $E_1 - E_2$ for *both* forward and reverse reactions. A consequence of this is that a catalyst which accelerates the forward reaction will also accelerate the reverse reaction.

If the reactants and catalyst are not in the same phase then the catalysis is said to be *heterogeneous*. For example, the reaction between carbon dioxide and hydrogen is greatly accelerated by the presence of finely divided platinum, the latter functioning as a catalyst.

$$CO_2 + H_2 \xrightarrow{\text{Pt.}} H_2O + CO$$

This acceleration is found to be dependent on the amount of catalyst surface available to the reactant materials, the rate increasing with increase in surface area. The mechanism by which this and similar reactions are catalysed heterogeneously cannot be described in quite such precise terms as is possible for reactions catalysed homogeneously. The sequence of events for the catalysed reaction between carbon dioxide and hydrogen is considered to involve three steps.

(a) *The adsorption of reactant molecules on the surface*

The process of adsorption[†] can be regarded somewhat loosely as reaction between the reactant molecules and the surface atoms of catalyst.

$$\left.\begin{array}{l} CO_2 + Pt \rightarrow CO_2(Pt) \\ H_2 + Pt \rightarrow H_2(Pt) \end{array}\right\} \text{ Adsorbed reactants}$$

As a result of interaction between the surface atoms of catalyst and reactant molecules there may be a weakening of particular bonds in the adsorbed molecules, thus making them more susceptible to reaction, i.e. they have become activated in the process of adsorption.

† Adsorption is the term used to describe the adhesion of molecules on the *surface* of a solid. Any increase in surface area will lead to an increase in adsorption.

(b) *Surface reaction*

Reaction occurs on the surface between the now activated molecules provided they occupy adjacent positions on the surface.

$$CO_2(Pt) + H_2(Pt) \rightarrow H_2O(Pt) + CO(Pt)$$

$\underbrace{\qquad\qquad\qquad\qquad}_{\text{Adsorbed reactants}}$ $\underbrace{\qquad\qquad\qquad\qquad}_{\text{Adsorbed products}}$

(c) *Desorption of products*

Product molecules break free from the surface, i.e. are desorbed, and continue their existence in the gaseous state.

$$H_2O(Pt) \rightarrow H_2O + Pt$$
$$CO(Pt) \rightarrow CO + Pt$$

Essentially the presence of a catalyst has provided (as before) an alternative route having an activation energy lower than that of the non-catalysed reaction. This decrease in activation energy accounts for the increase in reaction rate.

Exercises

13.1. The gas phase reaction.

$NO + O_3 \rightarrow NO_2 + O_2$ proceeds by a one-step conversion involving collisions between NO and O_3 molecules. If the concentration of both NO and O_3 are trebled, by what factor will the reaction rate change?

[*Ans.*: 9.]

13.2. In the gas phase reaction

$$CO + NO_2 \rightarrow CO_2 + NO$$

two possible collisions are shown below:

collision A   collision B

Which collision is more likely to be effective and lead to the formation of products?

[*Ans.*: Collision A.]

13.3. Molecules A and B react according to the equation

$$A + B \rightarrow C + D \qquad \Delta H = 50 \text{ kJ}$$

The activation energy for the reaction is 80 kJ. Plot the potential energy of the system as a function of the course of the reaction. What is the enthalpy change and activation energy for the reverse reaction

$$C + D \rightarrow A + B$$

What is the minimum value of activation energy for an endothermic reaction?

[*Ans.*: $\Delta H = -50 \text{ kJ}$; $E_{act} = 30 \text{ kJ}$; $E_{MIN} = \Delta H$.]

13.4. The rate of a certain reaction increases by a factor of three for every 10 K increase in temperature. By what factor does the rate increase for a temperature increase of 100 K?

[*Ans.*: Approx. 60,000.]

13.5. The hydrogenation of ethylene

$$C_2H_4 + H_2 \rightarrow C_2H_6$$

is immeasurably slow at normal temperatures but is greatly accelerated in the presence of finely divided nickel. That this is so is largely due to the metal's ability to strongly adsorb hydrogen which is incorporated on the metal surface as hydrogen atoms.

Write down a simple reaction sequence for the catalysed reaction and explain why the metal is used in a finely divided condition.

13.6. Finely divided platinum, suitable for use in surface catalysis, is prepared by precipitation from solution as minute spheres of average diameter 200 nm. Calculate the surface area per gram of platinum.

Density of platinum = 21·37 g cm^{-3}. (1 nm = 10^{-9} m.)

[*Ans.*: 1·41 m^2.]

CHAPTER 14

Chemical Equilibria

IT IS an experimentally observed fact that for many reactions the interacting substances are not completely consumed. Thus, if hydrogen and iodine are heated in a closed vessel, they will react only partially to produce hydrogen iodide.

$$H_2(g) + I_2(g) \rightarrow 2HI(g)$$

An essential requirement for this and indeed any reaction to occur is that the reacting molecules should collide. Now only a fraction of the collisions will result in the formation of hydrogen iodide molecules and as their number increases, the probability of a collision occurring between two hydrogen iodide molecules likewise increases. A fraction of these collisions will lead to the formation of hydrogen and iodine.

$$2HI(g) \rightarrow H_2(g) + I_2(g)$$

A stage is finally reached when no further change in overall composition with time is observed. The system is then said to be in a state of *dynamic equilibrium*. As a result of molecular collision the reaction continues but in such a way that for every molecule of hydrogen which reacts through collision with one molecule of iodine, two molecules of hydrogen iodide decompose elsewhere in the system. There is therefore an equality between the rate at which hydrogen iodide is formed and the rate at which it is decomposed. Such a reaction is sensibly called *reversible* and is represented in the chemical equation by the symbol $\rightleftharpoons$ as in

$$H_2(g) + I_2(g) \rightleftharpoons 2HI(g)$$

It is possible that all reactions are reversible—the problem is to select the conditions which serve to reveal the reversible character.

14.1. Equilibrium constants

Consider the general case of an equilibrium involving different numbers of reactant and product molecules.

$$aA + bB \cdots \rightleftharpoons cC + dD \cdots$$

It is found experimentally *that for any reversible reaction under equilibrium conditions, the ratio of the product of product concentrations to the product of reactant concentrations (each concentration being raised to a power equal to the coefficients in the balanced equation) is a constant characteristic of the reaction and dependent only on the temperature.* The above statement, referred to as the *law of chemical equilibrium,* is the foundation upon which an understanding of chemical equilibria is based.

The constant referred to in the above law is called the *equilibrium constant* and is related to the concentrations of reactants and products by the expression

$$K = \frac{[C]^c [D]^d \cdots}{[A]^a [B]^b \cdots}$$

where [ ] represents the equilibrium concentration (mol l^{-1}) of whatever component is included within the brackets.

The fact that the product concentrations appear in the numerator while reactant concentrations form the denominator is a purely arbitrary choice—obviously we must be consistent and this convention is the one universally accepted.

EXAMPLE 14.1. *Write down the equilibrium constant expression for each of the following reactions:*

$$H_2(g) + I_2(g) \rightleftharpoons 2HI(g)$$

$$2HI(g) \rightleftharpoons H_2(g) + I_2(g)$$

$$2SO_2(g) + O_2(g) \rightleftharpoons 2SO_3(g)$$

$$4NO(g) + 6H_2O(g) \rightleftharpoons 4NH_3(g) + 5O_2(g)$$

For the reaction

$$H_2(g) + I_2(g) \rightleftharpoons 2HI(g)$$

$$K = \frac{[HI]^2}{[H_2][I_2]}$$

The reaction written in this way implies that the starting materials are $H_2(g)$ and $I_2(g)$. If the starting material was instead HI(g), the equation would be more reasonably written as

$$2HI(g) \rightleftharpoons H_2(g) + I_2(g)$$

for which

$$K' = \frac{[H_2][I_2]}{[HI]^2} = \frac{1}{K}$$

For the reaction

$$2SO_2(g) + O_2(g) \rightleftharpoons 2SO_3(g)$$

$$K = \frac{[SO_3]^2}{[SO_2]^2[O_2]}$$

For the reaction

$$4NO(g) + 6H_2O(g) \rightleftharpoons 4NH_3(g) + 5O_2(g)$$

$$K = \frac{[NH_3]^4[O_2]^5}{[NO]^4[H_2O]^6}$$

The extent to which a reaction proceeds from left to right can be deduced from the magnitude of the equilibrium constant; if large, then the concentrations of at least one of the products must be high relative to those of the reactants, indicating that the position of equilibrium lies well over to the right. Conversely a small equilibrium constant implies that the reaction proceeds from left to right to only a small extent.

14.2. The equilibrium constant K_p

If a reaction occurs in the gas phase, it is convenient to express the concentrations of components present by their respective partial pressures. That this is justifiable can be easily demonstrated. For a gas A in a mixture

$$P_A = \frac{n_A RT}{V}$$

where

$$P_A = \text{partial pressure of A}$$

and

$$\frac{n_A}{V} = \text{molar concentration of A} = [A]$$

∴   At constant temperature $P_A \propto [A]$

The equilibrium constant for the reaction

$$aA + bB \cdots \rightleftharpoons cC + dD \cdots$$

is then written as

$$K_p = \frac{(P_C)^c (P_D)^d \cdots}{(P_A)^a (P_B)^b \cdots}$$

where P_A, P_B, P_C and P_D are the partial pressures of A, B, C and D respectively. It is advisable at all times to express partial pressures in atmospheres.

14.3. Relationship between K_p and K

Consider once more the general equation

$$aA + bB \cdots \rightleftharpoons cC + dD \cdots$$

in which, for a reaction occurring in the gas phase,

$$P_A = \frac{n_A RT}{V} = [A]RT$$

Similarly $P_B = [B]RT$, $P_C = [C]RT$, $P_D = [D]RT$

$$K_p = \frac{(P_C)^c (P_D)^d \cdots}{(P_A)^a (P_B)^b \cdots} = \frac{[C]^c [D]^d (RT)^{(c+d-a-b)}}{[A]^a [B]^b} = K(RT)^{\Delta n}$$

where $\Delta n = (c+d-a-b) = $ the change in the number of gaseous moles.

15*

In using the above expression care must be exercised to ensure numerical equality between the left- and right-hand sides of the equation. Thus if R takes the SI value of $8.314 \text{ J K}^{-1} \text{mol}^{-1}$, then pressures and concentrations must be expressed in N m^{-2} and mol m^{-3} respectively. If, however, as is more convenient, pressures and concentrations are expressed in atm and mol l^{-1} (mol dm^{-3}), then R assumes a value of $0.082 \text{ l atm K}^{-1} \text{mol}^{-1}$ (Section 4.3).

EXAMPLE 14.2. *For the decomposition of nitrogen tetroxide to nitrogen dioxide at* 318 K, *according to the equation*

$$N_2O_4(g) \rightleftharpoons 2NO_2(g)$$

$K_p = 0.6533$ *when partial pressures are expressed in atmospheres. What is the value of K at this temperature?*

$$K_p = K(RT)^{\Delta n} = 0.6533$$

$$\Delta n = +1, \quad \text{and} \quad R = 0.082 \text{ l atm K}^{-1} \text{mol}^{-1}$$

$$\therefore \quad K = \frac{0.6533}{0.082 \times 318} = 0.02505$$

14.4.   Heterogeneous equilibria

A heterogeneous equilibrium is one in which two or more phases are present. For example, the reaction between steam and red-hot coke involves both gaseous and solid phases

$$C(s) + H_2O(g) \rightleftharpoons CO(g) + H_2(g)$$

whence

$$K' = \frac{[CO] [H_2]}{[C] [H_2O]}$$

The concentration of a solid is numerically identical to its density when the latter is expressed in units of mole l^{-1}. Clearly the density of a solid is not dependent on the amount of solid present, and we can therefore combine this constant concentration with the original equilibrium constant,

i.e.
$$K'[C] = \frac{[CO][H_2]}{[H_2O]} = K$$

Thus, in any heterogeneous equilibrium, concentrations of pure solids do not appear in the equilibrium constant expression; nor, by a similar argument, do the concentrations of pure liquids.

14.5. Influence of concentration on the equilibrium position

Consider once more the reaction

$$C(s) + H_2O(g) \rightleftharpoons CO(g) + H_2(g)$$

for which

$$K = \frac{[CO][H_2]}{[H_2O]}$$

Suppose that to an equilibrium mixture more steam is added; if the temperature is kept constant, the value of the equilibrium constant must remain the same and it can only do so if some of the additional steam reacts with carbon to produce the appropriate amounts of carbon monoxide and hydrogen, thereby displacing the equilibrium position to the right. *Generally, when the concentration of one of the components making up an equilibrium system is increased the reaction promoted is that which tends to consume some of the added substance.*

The same result could be obtained by removing the products from the sphere of action. Their continued removal would cause more steam to react with carbon in order that the equilibrium constant should maintain its value, and in this way the reaction could be forced to completion.

EXAMPLE 14.3. *For the reaction*

$$H_2(g) + I_2(g) \rightleftharpoons 2HI(g)$$

$$K = 50 \cdot 25 \quad at \quad 717 \text{ K}$$

(i) *Calculate the number of moles of* HI *formed when two moles of* H_2 *and one mole of* I_2 *react at* 717 K *under a condition of constant volume, V.*

(ii) *Compare this result with that obtained when a further one mole of* H_2 *is introduced into the system.*

(i) Let $x =$ moles of H_2 undergoing decomposition.

The equilibrium concentrations are therefore as follows:

$$[H_2] = \frac{2-x}{V}, \quad [I_2] = \frac{1-x}{V}, \quad [HI] = \frac{2x}{V}$$

$$\therefore \quad K = \frac{[HI]^2}{[H_2][I_2]} = \frac{4x^2}{(2-x)(1-x)} = 50\cdot25$$

$$\therefore \quad 46\cdot25x^2 - 150\cdot75x + 100\cdot5 = 0$$

This is a quadratic equation of the form

$$ax^2 + bx + c = 0$$

the solution of which is

$$x = \frac{-b \pm \sqrt{b^2 - 4ac}}{2a}$$

Following this procedure

$$x = 2\cdot327 \quad \text{or} \quad 0\cdot9335$$

As there were only two moles of H_2 initially present, a value of $x = 2\cdot327$ is clearly ridiculous. Hence the number of HI moles present at equilibrium $= 2x = 1\cdot8670$.

(ii) Let $y =$ moles of H_2 undergoing decomposition.

The equilibrium concentrations are therefore as follows:

$$[H_2] = \frac{3-y}{V}, \quad [I_2] = \frac{1-y}{V}, \quad [HI] = \frac{2y}{V}$$

$$\therefore \quad K = 50\cdot25 = \frac{4y^2}{(3-y)(1-y)}$$

$$\therefore \quad y = 3\cdot383 \quad \text{or} \quad 0\cdot9645$$

The second value is the only permissible one, and so the number of HI moles present at equilibrium is $1 \cdot 9390$. It is evident from a comparison of the two results that increasing the concentration of HI forces the equilibrium to the right.

14.6. Influence of pressure on the equilibrium position

We have already seen that for the general equation

$$aA + bB \cdots \rightleftharpoons cC + dD \cdots$$

$$K_P = \frac{(P_C)^c (P_D)^d \cdots}{(P_A)^a (P_B)^b \cdots}$$

Now $P_A = x_A P$, $P_B = x_B P$, $P_C = x_C P$, $P_D = x_D P$ where P is the total pressure and x_A, x_B, x_C, and x_D are the mole fractions of A, B, C and D respectively.

$$\therefore \qquad K_P = \frac{x_C^c x_D^d}{x_A^a x_B^b} P^{\Delta n} = A\, P^{\Delta n} \quad \text{where} \quad A = \frac{x_C^c x_D^d}{x_A^a x_B^b}$$

This equation reveals the influence that pressure has on the position of equilibrium.

Consider a reaction characterized by a positive value of Δn. If the pressure acting on the system is increased, the A term must decrease if the value of the equilibrium constant is not to change. This decrease can only be achieved if C and D react to some extent (thereby reducing the numerator) to produce A and B (thereby lowering the denominator). Thus for such a reaction, an increase in pressure will favour the formation of reactant materials, the equilibrium position being displaced to the left. A similar argument leads to the conclusion that for a reaction having a negative value of Δn, an increase in pressure causes a shift of the equilibrium position to the right. For a reaction for which Δn is zero, a change in pressure has no effect on the equilibrium position, and this reflects itself in the equality between the terms K and K_P.

Summarizing, we can say that *an increase in pressure favours the formation of those products formed with a decrease in the number of gaseous molecules; a decrease in pressure favours the formation of those products formed with an increase in the number of gaseous molecules; a change of pressure has no effect on a system for which there is no change in the number of gaseous molecules.*

Dissociation equilibria provide a good illustration of a group of reactions for which Δn is not equal to zero. The breakdown, i.e. dissociation, of nitrogen tetroxide into nitrogen dioxide exemplifies this type of reaction.

$$N_2O_4(g) \rightleftharpoons 2NO_2(g)$$

For such a reaction, the fraction of one mole undergoing complete decomposition is referred to as the *degree of dissociation*, normally represented by α.

Thus, if a moles of N_2O_4 are initially present, then:

$$a\alpha = \text{moles of } N_2O_4 \text{ undergoing complete decomposition}$$
$$a - a\alpha = a(1-\alpha) = \text{moles of } N_2O_4 \text{ present at equilibrium}$$
and $$2a\alpha = \text{moles of } NO_2 \text{ present at equilibrium}$$

$\therefore$ $(a - a\alpha) + 2a\alpha = a(1+\alpha) = \text{total moles present at equilibrium}$

$\therefore$ $$P_{N_2O_4} = \frac{n_{N_2O_4}}{n_T} P = \frac{a(1-\alpha)}{a(1+\alpha)} P = \frac{1-\alpha}{1+\alpha} P$$

and $$P_{NO_2} = \frac{n_{NO_2}}{n_T} P = \frac{2a\alpha}{a(1+\alpha)} P = \frac{2\alpha}{1+\alpha} P$$

$\therefore$ $$K_P = \frac{(P_{NO_2})^2}{(P_{N_2O_4})} = \frac{4\alpha^2}{1-\alpha^2} P$$

EXAMPLE 14.4. *At 318 K and 1 atm, nitrogen tetroxide is 37·8% dissociated. To what extent would it be dissociated at 318 K and 2 atm?*

The degree of dissociation is given here as a percentage. This means that for every 100 moles of N_2O_4, 37·8 moles undergo dissociation. Hence the fraction of one mole dissociating (α) is 0·378.

$$K_P = \frac{4\alpha^2}{1-\alpha^2} P = \frac{4 \times 0·378^2 \times 1}{1 - 0·378^2} = 0·6533$$

The new condition of pressure does not alter the value of K_p. Hence

$$K_P = 0.6533 = \frac{8\alpha^2}{1-\alpha^2}$$

and

$$\alpha = 0.2742$$

An increase of pressure has, in agreement with our previous observation, reduced the value of α as a result of the equilibrium position being displaced to the left.

14.7. Influence of temperature on the equilibrium position

Changing the concentration of materials involved in an equilibrium or altering the pressure acting on the system can cause a displacement of the equilibrium position. There is, however, no consequent change in the value of the equilibrium constant unless the temperature changes.

By an argument beyond the scope of this book, it can be shown that the dependence of K_P on temperature is given by the equation

$$\frac{d \ln K_P}{dT} = \frac{\Delta H}{RT^2}$$

where R is the familiar gas constant and ΔH is the enthalpy change, i.e. the heat evolved (or absorbed) under a condition of constant pressure when the reaction proceeds *completely* from left to right.

The interesting feature of the above equation is that it indicates the influence of the sign and magnitude of the enthalpy term in determining the manner in which the equilibrium constant varies with temperature. Thus for an exothermic reaction, for which ΔH is negative, $d \ln K_P$ is negative when dT is positive. Hence K_P will decrease with increase in temperature and the equilibrium will be displaced in the direction which *absorbs* heat. If, on the other hand, ΔH is positive (endothermic reactions) then $d \ln K_P/dT$ will be positive, indicating that K_P will increase with increase in temperature, the equilibrium position again being displaced in the direction which absorbs heat. Conversely, a decrease in temperature (for which dT is negative) can be shown to displace the position of equilibrium in the direction which liberates heat. In short, *an increase in temperature favours the formation of those products formed with an absorption of heat, whilst a decrease in temperature favours the formation of those products formed with a liberation of heat.*

If it is desired to know values of K_P for a variety of temperatures, then the equation may be integrated between the limits of temperature T_1 and T_2. Thus, if $(K_P)_1$ and $(K_P)_2$ are the equilibrium constants corresponding to temperatures T_1 and T_2, then

$$\int_{(K_P)_1}^{(K_P)_2} d \ln K_P = \int_{T_1}^{T_2} \frac{\Delta H}{RT^2} dT$$

If we assume the temperature range to be not too large, ΔH can be regarded as a constant.

$$\therefore \qquad \ln \frac{(K_P)_2}{(K_P)_1} = \frac{\Delta H}{R} \int_{T_1}^{T_2} \frac{dT}{T^2} = \frac{\Delta H}{R} \left(\frac{T_2 - T_1}{T_1 T_2} \right)$$

$$\therefore \qquad \log \frac{(K_P)_2}{(K_P)_1} = \frac{\Delta H}{2 \cdot 303 R} \left(\frac{T_2 - T_1}{T_1 T_2} \right) \qquad \ln x = 2 \cdot 303 \log x$$

EXAMPLE 14.5. *For the reaction*

$$N_2(g) + O_2(g) \rightleftharpoons 2NO(g)$$

$K_p = 2 \cdot 31 \times 10^{-4}$ *at* 1900 K *and* $\Delta H = 90 \cdot 3$ *kJ* *mol*$^{-1}$. *What is the value of* K_p *at* 2300 K?

The value of ΔH given represents the heat absorbed (under a condition of constant pressure) when *one* mole of NO is produced from its elements; hence, when two moles are formed 180·6 kJ will be absorbed.

$$\therefore \qquad \log \frac{K_{2300}}{K_{1900}} = \frac{180,600}{2 \cdot 303 \times 8 \cdot 314} \left(\frac{2300 - 1900}{2300 \times 1900} \right) = 0 \cdot 863$$

$$\therefore \qquad \frac{K_{2300}}{K_{1900}} = 7 \cdot 30 \quad \text{and} \quad K_{2300} = 16 \cdot 9 \times 10^{-4}$$

This result is consistent with our previous observations in that the equilibrium position has been displaced in such a way as to increase the concentrations of those products formed with the absorption of heat.

14.8. Influence of catalyst on equilibrium position

In Chapter 13 we noted that in a reversible reaction, a catalyst which accelerates the forward reaction will also accelerate the reverse reaction. In fact, the two opposing rates change proportionally. The net effect is to hasten the establishment of equilibrium without altering the equilibrium concentrations or the equilibrium constant.

14.9. Industrial manufacture of ammonia

Ammonia is produced industrially by the reaction between nitrogen and hydrogen.

$$N_2 + 3H_2 \rightleftharpoons 2NH_3$$

This reversible reaction is characterized in the forward direction by a reduction in the number of gaseous moles ($\Delta n = -2$) and by the liberation of heat ($\Delta H = -92$ kJ). Consider the consequences of these characteristics with a view to predicting the conditions most likely to result in a maximum yield of ammonia.

(a) *Equilibrium considerations*

The equilibrium constant for the ammonia synthesis reaction is given by

$$K_P = \frac{(P_{NH_3})^2}{(P_{N_2})(P_{H_2})^3} = \frac{(x_{NH_3})^2}{(x_{N_2})(x_{H_2})^3}(P)^{-2} = A P^{-2}$$

The above relationship indicates the influence of pressure on the position of equilibrium. In this case, an increase in pressure must result in an attendant increase in the A term if the equilibrium constant is to maintain its value appropriate to a particular temperature. *High pressure will therefore favour the formation of ammonia.*

The influence of temperature on the equilibrium position can be deduced from the equation

$$\frac{d \ln K_P}{dT} = \frac{\Delta H}{RT^2}$$

When ΔH is negative (exothermic reaction) a decrease in temperature must lead to an increase in the value of K_p. *Thus the use of low temperatures will favour the formation of ammonia.*

Our predictions regarding the influence of pressure and temperature on the position of equilibrium are confirmed in the following graph (Fig. 14.1) in which experimental values of the equilibrium percentage of ammonia in a 3 : 1 hydrogen–nitrogen gas mixture are plotted against the pressure at a variety of temperatures.

FIG. 14.1. Equilibrium concentration of ammonia in the reaction between nitrogen and hydrogen.

(b) *Rate considerations*

Equilibrium considerations indicate *how far* we can expect the reaction to proceed to the right under certain conditions of pressure and temperature. Such information will not, however, tell us *how rapidly* the equilibrium will establish itself. Thus a decrease in temperature, although increasing the equilibrium percentage of ammonia, decreases the rate at which nitrogen and hydrogen react, so much so that at temperatures below 300°C the reaction would not be economically feasible. Nor would the situation be improved by using very high temperatures for although the rate would markedly increase (as a result of an increase in the proportion of effective collisions) the equilibrium percentage of ammonia would be decreased.

Circumstances such as this suggest the use of a catalyst, in this case an iron oxide. In this way, the activation energy for the reaction is decreased and the temperature which optimizes the rate and yield is in the region of 500°C. The pressures used vary from 100 to 1000 atm.

At present, all plants manufacturing ammonia use modifications of the original Haber process introduced at the turn of this century. The sources of nitrogen and hydrogen are many and varied. Two common sources are *producer gas* ($CO + N_2$) and *water gas* ($CO + H_2$).

To obtain gas containing the proper proportions of nitrogen and hydrogen, a gas "producer" alternately manufacturing producer and water gas is used (Section 12.5). These two gases are mixed in such proportions as to yield after suitable treatment nitrogen and hydrogen in the ratio of 1 : 3.

14.10. Producer gas

Producer gas (Section 12.5) is obtained when air is passed through a bed of white-hot coke, the sequence of reactions being

(1) $\quad C + O_2 \quad \rightarrow \quad CO_2 \qquad \Delta H = -394 \text{ kJ}$

(2) $\quad C + CO_2 \rightleftharpoons 2CO \qquad \Delta H = \quad 173 \text{ kJ}$

Reaction (1) proceeds to virtual completion; reaction (2), however, is reversible and an equilibrium is set up in the gas producer between the carbon, carbon monoxide and carbon dioxide. Irrespective of whether the producer gas is to be used as a fuel or as a source of nitrogen, it is desirable to have as little carbon dioxide present in the final mixture of gases as possible. We wish, in other words, to select conditions which will generate as high a ratio of carbon monoxide to carbon dioxide as possible. In view of the fact that the carbon monoxide is formed with the absorption of heat, its formation will therefore be favoured by employing high temperatures. Such a prediction is confirmed by the experimental data (Fig. 14.2). From these results it is clear that the temperature of the producer should not be allowed to fall below 1000°C if the proportion of carbon dioxide is to be kept to a minimum. (The condition of high tem-

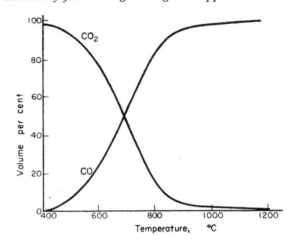

FIG. 14.2. Equilibrium concentration of gases in the reaction between carbon and oxygen.

perature has, of course, the added advantage of increasing reaction rate.) If this temperature is maintained, effectively all the carbon dioxide produced in reaction (1) is consumed in reaction (2) and the overall heat change is moderately exothermic.

$$C + O_2 \rightarrow CO_2 \qquad \Delta H = -394 \text{ kJ}$$
$$\underline{C + CO_2 \rightarrow 2CO \qquad \Delta H = 173 \text{ kJ}}$$
$$2C + O_2 \rightarrow 2CO \qquad \Delta H = -221 \text{ kJ}$$

14.11.   Water gas

The passage of steam through a bed of hot coke results in the production of water gas (Section 12.5). The reaction between steam and coke is reversible.

$$C + H_2O \rightleftharpoons CO + H_2 \qquad \Delta H = 131 \text{ kJ} \qquad (1)$$

Some carbon dioxide is always present due to the reaction

$$CO + H_2O \rightleftharpoons CO_2 + H_2 \qquad \Delta H = -41 \text{ kJ} \qquad (2)$$

As with producer gas, the carbon dioxide content should be as low as possible. This condition can be realized by maintaining a high temperature which in addition to producing high reaction rate will favour the formation of products in the endothermic reaction (1) and will suppress the formation of products in the exothermic reaction (2), the equilibrium position being shifted well over to the left.

The manner in which the composition of the system changes with temperature (Fig. 14.3) clearly confirms the need to maintain a temperature of at least 1000°C if the carbon dioxide content of the mixture is to be minimal.

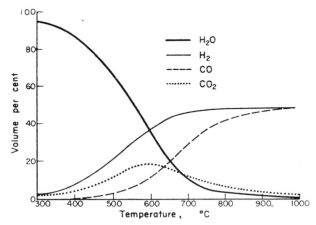

FIG. 14.3. Equilibrium concentration of gases in the reaction between steam and carbon. (Munro, L. A., *Chemistry in Engineering*, Prentice-Hall, New Jersey, 1964.)

14.12. Catalytic decomposition of water gas

In the production of water gas, a condition of temperature was chosen so as to minimize the extent to which the carbon monoxide was decomposed by steam

$$CO + H_2O \rightleftharpoons CO_2 + H_2 \qquad \Delta H = -41 \text{ kJ}$$

If it is desired to obtain hydrogen by this reaction (using water gas as a source of carbon monoxide) then conditions have to be chosen which facilitate the decomposition of carbon monoxide. Thus, while the yield of hydrogen will be clearly favoured by a low temperature (the reaction being exothermic) it will be unaffected by whatever pressure is used, there being no change in the number of gaseous moles ($\Delta n = 0$).

Although from a consideration of equilibrium factors a low temperature is ideal, such a condition is unsuitable from the point of view of reaction rate. Once more, a compromise must be made. In this case, a temperature of 450–500°C gives a reasonable rate with the help of a catalyst (ferric oxide and chromium oxide) together with a good equilibrium yield of hydrogen, which indeed can still further be improved by increasing the proportion of steam. This results, in accordance with the law of chemical equilibrium, in a displacement of the equilibrium position from left to right.

14.13. Steam reforming

The method used to obtain hydrogen as a feedstock for the manufacture of ammonia depends on the availability of suitable starting materials. Until the discovery of natural gas in the North Sea, hydrogen was obtained in the United Kingdom by the gasification of coal (Section 25.6) followed by the water gas reaction. However, the rising cost of coal compared with the relatively cheap North Sea gas encouraged the development of a process which could employ natural gas as a starting material. This process involves the so-called *steam reforming* of methane, the principal constituent of natural gas. The reaction is carried out by first passing a mixture of gas and steam over a nickel catalyst.

$$CH_4 + H_2O \rightleftharpoons CO + 3H_2 \qquad \Delta H = 206 \text{ kJ}$$

The endothermic character of the reaction suggests the use of a high temperature which in actual practice is in the range of 750–850°C.

In a second stage of the reforming there is added to the gas mixture (consisting mainly of steam, carbon monoxide, hydrogen and some residual methane) a supply of air, the amount of which is calculated to give the necessary N_2/H_2 ratio of 3 to 1 for subsequent ammonia synthesis. The relatively small amount of oxygen present in the air oxidizes a small amount of the gases, and is itself completely consumed.

The gas stream is then cooled to about 200°C in a separate vessel where the carbon monoxide is catalytically converted to carbon dioxide by the reaction

$$CO + H_2O \rightleftharpoons CO_2 + H_2 \qquad \Delta H = -41 \text{ kJ}$$

As already noted in Section 14.12, a low temperature displaces the equilibrium from left to right and at 200°C the conversion to carbon dioxide is almost complete.

The carbon dioxide is then almost entirely removed by absorption in aqueous potassium carbonate solution, but because even small amounts of carbon monoxide and carbon dioxide would "poison" the catalyst used in the ammonia synthesis, their concentration must be further reduced. This is ingeniously achieved through the reversal of the initial steam-reforming reaction

$$CO + 3H_2 \rightleftharpoons CH_4 + H_2O \qquad \Delta H = -206 \text{ kJ}$$

which when carried out at the lower temperature of about 350°C almost completely converts carbon monoxide to methane which, in small amounts, is not harmful to the catalyst. The carbon dioxide is similarly reduced to methane according to the equation

$$CO_2 + 4H_2 \rightleftharpoons CH_4 + 2H_2O \qquad \Delta H = -165 \text{ kJ}$$

Exercises

14.1. Under what conditions should carbon dioxide be stored in order to minimize the decomposition that can occur through the endothermic reaction

$$CO_2 \rightleftharpoons CO + O_2$$

[*Ans.*: High pressure, low temperature.]

14.2. Use the following information to deduce whether the reaction

$$A \rightleftharpoons B$$

is endothermic or exothermic in the forward direction

Temperature	Equilibrium constant
1000 K	10^{-4}
1200 K	10^{-5}

[*Ans.*: Exothermic.]

14.3. Show that for the homogeneous gas phase reaction

$$A(g) \rightleftharpoons 2B(g) + C(g)$$

the equilibrium constant K_p can be written as

$$K_p = \left[\frac{X_B^2 X_C}{X_A} \right] P^2$$

where P is the total pressure, and X_A, X_B, X_C are the equilibrium mole fractions of A, B and C respectively.

14.4. Gaseous formaldehyde dissociates according to the equation

$$H \cdot CHO(g) \rightleftharpoons H_2(g) + CO(g)$$

Show that the degree of dissociation α of formaldehyde is related to the total pressure P by the expression

$$\alpha = \sqrt{\frac{K_p}{K_p + P}}$$

14.5. 2·00 moles of A and 4·00 moles of B are placed in a 10 l container where they react according to the equation

$$A + 2B \rightleftharpoons 3C$$

The equilibrium concentration of A was found to be 0·16 mol l^{-1}. (a) What are the equilibrium concentrations of B and C? (b) What is the value of the equilibrium constant?

[*Ans.*: (a) 0·32, 0·12 mol l^{-1}; (b) 0·11.]

14.6. The equilibrium constant for the reaction

$$CO(g) + H_2O \rightleftharpoons CO_2(g) + H_2(g)$$

is 1·9 at 950 K. Show that an original equimolar mixture of CO and H_2O yields 29% of CO_2 (or H_2O) at equilibrium.

14.7. Ethanol is manufactured industrially by the gas phase hydration of ethylene

$$C_2H_4(g) + H_2O(g) \rightleftharpoons C_2H_5OH(g)$$

Deduce qualitatively the conditions of temperature and pressure necessary to produce the maximum *equilibrium* yield of ethanol.

What other factor must be considered in order to obtain the maximum *economic* yield?

Substance	$\Delta H_f / kJ\ mol^{-1}$
$C_2H_4(g)$	52·3
$H_2O(g)$	− 241·8
$C_2H_5OH(g)$	− 235·3

[*Ans.*: High pressure, low temperature.]

CHAPTER 15

Acids and Bases

15.1. Dissociation constants

Although some characteristics of acids and bases (e.g. bitter taste, corrosive action) are well known, these properties cannot be used to define the terms "acid" and "base". Perhaps the most widely used definition of these terms is that an acid is a substance which gives rise to hydrogen ions in solution while a base produces hydroxyl ions:

$$HNO_3 \rightarrow H^+ + NO_3^-$$
$$KOH \rightarrow K^+ + OH^-$$

Both the hydrogen and hydroxyl ions will be solvated, i.e. will be fairly strongly associated with one or more solvent molecules. In particular, the very small hydrogen ion H^+ is not very stable, and in aqueous solution will form a co-ordinate linkage with a water molecule to form a *hydronium ion* H_3O^+.

$$H^+ + O{\overset{\displaystyle H}{\underset{\displaystyle H}{<}}} \rightarrow \left[H-\overset{\displaystyle \overset{H}{|}}{O}-H \right]^+$$

On this basis the characteristic properties of acids are in fact the properties of the hydronium ion, and since the common acids are covalent molecules in the pure state, they do not then behave as acids. Pure anhydrous hydrogen chloride, for example, will have no effect on litmus.

Another point to note is that the formula of any compound gives little indication of its properties. If we consider the three materials phenol (C_6H_5OH), methanol (CH_3OH) and sodium hydroxide (NaOH), we find

236

that although each possesses an —OH group, phenol behaves as an acid, sodium hydroxide as a base, and methanol as neither.

Since acidic properties have been defined as the properties of the hydrogen (or hydronium) ion, it follows that the greater the concentration of hydrogen ions in solution, the more pronounced will be the acidity. If a molecule dissociates in solution to form hydrogen ions, the extent of dissociation will be governed by the equilibrium constant for the reaction:

$$HA \rightleftharpoons H^+ + A^+$$

$$K_a = \frac{[H^+][A^-]}{[HA]}$$

This equilibrium constant is called the *dissociation constant* of the acid.

In aqueous solution at 25°C the value of K_a for HCl is of the order of $1{\cdot}0 \times 10^7$, while K_a for acetic acid is $1{\cdot}8 \times 10^{-5}$. In aqueous solution, therefore, HCl is almost completely dissociated into ions and is said to be a *strong acid*, while acetic acid is only slightly dissociated and is described as a *weak acid*.

Similarly the strength of a base may be classified by the value of the equilibrium constant for the reaction

$$MOH \rightleftharpoons M^+ + OH^-$$

$$K_b = \frac{[M^+][OH^-]}{[MOH]}$$

Again we find *strong bases* (e.g. sodium hydroxide) with very large dissociation constants, and *weak bases* (e.g. ammonium hydroxide) which are only weakly dissociated.

15.2. Proton transfer theory

Although the definition of acids and bases outlined above is a very useful one, particularly for aqueous solutions, it has certain limitations. One of the most serious is that bases are limited to compounds which can give rise to hydroxyl ions. Furthermore, many compounds which cannot be classified as acids or bases on our simple definition exhibit acidic or basic properties in for example catalytic reactions, or in solvents other than water.

For these reasons the acid–base concept was broadened by Brønsted and Lowry, independently in 1923. They defined an acid as any substance which can give up a proton, while a base is any substance which can accept a proton. In other words, an acid is a *proton donor*, and a base is a *proton acceptor*. On this basis, the relationship between an acid and a base is given by the equation:

$$A \rightleftharpoons B + H^+$$

Acid   Base   Proton

Here the acid A and base B which differ from one another by a proton are described as a *conjugate pair*, i.e. the base B is said to be the conjugate base of the acid A.

We have already seen that a proton does not exist as such in solution, but combines with a water molecule to form the hydronium ion. In this case the water molecule has acted as a proton acceptor, that is, as a base. Whereas on the simple theory we would write

$$HNO_3 \rightleftharpoons H^+ + NO_3^-$$

$$CH_3CO_2H \rightleftharpoons H^+ + CH_3CO_2^-$$

$$HSO_4^- \rightleftharpoons H^+ + SO_4^{2-}$$

on the proton transfer picture these equations would be changed to

$$HNO_3 + H_2O \rightleftharpoons H_3O^+ + NO_3^-$$

Acid$_1$ Base$_2$ Acid$_2$ Base$_1$

$$CH_3CO_2H + H_2O \rightleftharpoons H_3O^+ + CH_3CO_2^-$$

Acid$_1$ Base$_2$ Acid$_2$ Base$_1$

$$HSO_4^- + H_2O \rightleftharpoons H_3O^+ + SO_4^{2-}$$

Acid$_1$ Base$_2$ Acid$_2$ Base$_1$

In each of these reactions water acts as a base. Furthermore, the nitrate, sulphate and acetate ions are classified as bases, because each is capable of accepting a proton to form its conjugate acid.

The proton transfer theory interprets the term "base" much more broadly than did the simple dissociation theory, which limited the definition to solutions of OH^- ions. From the few examples quoted above it can be seen that the OH^- ion is only one of a large number of bases, and while compounds such as NaOH are still described as bases, it is the OH^- ion which is the active constituent.

We have described how water can act as a proton acceptor, i.e. as a base:

$$H_2O + H^+ \rightleftharpoons H_3O^+$$
$$\text{Base} \qquad\qquad \text{Acid}$$

but it can also act as a proton donor, i.e. an acid

$$H_2O \rightleftharpoons H^+ + OH^-$$
$$\text{Acid} \qquad\qquad \text{Base}$$

These two equations may be combined to give

$$H_2O + H_2O \rightleftharpoons H_3O^+ + OH^-$$
$$\text{Acid}_1 \quad \text{Base}_2 \qquad \text{Acid}_2 \quad \text{Base}_1$$

In other words, water undergoes a self-ionization process and the acidity or basicity of any aqueous solution can be regarded as the result of disturbing this equilibrium by the addition of some other material. The well-known neutralization reaction of an acid with a base, once defined by the equation

$$\text{Acid} + \text{Base} \rightarrow \text{Salt} + \text{Water}$$

is simply the reaction between hydronium and hydroxyl ions, i.e. the reverse reaction written above.

It must be appreciated that water is by no means unique in its ability to undergo self-ionization, and that other solvents can act in this same way. Ammonia, for example, can act either as an acid, or as a base:

$$NH_3 \rightleftharpoons NH_2^- + H^+$$
$$\text{Acid} \quad\ \text{Base}$$

$$NH_3 + H^+ \rightleftharpoons NH_4^+$$
$$\text{Base} \qquad\qquad \text{Acid}$$

15.3. Lewis acids and bases

The acid–base concept can be broadened even further if the definition given by G. N. Lewis is adopted. He defined an acid as any substance which is capable of *accepting* an electron pair, and a base as any substance capable of *donating* an electron pair. Thus in the reaction

$$
H^+ + :N\!\div\!H \quad\rightleftharpoons\quad \left[H\!\div\!N\!\div\!H \right]^+
$$

H^+ is the acid and NH_3 the base because of the formation of a co-ordinate bond. Now, however, substances containing no hydrogen at all may also be classified as acids, as in the reaction

$$
F\!\div\!B + :N\!\div\!H \quad\rightleftharpoons\quad F\!\div\!B\!\div\!N\!\div\!H
$$

Here boron trifluoride is termed an acid because again a co-ordinate bond is formed.

While this concept can be extremely useful in considering, for example, certain organic reactions, it suffers from several disadvantages and need not concern us further here.

15.4. The strength of acids and bases

It has already been described (Section 15.1) how the strength of acids and bases may be given by the value of the dissociation constant K_a for the reaction

$$
HA \rightleftharpoons H^+ + A^-
$$

From what has been said in the previous section it is obvious that the greater the ability of the solvent to act as a base, the greater will be the strength of the acid. In other words, the equilibrium in the reaction

$$
HA + H_2O \rightleftharpoons H_3O^+ + A^-
$$

will be well over to the right-hand side. The effective strength of any acid, therefore, can be varied simply by changing the solvent.

Acetic acid, for example, is normally thought of as a weak acid, i.e. a substance which has a slight tendency to donate protons. It can, however, be made to act as a proton acceptor, although it will not act in this way as readily as does water. The extent of the formation of the $CH_3CO_2H_2^+$ ion

$$HA + CH_3CO_2H \rightleftharpoons CH_3CO_2H_2^+ + A^-$$
$$\text{Acid}_1 \qquad \text{Base}_2 \qquad\quad \text{Acid}_2 \qquad \text{Base}_1$$

depends very much on the strength of the acid HA.

If we consider the strengths of some of the common acids in aqueous solution we find that a relationship exists between the strength and structure of the acids. The common acids are of two main types, *hydro acids* (HCl, H_2S) and *oxy acids* (HNO_3, H_2SO_4).

In the hydro acid series, two generalizations can be made. Within any group in the periodic table, acid strength increases with increasing molecular weight. Thus acid strength increases in the group

$$HF < HCl < HBr < HI$$

Since it is the function of an acid to yield protons in solution this trend may seem somewhat anomalous when it is remembered that the H—F bond has far more ionic character than has the H—I bond. The important factor in this case is, however, the difference in size between the fluorine and iodine atoms. Because the iodine atom is so much larger than fluorine, the H—I bond is correspondingly weak; consequently a proton can be more easily abstracted from HI by the action of a base.

The second generalization is that within any horizontal period of the periodic table, acid strength increases as the electronegativity of the atom bonded to hydrogen increases. Thus in the series

$$NH_3 : H_2O : HF$$

HF is the strongest acid. This is because the change in size of the central atom in moving horizontally along a periodic series is not nearly so pronounced as that involved in moving down a group. Consequently the more polar the bond, the more readily may a proton be removed from the molecule.

Similarly the strength of an oxy acid is determined by both the electronegativity and the size of the central atom, the former usually being the dominant factor.

15.5.   The pH scale

It has already been explained in Section 15.2 that water undergoes a self-ionization process to form hydrogen and hydroxyl ions

$$H_2O + H_2O \rightleftharpoons H_3O^+ + OH^-$$

It is more convenient to write this equation as

$$H_2O \rightleftharpoons H^+ + OH^-$$

and simply assume that H^+ represents a hydrated proton. It is obvious from this equation that for each H^+ ion produced by the ionization of water, one OH^- ion is also produced. In other words, pure water contains an equal concentration of both H^+ and OH^- ions, and is said to be *neutral*.

The equilibrium constant for the ionization of water can be written

$$K = \frac{[H^+][OH^-]}{[H_2O]}$$

Since only a very small fraction of the water molecules ionize (at 25°C the concentration of both H^+ and OH^- ions is 1×10^{-7} mol l^{-1}) the concentration of undissociated water in the above equation may be taken as constant. By combining this constant concentration with the equilibrium constant a new constant K_w is obtained:

$$K[H_2O] = K_w = [H^+][OH^-]$$

where K_w is called the *ionic product* of water. In any aqueous solution, therefore, the product of the concentration of H^+ ions (mol l^{-1}) and the concentration of OH^- ions (mol l^{-1}) is a constant and has the value $1 \cdot 0 \times 10^{-14}$ at 25°C.

It follows from this that all aqueous solutions contain both H^+ and OH^- ions, even if the solution is strongly basic or acidic. For example, a 1 molar solution of HCl, assuming complete dissociation, has a H^+ concentration of $1 \cdot 0$ mol l^{-1}. Since $K_w = [H^+][OH^-] = 1 \cdot 0 \times 10^{-14}$ the concentration of OH^- ions is given by

$$[OH^-] = \frac{1 \cdot 0 \times 10^{-14}}{[H^+]}$$
$$= 1 \cdot 0 \times 10^{-14} \text{ mol } l^{-1}$$

Similarly, in a 1 molar solution of NaOH, the OH^- concentration is $1 \cdot 0$ mol l^{-1}; thus the H^+ ion concentration is:

$$[H^+] = \frac{1 \cdot 0 \times 10^{-14}}{[OH^-]}$$
$$= 1 \cdot 0 \times 10^{-14} \text{ mol } l^{-1}$$

From these simple examples we find that in ordinary solutions the H^+ ion concentration can vary from $1 \cdot 0$ to $1 \cdot 0 \times 10^{-14}$ mol l^{-1}. Since this range of concentration is too large to be convenient, the acidity of a solution is usually expressed on a logarithmic scale called the *pH scale*. The pH of a solution is defined:

$$pH = -\log_{10}[H^+] = \log_{10}\frac{1}{[H^+]}$$

Thus in the examples already considered in this section, the pH may be calculated:

(a) Pure water

$$[H^+] = 1 \cdot 0 \times 10^{-7} \text{ mol } l^{-1}$$
$$pH = -\log_{10}(10^{-7})$$
$$= -(-7)$$
$$= 7$$

(b) 1 molar HCl

$$[H^+] = 1 \cdot 0$$
$$pH = -\log_{10}(1 \cdot 0)$$
$$= 0$$

(c) 1 molar NaOH

$$[H^+] = 1 \cdot 0 \times 10^{-14}$$
$$pH = -\log_{10}(10^{-14})$$
$$= -(-14)$$
$$= 14$$

It will be noticed that a pH value of 7 signifies a neutral solution, a value of less than 7 an acid solution, and a value of greater than 7 an alkaline solution.

15.6.   Buffer solutions

There are occasions when it is essential to control the pH of a chemical system within a fairly narrow range, e.g. at certain stages in the manufacture of melamine–formaldehyde resins (see Chapter 28). This can be achieved by using a buffer solution, which is a solution whose pH is not appreciably altered by the addition of either H^+ or OH^- ions. Buffer solutions are usually mixtures of a weak acid and its salt.

The reason for the resistance to change in pH of a buffer solution can most readily be understood by considering a specific example. A mixture of acetic acid and sodium acetate can act as a buffer solution. If H^+ ions are added to such a system, the acetate ions will combine with the added H^+ ions to form undissociated acetic acid. Similarly, if OH^- ions are added, they will combine with the H^+ ions already in solution to form water, consequently some of the acetic acid molecules will then ionize to maintain the equilibrium. In neither case, therefore, will the pH of the system be altered to any extent.

Buffer solutions can be prepared to cover the whole pH range, but any given buffer solution is useful only over a range of about 2 pH units.

*

15.7.   Common acids and bases

Sulphuric acid (H_2SO_4), perhaps the most well-known mineral acid, is widely used in the manufacture of fertilizers, textiles and paper. It is itself a fairly cheap chemical, the basic raw material being sulphur or a sulphur containing ore such as FeS_2 (pyrites). The sulphur is easily converted to gaseous SO_2 which may then be oxidized to sulphuric acid.

One method of achieving this last oxidation step is the *lead chamber process*, where sulphur dioxide, oxygen and steam, together with oxides of nitrogen, are reacted in large lead-lined vessels. This is an example of a *homogeneous catalytic reaction*, where the nitrogen oxides catalyse the oxidation of the gaseous sulphur dioxide. The overall reaction may be represented:

$$SO_2 + H_2O \xrightarrow{\ NO_2\ } (NO)HSO_4 \xrightarrow{\ H_2O\ } H_2SO_4 + NO + NO_2$$

nitrosyl
sulphonic acid

It will be noticed that the oxides of nitrogen react in the early stages of the reaction to form an intermediate compound which subsequently hydrolyses to form sulphuric acid, at the same time regenerating the nitrogen oxide catalyst.

The lead chamber process has now been superseded by the *contact process*, which is more suitable for the preparation of high purity and over-strength acids. In this case the oxidation of SO_2 to SO_3 is achieved by means of a *heterogeneous catalyst*. The original catalyst used was platinized asbestos, but vanadium pentoxide has been found to be cheaper and is not so susceptible to "poisoning" by impurities. The overall reaction may be written:

$$SO_2 + O_2 \xrightarrow{\ V_2O_5\ } SO_3$$
$$SO_3 + H_2O \longrightarrow H_2SO_4$$

In practice it is found that the absorption of SO_3 in water is not very successful, but it is readily absorbed in sulphuric acid itself to yield pyrosulphuric acid, which can be converted to sulphuric acid by the addition of water.

$$SO_3 + H_2SO_4 \rightarrow H_2S_2O_7$$
$$H_2S_2O_7 + H_2O \rightarrow 2H_2SO_4$$

It is interesting to consider the contact process in terms of the principles of chemical equilibrium described in Chapter 14. We might expect that the reaction would be carried out at a high pressure since the formation of SO_3 from SO_2 and O_2 involves a reduction in volume. In fact the reaction is carried out at low pressure, the reason being that the rate of formation of SO_3 is determined by the rate of diffusion of the reactants to the catalyst surface through a strongly adsorbed layer of SO_3. Increasing the pressure does not accelerate this reaction. Furthermore, since the reaction

$$2SO_2 + O_2 \rightarrow 2SO_3$$

is exothermic, we expect that a low temperature will be necessary for maximum yield of product. Unfortunately the rate of reaction is found to be too slow at low temperatures and a compromise temperature of about 450°C is used.

Sulphuric acid is a strong dibasic acid, ionizing readily in aqueous solution in two stages:

$$H_2SO_4 \rightleftharpoons H^+ + HSO_4^-$$
$$HSO_4^- \rightleftharpoons H^+ + SO_4^{2-}$$

The first dissociation is complete, but the second occurs only to a limited extent. It is thus possible to make two types of salts from sulphuric acid: *acid salts* and *normal salts*.

$$H_2SO_4 + NaOH \rightarrow \underset{\substack{\text{Sodium} \\ \text{bisulphate}}}{NaHSO_4} + H_2O$$

Sodium bisulphate is described as an acid salt because it contains the HSO_4 ion which can itself dissociate to yield protons.

$$NaHSO_4 + NaOH \rightarrow Na_2SO_4 + H_2O$$

Apart from being used for the preparation of hydrochloric and nitric acids from their salts, sulphuric acid reacts with hydrocarbons to form another interesting series of acids called *sulphonic acids*.

$$\hexagon C_{10}H_{21} + H_2SO_4 \rightarrow HO_3S \hexagon C_{10}H_{21} + H_2O$$

These are monobasic acids, and their salts are important commercially as *detergents* (Chapter 29).

Concentrated sulphuric acid has a very strong affinity for water and indeed can form several hydrates, e.g. $H_2SO_4 \cdot H_2O$; $H_2SO_4 \cdot 2H_2O$. An obvious application of this property is the use of H_2SO_4 as a drying agent or as a dehydrating agent, e.g. dehydration of alcohols.

Another important property of sulphuric acid is its ability to act as an oxidizing agent in concentrated solution. Many non-metals are readily oxidized by means of sulphuric acid, e.g. C, S, Br^-, I^-.

$$C + 2H_2SO_4 \rightarrow 2SO_2 + CO_2 + 2H_2O$$

Furthermore, metals which lie below hydrogen in the electrochemical series, and which are therefore not attacked by dilute acids, can be readily oxidized by concentrated H_2SO_4.

$$Cu + 2H_2SO_4 \rightarrow CuSO_4 + 2H_2O + SO_2$$

Hydrochloric acid is the name commonly given to a solution of HCl in water, while pure HCl, which is a gas, is referred to simply as hydrogen chloride.

It has already been mentioned that HCl can be prepared by the action of sulphuric acid on metal chlorides, but it is also prepared as a by-product from the chlorination of organic materials, and this source is becoming increasingly important.

$$CH_3CH_3 + Cl_2 \rightarrow CH_3CH_2Cl + HCl$$

Apart from undergoing the usual reactions with metals, with the evolution of hydrogen, hydrochloric acid in concentrated solution can act as a *complexing agent*. A very powerful solvent can be made by mixing three parts of concentrated hydrochloric acid with one part concentrated nitric acid. This mixture, called *aqua regia*, owes its high solvent power to the combination of the oxidizing properties of nitric acid with the complexing properties of the hydrochloric acid. Metallic gold, for example, which is insoluble in either of the acids alone, dissolves in aqua regia to form the complex anion $[AuCl_4]^-$.

Nitric acid, like HCl, can be simply prepared by the action of sulphuric acid on metal nitrates, but the bulk of commercial nitric acid is obtained from the hydrolysis of NO_2, which is in turn obtained by the oxidation of ammonia.

The high temperature oxidation of ammonia can yield two main products, nitrogen or nitric oxide (NO).

$$4NH_3 + 3O_2 \rightarrow 2N_2 + 6H_2O$$
$$4NH_3 + 5O_2 \rightarrow 4NO + 6H_2O$$

In the presence of platinum metal as catalyst, the second reaction is greatly accelerated relative to the first, and nitric oxide is the main product. This very readily reacts with oxygen to form nitrogen dioxide which dissolves in water to give nitric acid:

$$2NO + O_2 \rightarrow 2NO_2$$
$$3NO_2 + H_2O \rightarrow 2HNO_3 + NO$$

Nitric acid is a strong monobasic acid, being completely dissociated in dilute aqueous solution. It reacts readily with metals and metal oxides to form nitrates which are generally more readily soluble than the salts of other acids.

In the majority of its reactions, nitric acid acts as an oxidizing agent. In its reactions with metals, the gaseous product evolved is not usually hydrogen, but oxides of nitrogen. For example, a fairly non-reactive metal like copper will reduce nitric acid to NO_2 or NO, depending on the concentration of the acid used:

$$Cu + 4HNO_3 \xrightarrow[HNO_3]{conc.} Cu(NO_3)_2 + 2NO_2 + 2H_2O$$
$$3Cu + 8HNO_3 \xrightarrow[HNO_3]{dil.} 3Cu(NO_3)_2 + 2NO + 4H_2O$$

More reactive metals can reduce the nitric acid even further, to yield ammonium salts:

$$4Zn + 10HNO_3 \rightarrow 4Zn(NO_3)_2 + NH_4NO_3 + 3H_2O$$

However, not all metals react readily with nitric acid. Some (e.g. Al, Cr, Fe) are rendered "passive" by the formation of non-reactive oxide films on the metal surface. One further important application of nitric acid is its use in nitrating certain organic materials (Section 22.5).

Sodium carbonate (Na_2CO_3) is probably the most important alkali metal salt in industrial chemistry, being the cheapest available alkali. It is widely used in the manufacture of, for example, soap, glass and sodium hydroxide. The main method of manufacture is by the *solvay process*, where CO_2 gas is bubbled through a solution of ammoniacal brine to form sodium bicarbonate which can readily be converted by heating to sodium carbonate. The reaction essentially involves the formation of HCO_3^- ions which will combine with the Na^+ ions in solution to form the sparingly soluble bicarbonate.

$$H_2CO_3 \rightleftharpoons H^+ + HCO_3^-$$
$$NH_3 + H^+ \rightleftharpoons NH_4^+$$
$$Na^+ + HCO_3^- \rightarrow NaHCO_3$$

Carbonic acid is a weak acid which does not ionize to any appreciable extent. However, the presence of NH_3, which can combine with protons to form the NH_4^+ ion, greatly increases the ionization of the weak acid. The bicarbonate is converted to carbonate simply by heating, at the same time yielding CO_2 which can be recycled.

$$2NaHCO_3 \rightleftharpoons Na_2O_3 + H_2CO + CO_2$$

It should be noted that the other gaseous reactant (NH_3) can also be recycled, since it is easily recovered by treating the NH_4Cl byproduct with lime.

$$2NH_4Cl + Ca(OH)_2 \rightarrow CaCl_2 + 2H_2O + 2NH_3$$

Sodium hydroxide is manufactured either by a *causticizing process* or by *electrolysis*. The former method involves heating a solution of Na_2CO_3 with a suspension of $Ca(OH)_2$, when $CaCO_3$ precipitates from solution

$$Na_2CO_3 + Ca(OH)_2 \rightleftharpoons 2NaOH + CaCO_3$$

Since both $Ca(OH)_2$ and $CaCO_3$ are solids, the equilibrium constant for the reaction is

$$K = \frac{[NaOH]^2}{[Na_2CO_3]}$$

from which it can be seen that the reaction is more complete in dilute solution. Since this would involve increased evaporation costs eventually intermediate concentrations must be used.

The second method of manufacture involves the electrolysis of brine solution using a carbon anode and mercury cathode. Chlorine is evolved at the anode, and the sodium generated at the cathode dissolves in the mercury to form an amalgam which can then react with water in a separate cell to form sodium hydroxide solution.

$$2Cl^- \rightarrow Cl_2 + 2e$$
$$Na^+ + 1e \rightarrow Na$$

Exercises

15.1. Define the following terms:

(i) hydronium ion;
(ii) dissociation constant;
(iii) strong acid;
(iv) weak base;
(v) buffer solution;
(vi) ionic product of water;
(vii) oxy acids.

15.2. Explain what is meant by:

(i) the proton transfer theory of acids and bases;
(ii) the concept of Lewis acids and bases.

15.3. For each of the following groups of acids indicate whether you would expect the acid strength to increase or decrease as we move along the sequence. Give reasons for your answers.

(i) H_2O; H_2S; H_2Se; H_2Te;
(ii) HIO_3; $HBrO_3$; $HClO_3$.

[*Ans.*: (i) increase; (ii) increase.]

15.4. (a) Explain what is meant by the pH scale.

    (b) Calculate the pH of a solution where the concentration of hydrogen ions is:

      (i) 0.03 mol l^{-1};

      (ii) 0.06 mol l^{-1};

      (iii) 0.015 mol l^{-1}.

[*Ans.*: (i) 1.52; (ii) 1.22; (iii) 1.82.]

15.5. The solubility of $Ca(OH)_2$ is 0.02 mol l^{-1}. Assuming that all of the dissolved material exists in solution in the form of Ca^{2+} and OH^- ions, calculate the pH of the solution.

[*Ans.*: 12.6.]

15.6. The pH of a sample of blood is 7.4. Calculate the concentration of hydrogen ions in the solution.

[*Ans.*: 4.0×10^{-8} mol l^{-1}.]

CHAPTER 16

Water Treatment

WATER is at the same time the most abundant and the most widely used chemical compound. Apart from being essential for life, it is necessary for a vast number of types of industry. It is, for example, the most common heat transfer agent and is used therefore as a coolant in condensers, cooling towers, etc. Other major uses are for steam generation and as a solvent. The overall importance of water may be appreciated when it is realized that the production of 1 ton of steel requires about 100 tons of water, while about 700 tons are required in the manufacture of 1 ton of paper.

16.1. Water sources

The primary source of all water is the sea, from which moisture is evaporated to be deposited later as rain. Sea water itself is sometimes used as a coolant, but because of the high salt content it generally cannot be used without extensive treatment. The large-scale purification of sea water is now becoming economically feasible and is discussed in some detail in Chapter 19.

For large-scale use, water may be obtained from one of two sources, *ground water*, i.e. water held underground in saturated zones of sand or porous rock, and *surface water*, i.e. rivers, lakes or artificial reservoirs. Water from these sources is never pure. Even rain water, the purest natural form of water, contains dissolved materials, e.g. dissolved oxygen, nitrogen and carbon dioxide, together with sulphur compounds if deposited over industrial regions. Once it reaches ground level even more impurities are gathered, depending on the nature of the terrain on which the water collects. Underground waters as obtained from wells are generally colourless and free of suspended material. They do, however, tend to have a fairly high concentration of dissolved solids, particularly calcium and magnesium bicarbonates and sulphates, and possibly iron salts. Surface waters can vary a great deal depending on locality. Moorland waters, for example, tend to be coloured brown and are often slightly acid due to the presence of weak organic acids derived from peaty ground. Lowland waters, on the other hand, tend to be colourless, but often contain suspended solids which settle only very slowly. Surface waters are almost inevitably contaminated by industrial or domestic effluents. It can readily be seen therefore that water may vary greatly according to its source, and will in general require treatment of some kind before it is suitable for industrial or domestic use.

16.2. Sterilization

Domestic water supplies may be put to various uses; the most important characteristics necessary are that it should be free of harmful bacteria and that it should be of pleasing appearance and taste.

Chemically the most common method of sterilization is *chlorination*. When chlorine is dissolved in water a chemical reaction takes place with the formation of hydrochloric acid, which is fully ionized, and hypochlorous acid, which dissociates to only a small extent.

$$Cl_2 + H_2O \rightleftharpoons H^+ + Cl^- + HOCl$$
$$\text{Hypochlorous acid}$$

$$HOCl \rightleftharpoons H^+ + OCl^-$$

These different species coexist in equilibrium, the relative amounts being determined mainly by the pH of the solution, and it is therefore immaterial whether the chlorine is introduced as such or as hypochlorite. In normal waters (pH 6–8) very little molecular chlorine exists and the bulk of the chlorine added is in the form either of undissociated hypochlorous acid or as the hypochlorite ion. It is generally believed that it is the undissociated acid which is the active species in the sterilizing reaction. The effectiveness of treatment depends both on the chlorine concentration used and on the time of contact, and different water supplies will of course require different doses. The ultimate test of effectiveness is a bacteriological examination of the water to ensure that sterilization is complete.

It is sometimes useful to add another type of chlorine-containing compound to the water supply, viz. *chloramines*, which may be formed by the action of either chlorine or hypochlorous acid on ammonia:

$$NH_3 + Cl_2 \rightleftharpoons NH_2Cl + H^+ + Cl^-$$
$$NH_3 + HOCl \rightleftharpoons NH_2Cl + H_2O$$

The compound NH_2Cl is called monochloramine, but compounds containing even more chlorine may be obtained:

$$NH_2Cl + HOCl \rightleftharpoons \underset{\substack{\text{Dichlor-}\\\text{amine}}}{NHCl_2} + H_2O$$

$$NHCl_2 + HOCl \rightleftharpoons \underset{\substack{\text{Trichlor-}\\\text{amine}}}{NCl_3} + H_2O$$

These reactions are all reversible and the chloramines decompose slowly in water to yield the active hypochlorous acid. They thus act as a reserve of chlorine which can kill organisms which develop after the initial treatment.

If it is necessary to over-chlorinate the supply initially, the excess chlorine must be removed because too high a concentration imparts both taste and colour to the water. This can be achieved fairly simply by passing the water through a bed of activated charcoal, or by treatment with suitable reducing agents, e.g. sulphur dioxide or sodium thiosulphate.

$$SO_2 + Cl_2 + 2H_2O \rightarrow H_2SO_4 + 2HCl$$
$$2Na_2S_2O_3 + Cl_2 \rightarrow Na_2S_4O_6 + 2NaCl$$

Other chemicals, e.g. permanganates, hydrogen peroxide, or ozone, can be used effectively to bring about sterilization, but are in general more expensive treatments and are not widely used.

16.3. Clarification

Natural waters usually contain an amount of suspended material, organic and inorganic, which must be removed. This can be achieved by a three-stage process of *settlement, coagulation* and *filtration.*

Settlement is brought about simply by prolonged storage of the water concerned. In general it is found that a storage time of around 24 hours removes about 90% of the suspended material, but the remaining 10% requires a different treatment. This is so because the particles suspended in the water are of different shapes and sizes. The rate of sedimentation of a spherical particle may be calculated from the equation:

$$v = \frac{2r^2(D-d)g}{9\eta}$$

where v = velocity of the particle,
   r = radius of the particle,
   D = density of the particle,
   d = density of water,
   η = viscosity of water,
   g = gravitational constant.

The rate of sedimentation is seen to be dependent upon the radius of the particle. Particles of diameter greater than 2 or 3×10^{-6} m will settle in less than 24 hours, but particles much smaller than this would take weeks or even years to settle out.

These extremely small particles less than 1×10^{-6} m in diameter are described as *colloidal* and are so small that they cannot be filtered out (Chapter 11). The most important property of material in this finely divided state is that it possesses a very high surface area and as a consequence is electrically charged due to the adsorption of ions on the surface. Since the small particles all carry charges of like sign, either positive or negative, they do not coalesce to form large aggregates because of the repulsive forces acting between them. In order to bring about coagulation, therefore, we must destroy these forces of repulsion. This can be achieved by adding highly charged ions to the water, these being adsorbed on the surface of the particles, thus neutralizing the existing charge so that coagulation may occur. Compounds commonly used as coagulants include aluminium sulphate, ferric sulphate and activated silica. Recently, synthetic polyelectrolytes have been used as coagulants or coagulating aids.

After the coagulation and settlement treatment the water is finally filtered. This may be carried out by passing the water through a bed of sand or anthracite. In a new filter the sand initially exerts a simple sieving action, but normally the growth of a biological "mat" of living organisms occurs in the surface layers of the filter bed and this helps to effect a further purification of the water.

16.4.   Water hardness

We know that natural waters contain a large number of dissolved impurities, and the presence of certain specific impurities leads to the water being termed *hard*. Hardness was originally defined as the "consumption" of soap by the water, due to the presence of calcium or magnesium ions which react with soap to form insoluble scums. Although other metal ions have this same property, e.g. iron, barium, manganese, these are rarely present in sufficient amount to cause any concern, and in practice the degree of hardness is determined by the amount of calcium or magnesium present.

Hardness may be of two types, *temporary* or *permanent*. Temporary hardness is caused by the presence of bicarbonates of magnesium or calcium and may be destroyed by boiling, when the corresponding carbonates precipitate from solution.

$$2HCO_3^- \xrightarrow{\text{heat}} CO_3^{2-} + H_2O + CO_2$$

$$Mg^{2+} + CO_3^{2-} \longrightarrow MgCO_3$$

Permanent hardness, which is not removed by boiling, is due to the presence of soluble salts of calcium or magnesium other than the bicarbonates, e.g. sulphates.

Measurement of hardness

Originally the degree of hardness of a specific water was determined by titration with a standard soap solution. Small amounts of the soap solution were added to a known volume of water, the mixture being shaken after each addition. The "endpoint" of the titration was taken as the point where a stable lather was produced. The *total hardness* was estimated as above, the permanent hardness by repeating the experiment using boiled water, and the temporary hardness by difference.

A more recent method involves titrating the water sample with a reagent which forms complexes with calcium and magnesium ions. The usual reagent is ethylenediaminetetraacetic acid (EDTA) which is normally used as the disodium salt:

$$HOOC-CH_2 \diagdown$$
$$\qquad\qquad N-CH_2-CH_2-N \diagup ^{CH_2-COOH}$$
$$Na^+{}^-OOC-CH_2 \diagup \qquad\qquad\qquad \diagdown CH_2-COO^-Na^+$$

This large anion *chelates* (i.e. complexes) the calcium and magnesium ions and effectively removes them from solution:

$$^-OOC-CH_2 \diagdown \quad \diagup CH_2-CH_2 \diagdown \quad \diagup CH_2COO^-$$
$$\qquad\qquad N \qquad\qquad\qquad N$$
$$CH_2 \diagup \qquad\qquad \diagdown \qquad\qquad \diagdown CH_2$$
$$CO\text{------}O \diagup Ca \diagdown O\text{------}CO$$

When all the calcium and magnesium ions have reacted with the EDTA, a suitable indicator changes colour.

Since different water samples are hard for different reasons, e.g. one may contain calcium bicarbonate while another magnesium sulphate, the degree of hardness is by convention described in terms of a concentration of calcium carbonate. Thus from a knowledge of the appropriate molecular weights we know:

$$50 \text{ g } CaCO_3 \equiv 12 \text{ g } Mg \equiv 42 \text{ g } MgCO_3 \equiv 81 \text{ g } Ca(HCO_3)_2$$

Unfortunately although convention decrees that the degree of hardness should be expressed in terms of calcium carbonate, the actual concentration units vary from country to country, and indeed within some countries. Probably the most useful method is to express the concentration as parts per million (ppm), since 1 ppm is equivalent to 1 mg per litre. Waters with a hardness of less than 100 ppm are generally considered to be *soft*, while very hard waters will have values in excess of 300 ppm.

Removal of hardness by precipitation

The most obvious method of removing dissolved salts from water is by distillation, and in certain circumstances this procedure may be economic. A detailed account of the principles involved in removing fairly high concentrations of dissolved salts from water by distillation and also by the use of membranes is given in Chapter 19. The two most widely used methods for the removal of hardness on a large scale are the *lime–soda process* and *ion exchange processes*.

Before we consider how the unwanted ions may be precipitated from solution, we must learn something about the factors which can affect the solubility of sparingly soluble salts. If we consider the situation when a sparingly soluble salt such as magnesium hydroxide is placed in water, we find that a very small amount of the material dissolves to form magnesium and hydroxyl ions in solution. The solution process rapidly stops, however, because the solution soon becomes saturated. At this point a state of equilibrium is attained which may be represented by the equation

$$Mg(OH)_2(s) \rightleftharpoons Mg^{2+}(aq) + 2OH^-(aq)$$

We can write an equilibrium constant for this reaction, and since we are considering a heterogeneous reaction (see Chapter 14) this would have the form

$$K = [Mg^{2+}][OH^-]^2$$

This constant is called the *solubility product* and is normally written K_{sp}. The value of this constant determines the extent to which magnesium hydroxide can dissolve at any given temperature. If the ionic product of the magnesium and hydroxyl ions exceeds the value of the solubility product then magnesium hydroxide will precipitate from solution. Similarly if the ionic product is less than the solubility product, more magnesium hydroxide will dissolve until the equilibrium value is attained.

It is easy to see the effect of pH on the solubility of magnesium hydroxide. If the pH is increased, the concentration of hydroxyl ions in solution is increased. Since the ionic product $[Mg^{2+}][OH^-]^2$ must remain a constant, it follows that the magnesium ion concentration must decrease, i.e. magnesium hydroxide will be less soluble the higher the pH of the solution.

The solubility product of calcium carbonate is derived in the same way, i.e.

$$CaCO_3(s) \rightleftharpoons Ca^{2+}(aq) + CO_3^{2-}(aq)$$
$$K_{sp} = [Ca^{2+}][CO_3^{2-}]$$

The effect of pH on the solubility of calcium carbonate is not immediately obvious from the equation given above. It must be remembered, however, that in solution the carbonate ion can undergo further reactions and a number of equilibria are possible, e.g.

$$H_2CO_3 \rightleftharpoons H^+ + HCO_3^-$$
$$HCO_3^- \rightleftharpoons H^+ + CO_3^{2-}$$

We can see, therefore, that when carbonate ions are introduced into a solution, some will react with water to form the bicarbonate ion, thus reducing the actual concentration of carbonate ions. This effect will be larger in acid than in neutral solutions. The concentration of carbonate ions in the solubility product expression refers only to the material existing in that actual form at equilibrium, i.e. if $1 \cdot 0$ mol l^{-1} of carbonate ion is introduced and under the experimental conditions half of this changes to bicarbonate, then the concentration of carbonate is only $0 \cdot 5$ mol l^{-1}.

TABLE 16.1.
SOLUBILITY PRODUCT VALUES

Salt	Solubility product at 18°C
$Mg(OH)_2$	9×10^{-12}
$MgCO_3$	1×10^{-5}
$Ca(OH)_2$	1×10^{-6}
$CaCO_3$	1×10^{-8}

We can now consider the precipitation of calcium and magnesium from hard water. The lime–soda process involves the addition of hydrated lime and sodium carbonate to the water in order to precipitate the unwanted ions. Temporary hardness is removed by lime:

$$Ca(OH)_2 + Ca(HCO_3)_2 \rightarrow 2CaCO_3 + 2H_2O$$
$$Ca(OH)_2 + Mg(HCO_3)_2 \rightarrow Mg(OH)_2 + Ca(HCO_3)_2$$

We find that the calcium ions are simply removed as insoluble calcium carbonate, while the magnesium ions precipitate as the hydroxide, but at the same time an equivalent amount of calcium ions is introduced to the solution. This in turn must be removed by the addition of more lime as shown in the former equation. The net result is that the complete removal of magnesium bicarbonate requires twice as much lime as does the removal of calcium bicarbonate. The reason for the different behaviour of the calcium and magnesium salts can be seen in Table 16.1 where we find that for calcium the carbonate is the more insoluble salt, but for magnesium the hydroxide is the more insoluble and will therefore precipitate first from solution.

Permanent hardness due to calcium is removed by the addition of the sodium carbonate, while any permanent hardness due to magnesium is removed by the lime as shown above.

$$CaSO_4 + Na_2CO_3 \rightarrow CaCO_3 + Na_2SO_4$$

It may seem paradoxical that hardness can be removed by adding to the solution a salt containing calcium, and it is true that care must be taken in treatment of this kind otherwise the water could finish with a higher degree of hardness than it had initially. Normally a slight excess of lime is added because the solubilities of both calcium carbonate and magnesium hydroxide decrease as the pH increases, but the hardness cannot be completely removed by this method.

Removal of hardness by ion exchange

The principle on which this method is based is that some materials, insoluble in water, are capable of absorbing certain ions from solution, at the same time liberating an equivalent number of different ions. If therefore we can find a material which will remove the ions that cause hardness, e.g. magnesium and calcium, while liberating harmless ions, e.g. sodium, then water softening is possible.

The earliest ion-exchange materials were naturally occurring inorganic compounds, *zeolites*, which are complex aluminosilicates. These are composed of large three-dimensional structures containing sodium ions rather loosely bound to the large interlocking lattice. When water containing calcium or magnesium ions is passed through a bed of this material, ion exchange occurs between the solution and the zeolite, as shown by the following equations (the symbol Z represents the complex zeolite):

$$\left.\begin{array}{c} Ca^{2+} \\ Mg^{2+} \end{array}\right\} + Na_2Z \rightleftharpoons \left.\begin{array}{c} CaZ \\ MgZ \end{array}\right\} + 2Na^+$$

The harmful calcium and magnesium ions are thus removed and an equivalent amount of sodium ions is introduced in their place. The exchange reactions are reversible and the zeolites can be regenerated by passing an excess of brine solution through the zeolite bed.

The purification of water by ion exchange can be taken a step further by using synthetic *cation* and *anion exchange* resins which exchange hydrogen and hydroxyl ions respectively. A typical cation exchange resin would consist of a highly cross-linked polymer (e.g. cross-linked polystyrene or phenolformaldehyde) containing acidic substituents such as the sulphonic acid grouping $Z-SO_3H$. When hard water is passed through a resin of this type exchange reactions analogous to the zeolite reactions occur:

$$\left.\begin{array}{c} Ca^{2+} \\ Mg^{2+} \end{array}\right\} + ZH_2 \rightleftharpoons \left.\begin{array}{c} CaZ \\ MgZ \end{array}\right\} + 2H^+$$

Again the exchange is reversible, thus allowing the original resin to be regenerated by treatment with acid.

Anion exchange resins are cross-linked polymer structures containing basic substituents such as quaternary ammonium groups $Z-NR_3^+$. In this case exchange reactions occur with the liberation of hydroxyl ions:

$$2ZNR_3^+OH^- + SO_4^{2-} \rightleftharpoons (ZNR_3^+)_2SO_4^{2-} + 2OH^-$$

If water is treated with both a cation and an anion exchanger, not only is softening achieved but all mineral ions are removed from solution, leaving a water of exceptionally high purity.

16.5.   Boiler feed water

Water which is to be used in modern high pressure boilers must be very carefully treated because of the extremes of temperature and pressure encountered in day to day working. The two main problems to be overcome are the prevention of harmful scale formation and the prevention of corrosion.

It might be thought that if the water were softened as described in the previous section no problem of scale formation would arise. It must be remembered, however, that although water may be demineralized by synthetic ion exchangers this process is expensive to apply on a large scale to hard waters, and the precipitation processes do leave a certain amount of *residual hardness*. Although only a small amount of material remains in solution, two processes occur in boiler operation to bring about scale formation, viz. the solubility of most scale forming salts decreases with increasing temperature, and the rate of evaporation in a modern boiler can be very high. In either case the solubility product is eventually exceeded and scale deposition may occur.

One method of internal treatment is to add sodium carbonate to the boiler water. In this case, if the concentration of calcium ions builds up due to evaporation, calcium carbonate is precipitated as a sludge in the boiler. It appears at first sight that this is precisely what we wanted to avoid, but the carbonate is precipitated as a sludge which is easily removed from the boiler, and not as a highly cohesive film as would occur with, for example, calcium sulphate under the same conditions.

At this point we encounter conflicting effects of scale prevention and protection against corrosion. Sodium carbonate cannot be added to high pressure boilers because the carbonate ion reacts under these conditions to yield hydroxyl ions:

$$CO_3^{2-} + H_2O \rightleftharpoons 2OH^- + CO_2$$

High local concentrations of hydroxyl ions can lead to *caustic embrittle-ment*, a type of corrosion to which steel is susceptible, where cracking occurs in areas of high stress, e.g. around rivet holes. It is therefore dangerous to use the sodium carbonate treatment in high pressure boilers.

The addition of phosphates in the correct concentrations can solve many of the problems posed by the previous method. Calcium phosphate is quite insoluble and is precipitated in the form of a sludge which is easily removed from the boiler. Unlike the carbonate ion, the phosphate is not attacked by water. A series of phosphate salts may form in solution depending on the pH, e.g.

$$H_2PO_4^- \rightleftharpoons HPO_4^{2-} \rightleftharpoons PO_4^{3-}$$
$$\underrightarrow{\text{increasing pH}}$$

Indeed, mixtures of phosphates act as buffer solutions (Section 15.6), thus helping to maintain the boiler water at a fixed pH.

Although simple phosphates may be added to the boiler water supply, it is sometimes more satisfactory to use polyphosphates with the general formula $(NaPO_3)_n$. These have a twofold action in that they can be hydrolysed to form the simple phosphate ions discussed above, and have the property of forming soluble complexes with calcium ions, thus preventing the deposition of calcium salts in the boiler feed lines.

Exercises

16.1. Define the following terms:

    (i) ground water;
    (ii) surface water;
    (iii) solubility product;
    (iv) ion exchange;
    (v) zeolites;
    (vi) caustic embrittlement.

16.2. Write notes on:

    (i) water clarification;
    (ii) water sterilization.

16.3. (a) Define what is meant by the terms: (i) temporary hardness, (ii) permanent hardness.

   (b) Explain how the hardness of water may be measured and how it may be removed.

16.4. The solubility product of $Ca(OH)_2$ is 1.0×10^{-6}. Would you expect $Ca(OH)_2$ to precipitate when:

   (i) 2.0 l of a solution containing 0.05 mol l^{-1} of $CaCl_2$ is mixed with 0.5 l of a solution containing 0.5 mol l^{-1} of NaOH?

   (ii) 2.0 l of a solution containing 0.01 mol l^{-1} of $CaCl_2$ is mixed with 0.5 l of a solution containing 0.5 mol l^{-1} of NeOH?

[*Ans.*: (i) Yes; (ii) No.]

CHAPTER 17

Separation Techniques: I, Distillation

DISTILLATION is the process by which vapour is first formed from a liquid (through boiling), then condensed back to liquid (by cooling). The technique is used extensively in industry to achieve a separation of volatile from non-volatile material (as in the desalination of sea-water), and in the separation of a mixture of volatile components (as in the fractionation of crude petroleum).

17.1. Boiling-point diagrams

To acquire an understanding of the distillation process, let us turn our attention to the simplest type of binary (two-component) liquid mixture, namely, one in which the two components are miscible in all proportions. If the two components A and B are of a similar chemical constitution the system will deviate only slightly from ideality, and will have a boiling point intermediate between those of the pure components. The variation of boiling point with composition is shown by the lower curve L in Fig. 17.1, while the upper curve V gives the composition of vapour in equilibrium with liquid at the boiling point of the latter. The horizontal lines connecting the L curve and V curve give the compositions of liquid and vapour in equilibrium with each other at the appropriate temperature. Such a line is called a tie-line.

The reader should find it helpful in the understanding of the above diagram to consider the fate of a liquid mixture of composition l_1 as it is heated at constant pressure from temperature T_0 to temperature T_4. Until temperature T_1 is reached, the system is in the liquid state. At temperature T_1 the liquid boils, producing vapour of composition v_1.

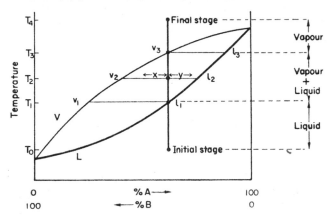

FIG. 17.1. Boiling point–composition diagram.

This vapour contains a higher proportion of the more volatile component B than does the liquid with which the vapour is in equilibrium. As the temperature is further increased to T_2, a number of changes occur in the system. Firstly, the composition of the liquid changes along the L curve to l_2; concurrent with this is the change in vapour composition along the V curve to v_2. Secondly, the relative amounts of liquid and vapour change. Whereas at T_1 the amount of vapour v_1 is infinitesimally small, at T_2 the amount of vapour v_2 is comparable to the amount of liquid present. The relative amounts of the two phases can be determined from the distances x and y. Thus, for a system having an overall composition l_1,

$$\frac{\text{amount of liquid phase}}{\text{amount of vapour phase}} = \frac{x}{y}$$

As the temperature continues to increase, the above ratio becomes smaller and smaller, until at the temperature T_3 it reduces to zero, thus indicating that the system is now entirely in the vapour state. No further change of state occurs upon heating to the temperature T_4.

17.2. Fractional distillation

From our remarks so far, we have established that partial vaporization of a binary liquid mixture yields a vapour which contains a higher proportion of the more volatile component than does the original liquid mixture; as a result a separation of the two components is made possible. The technique employed to effect this separation is that of *fractional distillation*. By this method, the liquid is boiled and the vapour produced is condensed to yield a distillate which is richer in the more volatile component than the original mixture. When this operation is performed a sufficiently large number of times a separation is achieved. Fortunately the process can be carried out automatically through the use of a *fractionating column*, the most common example of which is the "bubble-cap" type (Fig. 17.2). The unit shown is designed to operate continuously with the liquid being fed into the column at a rate just sufficient to balance the material loss from the column. Once such a column has been operating for some time, a steady-state condition is attained, one characteristic of which is that a temperature gradient is established, the column being hot at the bottom, gradually cooling upwards.

Let us now consider the fate of a liquid feed of composition 50% heptane in octane as it is continuously introduced into the column at plate 4. Let the temperature of plates 1, 2, 3, ... be $T_1, T_2, T_3, \ldots$, etc.

FIG. 17.2. Fractionating column.

The feed introduced into plate 4 will assume a temperature T_4, at which temperature liquid of composition l_4 will be in equilibrium with vapour of composition v_4 (Fig. 17.3(a)). This vapour in rising up through the column is obliged to bubble through the liquid in plate 5—in so doing its temperature drops to T_5 and liquid of composition l_5 condenses out. The vapour, now of composition v_5, rises, bubbles through liquid in plate 6 and partially condenses to liquid of composition l_6. The step-wise progression continues until at plate 7 the departing vapour is very nearly

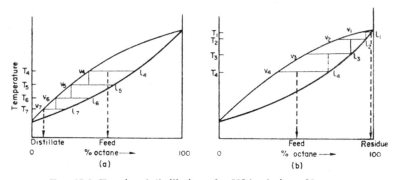

FIG. 17.3. Fractional distillation of a 50% solution of heptane in octane.

pure heptane. *Thus the vapour in rising up through the column experiences a continual enrichment in the more volatile component.*

Attention must now be drawn to the downward flow of liquid in the column (Fig. 17.3(b)). At plate 4, liquid of composition l_4 accumulates and overflows to plate 3 via the downspout. At the higher temperature T_3 partial vaporization of the more volatile component leaves a liquid of composition l_3 which in turn overflows to plate 2 at which the composition of liquid is l_2. Finally a residue of almost pure octane is obtained, i.e. *the liquid in flowing down the column has become enriched in the less volatile component.* Thus, if a fractionating column has sufficient plates to allow efficient contact between ascending vapour and descending liquid, a complete separation of a liquid mixture of the type considered can be achieved.

17.3. Petroleum refining

In the distillation considered above, the feed was a simple two-component mixture. When, however, a liquid mixture containing more than two components is being considered, a complete separation of each component becomes impossible if only one fractionating column is used. One particularly complex liquid mixture is crude petroleum, the first stage in the refining of which is to separate the mixture by distillation into fractions, each of which boils within a certain predetermined range. The petroleum is first heated to about 350°C, at which temperature it is almost completely vaporized. The vapour is then fed into the fractionating column (Fig. 17.4), where the pressure is maintained at a level slightly above atmospheric. By the mechanism already discussed, the more volatile materials condense in the upper plates of the column whilst the less volatile materials collect in the lower plates. The various fractions are continuously tapped off and in some cases processed by further distillation. Whereas the gasoline, kerosine and gas oil fractions can be

Gas	$C_1 - C_5$
Gasoline	$C_5 - C_{12}$
Kerosine	$C_{12} - C_{16}$
Gas oil	$C_{15} - C_{16}$
Fuel oil	$> C_{16}$
Lubricating oil	$> C_{20}$
Vapour feed	
Steam	
Residue: pitch, tar	$C_{36} - C_{90}$

Fig. 17.4. Petroleum fractionation.

distilled at near atmospheric pressure, the lubricating oil fraction is distilled under vacuum, thus reducing the boiling points of the oils in this fraction. This procedure is followed in order to avoid thermal decomposition of the oil, which becomes serious at temperatures in excess of about 400°C. To minimize still further the possibility of decomposition, vacuum distillation is frequently combined with steam distillation, which, for reasons to be discussed shortly, also has the effect of depressing the boiling point.

17.4.   Constant boiling mixtures

Returning once more to the discussion of two component liquid mixtures, it must be mentioned that very marked deviations from ideal behaviour can arise. Such behaviour can be expected if the attraction between dissimilar molecules is either much greater or less than that between similar molecules; the former is illustrated by the system HNO_3–H_2O while the C_2H_5OH–H_2O system exemplifies the latter. Figure 17.5(a) and (b) shows the type of boiling point diagrams for the

FIG. 17.5. Boiling point–composition diagrams for systems exhibiting a maximum and a minimum boiling point.

above liquid pairs. An examination of Fig. 17.5(a) and (b) reveals that liquid mixtures corresponding in composition to points M and N, on boiling, yield vapours of exactly the same composition M and N. Such mixtures therefore boil **unchanged and have** constant boiling points—

they are called *azeotropes* (which means to boil unchanged). Distillation of such mixtures cannot result in a separation of the two components. Thus if a nitric acid solution of composition 100–38% HNO_3 is distilled, the distillate obtained is pure nitric acid, whilst the residue is the azeotropic mixture. Alternatively if the composition of the original solution was more dilute than that of the azeotropic mixture, the resulting distillate would be pure water. The products obtained upon distillation of HNO_3–H_2O and C_2H_5OH–H_2O mixtures are summarized in Table 17.1.

TABLE 17.1. DISTILLATION PRODUCTS OF HNO_3–H_2O
AND C_2H_5OH–H_2O MIXTURES

Original solution	100–38% HNO_3	38–0% HNO_3	100–96% C_2H_5OH	96–0% C_2H_5OH
Residue	Azeotrope	Azeotrope	C_2H_5OH	H_2O
Distillate	HNO_3	H_2O	Azeotrope	Azeotrope

17.5. Steam distillation—completely immiscible liquids

If two liquids are completely immiscible in each other there will be virtually no attractive interaction between the dissimilar molecules. As a result, the physical properties of each will remain unaltered even when mixed; they will exert their own vapour pressures independently. Thus, if P_A° and P_B° are the vapour pressures of the pure liquids A and B at a temperature T, then the partial pressures of A and B in the vapour phase above the liquid mixture will likewise be P_A° and P_B° respectively at temperature T. The total pressure P exerted by the vapour mixture will then be given by Dalton's law, viz.

$$P = P_A^\circ + P_B^\circ$$

This expression contains no terms which relate to the composition of the original liquid mixture, and so it follows that the total pressure is dependent only on the temperature and not at all on the relative amounts of the two liquids present. The total pressure at any temperature can be easily obtained by plotting the sum of the vapour pressures against the temperature (Fig. 17.6).

Now when any liquid system is heated under normal atmospheric pressure, it will boil when its vapour pressure equals that of the confining atmosphere, viz. 1 atm. An inspection of the graphs in Fig. 17.6 shows that this condition is satisfied at temperature T_M for the mixture, at temperature T_A for liquid A and at temperature T_B for liquid B. It is thus

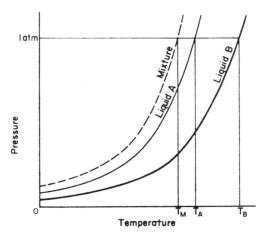

FIG. 17.6. Variation of vapour pressure with temperature for a liquid mixture made up of two completely immiscible liquids.

seen that *a liquid mixture comprising two immiscible liquids will boil at a temperature lower than the boiling point of either liquid and will continue to boil at that temperature (irrespective of its composition) provided the external pressure remains constant.*

The composition of the vapour produced from the boiling liquid also remains constant as the following procedure shows:

Let P_A° and P_B° be the vapour pressures of liquids A and B at the boiling point of the liquid mixture T_M. Then from Dalton's law

$$P_A^\circ = x_A P \quad \text{and} \quad P_B^\circ = x_B P$$

where x_A and x_B are the mole fractions of A and B in the vapour, and P is the total pressure = confining pressure.

$$\therefore \quad \frac{P_A^\circ}{P_B^\circ} = \frac{x_A}{x_B} = \frac{n_A}{n_B} = \frac{w_A}{M_A} \bigg/ \frac{w_B}{M_B}$$

$$= \frac{w_A}{(w_A + w_B)M_A} \bigg/ \frac{w_B}{(w_A + w_B)M_B}$$

$$= \frac{W M_B}{(1 - W)M_A}$$

where W = weight fraction of A in the distillate. Now M_A and M_B, being the molar masses of A and B respectively, are constants; so also are P_A° and P_B° provided the temperature remains constant throughout the distillation as indeed it does. It therefore follows that the distillate composition expressed by W must likewise remain constant.

The term *steam distillation* is used when one of the two immiscible liquids is water, and the technique is employed in the distillation of the heavy oil fractions in petroleum refining in order to avoid the serious decomposition that would occur if the distillation took place at normal pressure. Steam distillation is thus seen to be, in a sense, equivalent to distillation *in vacuo*. The following example emphasizes this point.

EXAMPLE 17.1. *A heavy oil (mol. wt. = ca. 268) on steam distillation boils at a temperature $t°C$ (where $t < 100$) under normal atmospheric pressure. The composition of the distillate is 99% (by weight) of water. Calculate the pressure necessary to distil off the oil at the same temperature $t°C$ using the technique of vacuum distillation.*

$$\frac{P_{H_2O}^\circ}{P_{HC}^\circ} = \frac{W M_{HC}}{(1 - W)M_{H_2O}}$$

where $P_{H_2O}^\circ$ = vapour pressure of water at $t°C$,
$\quad P_{HC}^\circ$ = vapour pressure of the hydrocarbon at $t°C$,
$\quad M_{H_2O}$ = molar mass of water,
$\quad M_{HC}$ = molar mass of the hydrocarbon,
$\quad W$ = weight fraction of water in the distillate.

Now P = total pressure
$\quad$ = atmospheric pressure
$\quad$ = $P_{H_2O}^\circ + P_{HC}^\circ$

$$\therefore \quad \frac{P - P_{HC}^{\circ}}{P_{HC}^{\circ}} = \frac{W M_{HC}}{(1-W) M_{H_2O}}$$

$$\therefore \quad P_{HC}^{\circ} = \frac{(1-W) M_{H_2O}\, P}{W M_{HC} + (1-W) M_{H_2O}}$$

$$= \frac{(1-0 \cdot 99) \times 18 \times 760}{0 \cdot 99 \times 268 + 0 \cdot 01 \times 18}$$

$$= 5 \text{ mmHg}$$

Thus at $t°$C, the heavy oil exerts a vapour pressure of 5 mmHg. Consequently, if it were to be distilled under vacuum at $t°$C, the external pressure would have to be reduced to 5 mmHg.

Exercises

17.1. The vapour pressures of oxygen and nitrogen at 83 K are 45·2 and 186·8 kPa respectively. What is the composition of a liquid mixture of oxygen and nitrogen which boils at 83 K at normal atmospheric pressure of 101 kPa? What is the composition of vapour initially produced from the boiling liquid? (1 N m^{-2} = 1 Pa.)

[*Ans.*: 60·6 mol% O_2; 27·1 mol% O_2.]

17.2. At 85 K the vapour pressures of oxygen and nitrogen are 58·0 and 232·8 kPa respectively. Repeat the above calculation (exercise 17.1) at 85 K and 101 kPa pressure.

[*Ans.*: 75·4 mol % O_2; 43·3 mol % O_2.]

17.3. At the standard pressure of 101 kPa (1 atm), oxygen boils at 90·0 K while nitrogen boils at 77·4 K. Use this information along with the data calculated in exercises 17.1 and 17.2 to plot a boiling point–composition diagram for this system.

17.4. Liquid A (b.p. 61°C) and liquid B (b.p 65°C) form an azeotrope boiling at 53°C and containing 87% by weight of A.

(a) Sketch the boiling point–composition diagram.

(b) Which component can be obtained pure by fractional distillation of a 40% by weight mixture?

(c) Calculate the maximum per cent of the pure component that can be recovered by carrying out the distillation in (b).

[*Ans.*: (b) B as residue; (c) 54% by wt. of B.]

17.5. Solid camphor ($M = 152$ g mol^{-1}) sublimes readily and being immiscible in water can be steam distilled. Use the following data to estimate the boiling point of the mixture at 760 mmHg (1 atm):

	Temp. (°C)		
	95	100	105
v.p. of water/mmHg	634	760	906
v.p. of camphor/mmHg	18	22	28

[*Ans.*: 99·2°C.]

17.6. What weight of steam would be required to transfer 1 kg of camphor into the distillate in the steam distillation of exercise 17.5?

[*Ans.*: 4·2 kg.]

CHAPTER 18

Separation Techniques: II, Extraction

DISTILLATION, considered in Chapter 17, can be thought of as an extraction technique. As such it relies on and exploits differences in the boiling points of liquids. However, in a number of situations distillation is an ineffective separation method. This is so if the components making up the mixture have similar boiling points or if azeotropes are formed or if excessive decomposition occurs even at very low pressures. In such an event, solvent extraction (in which solubility differences are exploited) is normally attempted. This technique is one whereby a solute is transferred from one liquid in which it is initially dissolved into another liquid immiscible in the first.

18.1. The distribution law

When a solute is added to two immiscible liquids, the solute distributes itself between the two solvents in such a manner as to make the ratio of its concentration in the two solvents a constant.[†] This is a statement of the *distribution law,* the ratio itself being called the *distribution coefficient (K).* The latter has a value characteristic of the system in question and ideally is dependent only on the temperature.

Thus, if a solute S is added to two immiscible solvents A and B, then

$$K = \frac{[S]_A}{[S]_B}$$

where $[S]_A$ and $[S]_B$ represent the equilibrium concentrations of solute in solvents A and B respectively.

[†] This is true provided the solute is in the same molecular state in each solvent.

If an excess of solute is added both solvents must become saturated simultaneously in order that K should maintain its value appropriate to a particular temperature. Consequently the distribution coefficient is equivalent to the ratio of the saturated solubilities of solute in each of the two solvents.

18.2. Batch extraction

In extraction work, the two solvents are referred to as the *extraction solvent* and the *raffinate solvent*, the latter being the one in which solute is initially dissolved. At any stage in the extraction, the phase rich in the extraction solvent is termed the *extract phase*, while the raffinate-rich phase is labelled the *raffinate phase*.

In batch extraction, the extraction solvent is mixed with the raffinate phase until the equilibrium amount of solute is present in each case, after which the two phases are separated. The operation can then be repeated using a fresh batch of extraction solvent for each single extraction.

Suppose then that we have W g of solute S dissolved in x litres of raffinate solvent. To this there is added y litres of extraction solvent. After mixing, the solute will be distributed between the two solvents and this distribution will be characterized by a particular value of the distribution coefficient which will then be given by:

$$K = \frac{[S]_{\text{Raffinate solvent}}}{[S]_{\text{Extraction solvent}}}$$

Let W_1 g = weight of solute remaining in the raffinate phase and $(W - W_1)$ g = weight of solute transferred to the extract phase.

$$K = \frac{W_1/x}{(W - W_1)/y}$$

$$\therefore \quad W_1 = W \frac{Kx}{Kx + y}$$

The two layers are separated and to the raffinate phase there is added a fresh portion of y litres of extraction solvent. Again, on mixing, solute will pass from one phase into the other until the two concentrations are such that K assumes its appropriate value. Thus if W_2 g is the weight of solute remaining in x litres of the raffinate phase after two extractions, $(W_1 - W_2)$ g is the weight of solute transferred to the second portion of extraction solvent. In this case

$$K = \frac{W_2/x}{(W_1 - W_2)/y}$$

$$\therefore \quad W_2 = W_1 \frac{Kx}{Kx + y}$$

$$\therefore \quad W_2 = W \left(\frac{Kx}{Kx + y} \right)^2$$

For n extractions, the weight W_n of solute remaining in the raffinate phase is given by

$$W_n = W \left(\frac{Kx}{Kx + y} \right)^n$$

The process is described in the form of a flow diagram in Fig. 18.1.

FIG. 18.1. Batch extraction.

EXAMPLE 18.1. *An aqueous solution of acetic acid was contaminated with small amounts of impurities and it was considered desirable to extract the acid with chloroform from which the acid could later be recovered by distillation. If the concentration of acid in the aqueous solution was initially 5 g l^{-1}, calculate the minimum number of extractions required to reduce the* *concentration below* $0 \cdot 5$ *g* l^{-1}, *using in each extraction* 100 *cm*^{3} *of aqueous solution and* 1 *l of chloroform.*

$$W_n = W\left(\frac{Kx}{Kx+y}\right)^n$$

$\therefore$
$$n = \frac{\log Wn/W}{\log Kx/(Kx+y)}$$

$$K = \frac{[\text{Acetic acid}]_{H_2O}}{[\text{Acetic acid}]_{CHCl_3}} = 26 \cdot 4$$

Weight of acid in 100 cm^3 of aqueous solution $= 0 \cdot 5$ g $= W$

Weight of acid in 100 cm^3 of aqueous solution $= 0 \cdot 05$ g after n extractions

$\therefore$
$$n = \frac{\log 0 \cdot 05/0 \cdot 5}{\log 26 \cdot 4 \times 0 \cdot 1/26 \cdot 4 \times 0 \cdot 2 + 1} = \frac{\log 0 \cdot 1}{\log 0 \cdot 7252} = 7 \cdot 163$$

Therefore eight separate extractions are necessary to reduce the concentration of acid below $0 \cdot 5$ g l^{-1}.

18.3. Extraction efficiency

A measure of extraction efficiency is given by the ratio of weight of solute extracted to the initial weight of solute,

i.e.
$$\text{per cent extraction} = \frac{100(W - W_n)}{W} = 100\left(\frac{1 - W_n}{W}\right)$$

For a single extraction, where $n = 1$

$$\text{per cent extraction} = 100\left[1 - \left(\frac{Kx}{Kx+y}\right)\right] = \frac{100y}{Kx+y}$$

EXAMPLE 18.2. *The distribution coefficient of lactic acid between water and chloroform at 25°C is*

$$K = \frac{[\text{Lactic acid}]_{H_2O}}{[\text{Lactic acid}]_{CHCl_3}} = 49 \cdot 26$$

What percentage of lactic acid will be extracted from 50 cm³ of an aqueous solution of the acid by shaking with 1 l of chloroform?

For $n = 1$, $x = 0.05$ l, $y = 1$ l.

$$W_1 = W\left(\frac{49.26 \times 0.05}{49.26 \times 0.05 + 1}\right)$$

$$= 0.7112W$$

$\therefore$ per cent extraction $= \dfrac{100(W - W_1)}{W}$

$$= \frac{100(W - 0.7112W)}{W} = 28.9\%$$

or more directly:

per cent extraction $= \dfrac{100y}{Kx + y} = \dfrac{100 \times 1}{49.26 \times 0.05 + 1} = 28.9\%$

In the equation

$$W_n = W\left(\frac{Kx}{Kx + y}\right)^n$$

y and n are related to each other by the simple equation

$$y \times n = V$$

where $V =$ the total volume of extraction solvent.

We must now establish the relationship between extraction efficiency and the number of extractions. Thus for n extractions

$$\text{per cent extraction} = 100\left(1 - \frac{W_n}{W}\right)$$

$$= 100\left[1 - \left(\frac{Kx}{Kx + y}\right)^n\right]$$

$$= 100\left[1 - \left(\frac{Kx}{Kx + V/n}\right)^n\right]$$

$$= 100\left[1 - \left(\frac{1}{1 + V/Kxn}\right)^n\right]$$

The binomial expansion of $(1 + V/Kxn)^n$ produces the following series:

$$\left(1 + \frac{V}{Kxn}\right)^n = 1 + n\left(\frac{V}{Kxn}\right) + \frac{n(n-1)}{2}\left(\frac{V}{Kxn}\right)^2 + \cdots \left(\frac{V}{Kxn}\right)^n$$

When $n = 1$, $\quad \left(1 + \frac{V}{Kxn}\right)^n = 1 + \frac{V}{Kx}$

Now this quantity is equivalent to the first two terms in the general series. From this it follows that

$$\text{for} \quad n = 1, \quad 1 + \frac{V}{Kx} < \left(1 + \frac{V}{Kxn}\right)^n, \quad \text{for} \quad n = 2, 3, 4; \ldots$$

Hence the extraction efficiency increases with the number of extractions. In other words, given a total volume V of extraction solvent, greater efficiency is obtained when n is made large and y small.

EXAMPLE 18.3. *Using the data given in Example 18.2, calculate the percentage of lactic acid which would have been extracted from 50 cm³ of an aqueous solution of the acid if two extractions had been performed with 500 cm³ of chloroform per extraction.*

For $n = 2$, $\quad x = 0.05$ l, $\quad y = 0.5$ l.

$$W_2 = W\left(\frac{49.26 \times 0.05}{49.26 \times 0.05 + 0.5}\right)^2$$

$$= 0.6911W$$

$$\text{per cent extraction} = 100\left(\frac{W - W_2}{W}\right) = 30.9\%$$

Comparison of this figure with the corresponding figure obtained in Example 18.2 verifies the fact that extraction efficiency increases with the number of extractions.

18.4.   Continuous countercurrent extraction

Batch extraction is a technique commonly employed in the laboratory but it finds little industrial application because of the inefficient use of the extraction solvent. The overall efficiency of the process can be greatly improved by the technique of *continuous countercurrent extraction*. By this method, the extraction solvent is made to flow continuously in one direction, the raffinate phase flowing continuously in the reverse direction. Now whereas in batch extraction mixing and settling take place in the same vessel, two separate vessels (referred to as the mixer and settler) are required when the flow of the liquids is continuous. Thus, in the two-stage countercurrent extraction shown in Fig. 18.2, fresh extraction

FIG. 18.2. Continuous countercurrent extraction.

solvent is used in the second extraction, the extract then being used in the first stage. For clarity only two stages have been considered; usually more stages are necessary. Although the extraction efficiency of this technique is high, the floor area required for such a multistage extraction is likewise high. This disadvantage is eliminated by using vertical tower extractors, one of the most modern of which is shown in Fig. 18.3.

The process is basically a continuous countercurrent extraction with the more dense liquid being fed in at the top of the column, the less

FIG. 18.3. Vertical tower extractor.

dense liquid entering at the bottom. There is a continuous transfer of solute material (which can be initially contained in either the light or heavy liquid) as one rises up and the other flows down the column. The column itself is essentially an assembly of mixers and settlers, the mixing being provided for by a rotating stirrer, while settling is achieved through the use of wire mesh.

18.5. Industrial applications

The technique of solvent extraction is most extensively used in the petroleum industry. For example, in the refining of the lubricating oil fractions the constituents of the oil are distributed between two immiscible solvents, e.g. propane and a phenolic mixture of phenols, cresols and xylenols. The type of plant used is similar to that shown in Fig. 18.2 in that it is made up of a series of mixer-settlers. However, in this case the oil is fed not into the first mixer, but into an intermediate one. The process is shown in its most simplified form in Fig. 18.4. Propane is introduced at one end of the series of mixer-settlers and, as it moves along the series, preferentially dissolves out the alkanes. In the meantime the phenolic solvent, introduced at the other end of the series and flowing

FIG. 18.4. Refining of the lubricating-oil fractions.

countercurrently to the propane, dissolves out the cycloalkanes and aromatics.

Another application is the removal of aromatics from kerosine using liquid sulphur dioxide. In this case the liquid sulphur dioxide is fed into the top of a column with the less dense kerosine entering at the bottom. The kerosine in floating to the top of the column makes intimate contact with a countercurrent flow of liquid sulphur dioxide and, in so doing, transfers its aromatic constituents to sulphur dioxide.

Although being particularly effective in removing aromatics, sulphur dioxide suffers from the disadvantage that it is a gas at normal pressures. Consequently the extraction unit must be designed to operate under pressure. To overcome this other less volatile solvents are sometimes used.

In addition to extracting aromatic compounds, sulphur dioxide is also effective in reducing the sulphur content of oil fractions. It is, however, very much less efficient than liquid hydrogen fluoride which is now commonly used for this purpose.

Exercises

18.1. The distribution coefficient of iodine between water and ether is

$$K = \frac{[I_2]_{H_2O}}{[I_2]_{ether}} = 5.0 \times 10^{-3}$$

What volume of ether would be necessary to remove 99.9% of the iodine from 1000 cm³ of an aqueous solution?

[*Ans.*: 5000 cm³.]

18.2. What *total* volume of ether would be necessary to remove 99·9% of the iodine from 1000 cm³ of an aqueous solution in three separate extractions using equal volumes of ether for each extraction?

[*Ans.*: 135 cm³.]

18.3. 100 cm³ of a solution containing $1·0 \times 10^{-3}$ g of impurity (I) is to be purified by extraction with a total volume of 1000 cm³ of extracting solvent that dissolves only the impurity. How many extractions will be required to reduce the amount of impurity to a level not exceeding $1·0 \times 10^{-6}$ g?

$$K = \frac{[I]_{\text{raffinate}}}{[I]_{\text{extract}}} = 0·50$$

[*Ans.*: 4.]

18.4. Substance X contains a 10% by weight of impurity Y. It is desired to reduce the level of impurity by extraction of an aqueous solution of X (containing Y) with an equal volume of ether. Given the two following distribution coefficients

$$K_x = \frac{[X]_{H_2O}}{[X]_{\text{ether}}} = 3·0 \quad \text{and} \quad K_y = \frac{[Y]_{H_2O}}{[X]_{\text{ether}}} = 1·0$$

calculate the percentage impurity in the aqueous phase after one extraction.

(*Hint*: imagine that the volume of aqueous solution selected is that which contains 100 g of solute, i.e. 90 g X and 10 g Y.)

[*Ans.*: 6·9%.]

18.5. Use the information in exercise 18.4 to calculate the percentage impurity in the aqueous phase after a second extraction with the same volume of ether.

[*Ans.*: 4·7%.]

18.6. Use the information in exercise 18.4 to determine the number of ether extractions required to reduce the impurity level to a value not exceeding 1%.

[*Ans.*: 6.]

CHAPTER 19

Separation Techniques: III, Desalination

THE processes of evaporation and condensation occur on a vast scale over the earth's surface. Where the surface is covered by water, evaporation occurs and the water vapour produced is carried by winds over land where it condenses as rain or snow.

The advent of the industrial revolution brought about a concentration of population in the rapidly growing cities and necessitated the construction of large reservoirs in order to trap condensed water. Today, however, the demand for fresh water is increasing at such a rate that these measures are becoming more and more inadequate. Even in the United Kingdom, with its temperate climate, water shortages are not uncommon. The situation in many countries is more critical than this; deserts must be reclaimed, but in a country where the rainfall is slight and the nearest freshwater supply is perhaps a thousand miles distant, this is no mean task. A partial solution is provided by the improvement and extension of existing trapping facilities, but in recent years much research and development has been concentrated on the sea as a source of fresh water. Unfortunately, the sea contains about 3·5% of salts (mostly sodium chloride with much lesser amounts of magnesium chloride, and magnesium and calcium sulphates). Small though this figure may seem, it is intolerable from the standpoint of human or agricultural needs—it is indeed too great by a factor of about 70.

A large number of processes, designed to extract fresh water from the sea and referred to collectively as *desalination processes*, have been suggested. Of these, distillation and freezing processes and reverse osmosis are the furthest advanced and our discussion will be restricted to these techniques.

19.1. Multistage flash distillation

The separation of fresh water from the sea is a simple enough matter, easily accomplished by, for example, distillation. The problem, however, is not *how* the separation can be achieved but *how efficiently* it can be done. A simple distillation, such as is shown in Fig. 19.1, would produce fresh water as distillate but would do so most inefficiently in that (i) the process is batch as distinct from continuous, and (ii) there is no recovery of the latent heat released when the vapour condenses in the condenser.

FIG. 19.1. Simple distillation.

If the technique of distillation is to be employed to obtain fresh water from the sea, then in order to be economically viable, there must be a recovery of this latent heat and at the same time the plant must be designed to allow a continuous flow of sea water. These requirements are particularly well satisfied in the technique referred to as *multistage flash distillation*.

By this method, sea water at a certain temperature is introduced into a vessel (the flash chamber) in which the pressure is maintained at a value less than that corresponding to the vapour pressure of the sea water. This condition results in rapid evaporation, the vapour being said to "flash" off. Now before considering the multistage process, let us examine the essential features of a single-stage flash distillation, as shown in Fig. 19.2. A sea-water feed enters the condenser tubes and, immediately prior to its entry into the flash chamber, is heated by steam to a temperature T at which its vapour pressure is P. It then enters the flash

FIG. 19.2. Single-stage flash distillation.

chamber where the pressure is maintained at P_1 (where $P_1 < P$). This condition causes a fraction of the liquid to flash off and because it is only the more energetic water molecules which leave the solution, the latter will experience a cooling effect, the temperature dropping to T_1. In the meantime, the flashed vapour condenses on the condenser coils, thus releasing its latent heat of condensation, which, by heat transfer through the walls of the condenser, is absorbed by the sea water.

A number of stages constitute the multistage process, which is outlined in a simplified form in Fig. 19.3. As can be seen, the sea-water feed entering the condenser in the nth flash chamber is preheated by condensing vapour; this recovery of energy is repeated in each subsequent condenser.

FIG. 19.3. Multistage flash distillation.

The temperature is then raised, as before, by steam and this is followed by the flashing off of vapour in successive chambers. The salt water, as it moves along from one chamber to the next, cools continually and also experiences an increase in salt concentration. Because of this cooling effect the pressure in each successive chamber must progressively decrease in order that the pressure within the chamber is less than the vapour pressure of the feed to that chamber. The vapour, condensing out as a freshwater distillate, could be withdrawn from each chamber. It is, however, more efficient to pump the distillate from one chamber to the next, for it also is cooled by flashing and, in this way, the preheating of the brine is made more appreciable.

The overall efficiency of a multistage flash distillation unit is largely dependent on the flashing temperature range $T-T_n$. Whilst the lowest temperature T_n is virtually fixed by that of the sea-water feed, the higher temperature T is determined by that of the steam supply. Thus if high pressure steam is used, sea water could enter the first flash chamber at temperatures in excess of 100°C, thereby increasing efficiency. There are, however, inherent disadvantages in the use of temperatures much higher than this, the most serious of these being an increase in corrosion and the formation of scale (see Chapter 16). Consequently, present practice is to keep the high temperature T below 110°C. However, looking to the future, it seems likely that higher temperatures will be employed as corrosion and scale difficulties are overcome through the use of resistant materials and improved techniques.

19.2. Freezing processes

Distillation is not the sole technique which can bring about the separation of fresh water from the sea. The freezing of sea water separates out ice leaving the salts in solution. Once more, however, the efficiency of the process is of paramount concern. Two freezing processes, each based on a common principle, will be discussed. These are the *vacuum freeze process* and the *secondary refrigerant process*.

Vacuum freeze process

By this method pre-cooled sea water at about 1°C is fed into a chamber (the freezer) where the pressure is maintained at a value less than that corresponding to the vapour pressure of freezing sea water (about 3 mmHg) in order that some water may flash off and produce a consequent cooling effect, sufficient to freeze out some ice (Fig. 19.4). Because of the low temperature and consequent low value of vapour pressure, the freezer chamber must operate under high vacuum. The ice–brine mixture

FIG. 19.4. Vacuum freeze process.

is pumped to a separator, where it is both separated and washed with fresh water, the washing being necessary to remove brine trapped between the ice crystals. Concurrently the water vapour, initially produced by flashing from the freezer, is compressed and condensed on washed ice contained in the melter. Thus at this stage water is produced by condensation of vapour together with the melting of ice.

Secondary refrigerant process

Whereas the freezing of sea water is achieved in the vacuum freeze process by evaporating a portion of the water under high vacuum, this process accomplishes the freezing by vaporizing another liquid immiscible in water. The vapour of this second liquid is then compressed and used to melt the ice. This second liquid is referred to as the secondary refrigerant. The advantage of using this technique is that a secondary refrigerant can be chosen which boils under normal atmospheric pressure in the region of 0°C. *n*-Butane (b.p. 0°C) is ideal as a secondary refrigerant.

The process operates as follows. Liquid butane, introduced into the freezer, evaporates to a considerable extent; the energy required for this evaporation is extracted from the sea water which cools and produces, as before, an ice–brine mixture which is transferred to the separator and then on to the melter (Fig. 19.5). Simultaneously the butane vapour removed from the freezer is compressed and condensed on the ice in

Fig. 19.5. Secondary refrigerant process.

the melter, thus regenerating liquid butane (which is then returned to the freezer) and producing fresh water (a small fraction of which is returned to the separator to wash the ice crystals).

An advantage of freezing processes over distillation techniques is their low energy requirement. However, practical difficulties associated with the separation and washing stages have up till now prevented the commercial exploitation of the processes although fairly large pilot plants are in operation.

19.3. Reverse osmosis

The phenomenon we now call *osmosis* was first noted by the Abbé Nollet in 1748. He observed that when wine was stored in a vessel sealed off with an animal membrane, such as a bladder, and immersed in water, the bladder swelled up and finally burst. This, we now know, is a result of the bladder permitting water (solvent) but not alcohol (solute) to pass through it. A similar effect can be seen when prunes or raisins swell and subsequently burst when immersed in water. Osmosis can thus be defined as *the spontaneous flow of solvent from a dilute to a more concentrated solution via a membrane which allows solvent but not solute molecules to pass through it.* Such a membrane is said to be *semipermeable.*

In Fig. 19.6, a tube fitted with a semipermeable membrane and containing a salt solution is partly immersed in water so that initially the two levels outside and inside the tube are the same. Water will tend to

$P = \pi$ $P = 0$ $P > \pi$
(a)   (b)   (c)

Fig. 19.6. Osmotic flow of solvent.

flow from the beaker into the tube through the semipermeable membrane. This flow can be prevented by applying a pressure of such a magnitude as to maintain the equality of the two levels. This excess pressure required to prevent osmosis is called the osmotic pressure, and is denoted by the symbol π. Now we can vary the pressure acting on the solution by incorporating a piston into the tube. Thus, if the flow of water is to be prevented, the excess pressure P acting on the piston must be equal to the osmotic pressure of the solution, π (Fig. 19.6(a)). If P is less than π, the piston will be pushed upwards as a result of the spontaneous flow of water into the tube via the semipermeable membrane. If no excess pressure is applied to the solution then water will enter the tube until the hydrostatic pressure exerted by the column of solution becomes equal to the osmotic pressure (Fig. 19.6(b)). Finally, if a pressure in excess of π is applied to the solution, water will pass *out of* the solution (Fig. 19.6(c)). This reversal of the normal spontaneous flow of solvent is referred to as *reverse osmosis*.

As a technique for desalination, reverse osmosis has been under development for only a short time, and although the process is basically attractive (on account of its simplicity and low-energy requirements), its commercial success will depend upon the production of membranes which are both cheap and durable. A solution to the problem is proving elusive and at the present stage of development the essential technological breakthrough has not yet been made.

Sea water has an osmotic pressure of about 25 atm and, if reverse osmosis is to occur, a pressure in excess of this must be applied on the solution side of the membrane (cellulose acetate). In practice it is found that much higher pressures than this must be developed, and this necessitates the use of a membrane support. Fibre glass, on account of its low cost, is particularly suitable for this purpose. At present, development of the technique has not passed beyond the pilot plant stage and is, as yet, only successful in the recovery of fresh water from a brackish source, i.e. one in which the salt content ranges from 0·1 to 1 % by weigth.

Exercises

19.1. Describe the essential features of multistage flash distillation.

19.2. What factor principally determines the overall efficiency of a multistage flash distillation? What is the advantage and disadvantage of employing high pressure steam to heat the sea water prior to its entry into the flash chamber?

19.3. Describe the two freezing processes used in desalination.

19.4. Define osmosis and osmotic pressure. How is the normal, spontaneous flow of solvent reversed?

19.5. The osmotic pressure π of sea water is 14·6 atm at 298 K. Given that π is inversely proportional to the volume containing a fixed amount of solute, determine the minimum excess of pressure which must be applied to 1000 cm^3 of sea water in order to transfer 100 cm^3 of water from the salt solution to pure water via a membrane permeable only to water.

[*Ans.*: 16·2 atm.]

19.6. The work W required to transfer a volume of solvent ΔV from solution to pure solvent via a semipermeable membrane is given by

$$W = \pi \, \Delta V$$

Calculate the energy requirement for the transfer of solvent in exercise 19.5. (1 atm = $1·013 \times 10^5$ N m^{-2}.)

[*Ans.*: 164 J.]

CHAPTER 20

Organic Chemistry—Basic Concepts

20.1.   What is organic chemistry?

Organic chemistry is defined as *the chemistry of carbon compounds.* Other elements occur frequently—notably hydrogen, which is seldom absent, and to a rather lesser extent oxygen, nitrogen, sulphur and the halogens. The appearance in organic compounds of elements outside of this group is comparatively rare.

It is striking that one major branch of chemistry is devoted to the compounds of only one of the 103 elements now placed in the periodic table. The reasons for this we shall presently examine.

20.2.   Chemical bonding in carbon compounds

Carbon exhibits a valency of four. Moreover, the valence links formed by carbon are essentially covalent. The covalent nature of organic compounds results in minimal molecular interaction due to electrical charge, and most of these materials are gases, liquids or low melting solids. Organic compounds are usually non-electrolytes and generally insoluble in water or other ionizing solvents. Valence type controls physical properties in these and other ways.

A given carbon atom may be single-bonded to other monovalent elements, as in methane or methyl chloride, or it may form multiple bonds with multivalent elements, as it does in formaldehyde and hydrogen cyanide.

$$\begin{array}{c} H \\ | \\ H-C-H \\ | \\ H \end{array} \qquad \begin{array}{c} H \\ | \\ H-C-Cl \\ | \\ H \end{array} \qquad \begin{array}{c} H \\ \diagdown \\ H \diagup \end{array} C{=}O \qquad H-C{\equiv}N$$

Methane   Methyl chloride   Formaldehyde   Hydrogen cyanide

In being tetravalent carbon is no different from other elements in group IV of the periodic table. It is capable of forming a uniquely large number of compounds (and so carbon chemistry is a uniquely large subject) because carbon has a great tendency to combine, not only with other elements, but also with itself.

Thus carbon in combination with hydrogen is potentially capable of forming, in addition to methane, a whole series of related *hydrocarbon* compounds such as ethane, propane, butane, and so on.

$$\begin{array}{cc} H & H \\ | & | \\ H-C-C-H \\ | & | \\ H & H \end{array} \qquad \begin{array}{ccc} H & H & H \\ | & | & | \\ H-C-C-C-H \\ | & | & | \\ H & H & H \end{array} \qquad \begin{array}{cccc} H & H & H & H \\ | & | & | & | \\ H-C-C-C-C-H \\ | & | & | & | \\ H & H & H & H \end{array}$$

Ethane   Propane   Butane

The extension of carbon chains by the formation of *saturated* carbon–carbon single bonds is a well-established phenomenon of organic chemistry. Carbon atom may also be bonded to carbon atom by *double* or *triple* valence links. The parent hydrocarbon compounds with these features are the gases ethylene and acetylene. Structures containing double or triple bonds are termed *unsaturated*.

$$\begin{array}{cc} H & H \\ | & | \\ C{=}C \\ | & | \\ H & H \end{array} \qquad H-C{\equiv}C-H$$

Ethylene   Acetylene

In organic compounds typified by methane, ethylene and acetylene the carbon skeleton is in the form of an open chain. Open-chain compounds as a class are known as *aliphatic* compounds. Carbon skeletons can occur not only as open chains but also as rings. Thus *alicyclic* compounds (known also as *naphthenes*) such as the hydrocarbons cyclopropane and cyclohexane provide further examples of the versatility with which carbon atoms bond to one another. *Aromatic* compounds (discussed in Chapter 22) constitute a second very important class of ring structures in organic chemistry. The hydrocarbon benzene, C_6H_6, is the parent member of this class.

$$H_2C \overline{\qquad} CH_2$$
$$\diagdown C \diagup$$
$$H_2$$
Cyclopropane

$$\overset{H_2}{\underset{}{C}}$$
$$H_2C \diagup \diagdown CH_2$$
$$H_2C \diagdown \diagup CH_2$$
$$\underset{H_2}{C}$$
Cyclohexane

20.3.   Structural isomerism

There is for all but the simplest organic structures more than one way of arranging the atoms involved while still satisfying valence require-ments. Thus in the hydrocarbon series methane, ethane, propane, butane and beyond the first three are described by one unambiguous structural representation. For the fourth member, butane C_4H_{10}, however, two satisfactory structures, shown below, can be written out. These are most commonly called *normal-* or *n*-butane (the prefix *normal* being used to indicate a straight-chain arrangement of carbon atoms) and isobutane.

$$H - \underset{\underset{H}{|}}{\overset{\overset{H}{|}}{C}} - \underset{\underset{H}{|}}{\overset{\overset{H}{|}}{C}} - \underset{\underset{H}{|}}{\overset{\overset{H}{|}}{C}} - \underset{\underset{H}{|}}{\overset{\overset{H}{|}}{C}} - H$$
n-Butane

$$\overset{H}{\underset{|}{}}$$
$$H - C - H$$
$$H - \underset{\underset{H}{|}}{\overset{\overset{|}{}}{C}} - \underset{\underset{H}{|}}{\overset{\overset{|}{}}{C}} - \underset{\underset{H}{|}}{\overset{\overset{|}{}}{C}} - H$$
Isobutane

The property whereby a single molecular formula may represent more than one structure is known as *structural isomerism*. Butane can exist as one of two structural isomers. The next member of this particular hydrocarbon series, pentane C_5H_{12}, can exist as one of three structural isomers. Beyond this point, as the number of carbon atoms in a given

n-Pentane

Isopentane

Neopentane

molecular formula increases, so too at a rapid rate does the number of possible structural isomers. Thus for the hydrocarbon pentacosane, $C_{25}H_{52}$, over 36 million structural variations are possible.

Structural isomers are quite distinct chemical compounds. In a group of isomers like pentanes, however, there are only minor differences in chemical behaviour between isomers, as we might expect from the similar carbon–carbon and carbon–hydrogen valence bonds in all three compounds. Structural isomerism, however, often tends to lead to very major differences in chemical behaviour. Thus isomeric arrangements of the molecular formula C_3H_6O, shown below, contain in part quite different types of valence bonds from one another, and the two compounds in question, propionaldehyde and methyl vinyl ether, are easily distinguished chemically from one another.

20

Propionaldehyde

Methyl vinyl ether

20.4. The homologous series

Table 20.1 lists the first eight members of the hydrocarbon family of compounds called alkanes. Only the molecular formulae are given, but remember that C_4H_{10} and higher formulae each represent more than one isomer.

TABLE 20.1. ALKANES

Name	Molecular formula
Methane	CH_4
Ethane	C_2H_6
Propane	C_3H_8
Butane	C_4H_{10}
Pentane	C_5H_{12}
Hexane	C_6H_{14}
Heptane	C_7H_{16}
Octane	C_8H_{18}

A series of this sort is called an *homologous series*. Inspection shows that each member differs from its predecessor by a constant amount (i.e. CH_2, or a molecular weight difference of 14, in our example). The above series can be expressed by the general formula C_nH_{2n+2}. Each individual member of the series is called an *homologue*.

Any homologous series shows a steady change in physical properties such as boiling point or specific gravity as the series is ascended. More important, all members of a given series exhibit very similar chemical behaviour. This permits the apparent complexity of organic chemistry, therefore, to be reduced to a study of a relatively small number of homologous series.

20.5. Functional groups

In most organic compounds there is present an atom or group of atoms whose characteristic chemical reactivity can be used as a basis for classification. This particular active point of a molecule is called the *functional group*. In a given homologous series of compounds, each member has present the same functional group.

In the alkane *n*-butane carbon–hydrogen and carbon–carbon bonds may be broken in the course of chemical reaction and each of these sites is a potential functional group. In the hydrocarbon 1-butene, however, the carbon–carbon double bond is much more active than any other bond in the molecule, and is by definition the functional group. Again, if one of the hydrogen atoms in *n*-butane is replaced by a bromine atom, as in 2-bromobutane, subsequent chemical reactions are found to involve in the main the carbon–bromine bond: this site in the molecule is the chemically dominant one, and is therefore the functional group.

$$CH_3 \cdot CH_2 \cdot CH_2 \cdot CH_3 \qquad CH_2 = CH \cdot CH_2 \cdot CH_3$$

$$\text{\textit{n}-Butane} \qquad\qquad\qquad \text{1-Butene}$$

$$\overset{\displaystyle Br}{\overset{\displaystyle |}{CH_3 \cdot CH \cdot CH_2 \cdot CH_3}}$$

$$\text{2-Bromobutane}$$

20.6. Hydrocarbon radicals

Every organic structure may be regarded as having been derived from a hydrocarbon by substitution of the relevant functional group in place of one or more hydrogen atoms.

The fundamental position of a hydrocarbon homologous series can be demonstrated by reference to the alkane series (Table 20.1). If in the alkanes one hydrogen atom is replaced by one chlorine atom, the pattern shown in Table 20.2 results.

The monochlorinated alkanes obviously constitute an homologous series, the *alkyl chlorides*, fitting the general formula $C_n H_{2n+1} Cl$.

20*

TABLE 20.2. MONOCHLORINATED
ALKANES

Name	Molecular formula
Methyl chloride	CH_3Cl
Ethyl chloride	C_2H_5Cl
Propyl chloride	C_3H_7Cl
Butyl chloride	C_4H_9Cl
	$\vdots$
	$C_nH_{2n+1}Cl$

If instead of chlorine we were to substitute some other monovalent atom or group X in the alkanes a series of general formula $C_nH_{2n+1} \cdot X$ would be obtained. The substituent group, which is in practice the functional group, may be complex, and may itself contain a carbon atom; e.g. substitution of the group would yield the series of aliphatic *aldehydes*

$$-C\underset{H}{\overset{O}{\diagup}}$$

whose general formula is therefore $C_nH_{2n+1} \cdot CHO$.

The hydrocarbon residues present in these substituted alkanes are called *alkyl radicals*—the methyl radical from methane, the ethyl radical from ethane, and so on. They are denoted by the general symbol R; hence the general formula commonly used for an alkyl chloride is $R \cdot Cl$.

Alkanes, then, form the basis of a wide selection of different organic aliphatic compounds classified, according to functional group, in homologous series. A parallel relationship exists in aromatic chemistry. The precursor of all aromatic compounds, benzene, C_6H_6, may have one or more of its hydrogen atoms replaced by other atoms or groups. Thus the first member of an aromatic homologous series can be represented as $C_6H_5 \cdot X$; the hydrocarbon radical C_6H_5 is called the phenyl radical. Radicals derived from aromatic systems in general are called *aryl radicals*, and are denoted by the symbol Ar.

20.7.   Radical isomerism

The monochlorination of either methane or ethane yields in each case one product only; propane, however, can be monochlorinated to give one of two structural isomers, *n*-propyl chloride or isopropyl chloride:

$$CH_3 \cdot CH_2 \cdot CH_2Cl \qquad CH_3 \cdot \overset{\overset{\displaystyle Cl}{|}}{C}H \cdot CH_3$$

  n-Propyl chloride   Isopropyl chloride

The propyl radical can therefore exist in one of two isomeric forms. In the case of butane there are two hydrocarbon isomers (*n*-butane and isobutane) from each of which can be derived two radical isomers. The four possible butyl chlorides are shown below.

$$CH_3 \cdot CH_2 \cdot CH_2 \cdot CH_2Cl \qquad CH_3 \cdot \overset{\overset{\displaystyle Cl}{|}}{C}H \cdot CH_2 \cdot CH_3$$

    n-Butyl chloride   sec-Butyl chloride

$$\begin{matrix} CH_3 \diagdown \\ \qquad\quad CH \cdot CH_2Cl \\ CH_3 \diagup \end{matrix} \qquad CH_3 \cdot \overset{\overset{\displaystyle CH_3}{|}}{\underset{\underset{\displaystyle CH_3}{|}}{C}}Cl$$

      Isobutyl chloride   tert-Butyl chloride

The prefix *secondary* or *sec* preceding a radical name indicates attachment of a substituent group to a carbon atom to which is also attached one hydrogen atom; iso- refers to the occurrence of the hydrocarbon fragment $(CH_3)_2CH-$ in the structure; and *tertiary-* or *tert-* indicates attachment of the substituent group to a carbon atom carrying no hydrogen atoms. The terms are widely used in organic nomenclature. Systematic methods of naming organic compounds are also in use: e.g. if the longest single carbon chain in an aliphatic compound is numbered, the position of substituent groups can readily be indicated. Examples of such a system are 1-butene and 2-bromobutane (p. 299). Isobutyl chloride is alternatively named 1-chloro-2-methylpropane; tert-butyl chloride becomes 2-chloro-2-methylpropane.

20.8.   Raw material sources

Raw materials for the organic chemical industry may be obtained from a number of naturally occurring resources.

Petroleum oil (including natural gas) has become the major source of organic raw materials within the last four decades. It is essentially a complex mixture of aliphatic and alicyclic hydrocarbon structures, but aromatic compounds can also be synthesized from petroleum by secondary processes. As oil becomes less available and more expensive towards the end of the twentieth century its dominant role in the chemical industry will weaken.

Coal has in the past been a rich source of aromatic compounds. Less conveniently, aliphatic structures can also be obtained. Coal deposits are much more extensive than those of petroleum oil, and are widely spread, although largely restricted to the northern hemisphere. It is likely, therefore, that the eminence which coal had at the beginning of the industrial revolution in the eighteenth century will be repeated as petroleum supplies dwindle.

Other naturally occurring organic raw materials include peat (which is geologically related to coal), tar sands (which are related to petroleum and contain bitumen) and oil shales. Huge quantities of organic material are bound up in these sources but no extensive use has as yet been made of them because of high recovery and processing costs.

Exercises

20.1. Write out structural formulae for the following compounds:

> 1,3-dichlorobutane;
> 2-pentene;
> isopropyl bromide;
> *n*-hexane;
> cycloheptane;
> diethyl ether.

20.2. How many structural isomers exist for the following molecular formulae:

C_4H_8;
$C_3H_6Cl_2$;
$C_4H_8Br_2$?

20.3. Give an example of each of the following:

an homologue;
an alkyl radical;
an alicyclic compound;
a functional group.

CHAPTER 21

Hydrocarbons: I, Aliphatic Compounds

ALIPHATIC hydrocarbons are divided into three families: alkanes (or paraffins), alkenes (or olefins) and alkynes (or acetylenes). Alkanes are abundantly available from natural sources: this is not the case with the industrially important alkenes and alkynes, which have to be obtained by synthesis.

ALKANES

21.1. Alkanes from petroleum

Alkanes are obtained from petroleum by fractional distillation, those of particular use to the chemical industry being the early members which up to the C_4 fraction exist as gases. Liquid petroleum contains a large quantity of these gases, perhaps as much as 200 times its own volume.

Natural gas is also a rich source of the lower alkanes, particularly methane, which is always by far the largest constituent. North Sea gas, for example, contains upwards of 90% methane.

Relatively pure single components can be obtained from natural gas or petroleum by extensive refining. Thus propane and butane, liquefied and stored under pressure, are available to the commercial and domestic market. (These products are also obtained from the breakdown of higher petroleum fractions.) The liquid fractions from petroleum, however, are not normally separated into pure components. The early liquid fractions are widely marketed as solvents, and for this purpose a hydrocarbon mixture is quite suitable. Hence "40–60" petroleum ether is a hydrocarbon mixture distilling between 40°C and 60°C, and is essentially the

C_5–C_6 fraction. Two "40–60" petroleum ethers may differ greatly in their detailed chemical analysis because of differences in source, but their solvent powers will be very similar.

21.2. Reactions of alkanes

Alkanes are not very reactive. Being saturated and having no labile functional group, they are chemically stable, at least under mild conditions. Some of their more important reactions include:

(a) *Substitution reactions*

The substitution of hydrogen by other elements in an alkane can occur progressively. Hydrocarbon chlorination illustrates this reaction.

$$CH_4 \xrightarrow{Cl_2} CH_3Cl \xrightarrow{Cl_2} CH_2Cl_2 \xrightarrow{Cl_2} CHCl_3 \xrightarrow{Cl_2} CBI_4$$

Meth- ane	Methyl chloride	Methylene chloride	Chloro- form	Carbon tetra- chloride

(b) *Cracking reactions*

The decomposition of a raw material often leads to useful products. If the decomposition is induced by high temperature, of the order of 400°C and above, the reaction is described as *pyrolysis*, or *carbonization*, or *cracking*. This last expression is the one most commonly used in referring to the thermal decomposition of alkanes as practised in the petrochemical industry.

When alkanes are cracked, multiple reactions occur, and reaction variables such as temperature, catalyst and contact time have to be optimized for a given result. Three types of reaction can be distinguished.

(i) Alkenes, outstandingly important starting materials for further synthesis, are produced along with hydrogen. Thus ethane cracking yields ethylene.

$$CH_3 \cdot CH_3 \rightarrow CH_2{=}CH_2 + H_2$$

Ethane	Ethylene

(ii) Smaller molecules are produced by breaking carbon–carbon bonds. If this occurs, alkenes again appear in the cracked products.

$$CH_3 \cdot CH_2 \cdot CH_3 \rightarrow CH_2 {=\!=} CH_2 + CH_4$$

Propane   Ethylene   methane

(iii) Rearrangement, or isomerization, takes place to give, for example, branched chain isomers from *n*-alkanes.

$$CH_3 \cdot CH_2 \cdot CH_2 \cdot CH_3 \rightarrow CH_3 \cdot \overset{\displaystyle CH_3}{\overset{\displaystyle |}{CH}} \cdot CH_3$$

n-Butane   Isobutane

Extensive industrial use is made of these reactions, all of which tend to occur in an increasingly complex fashion as the average chain length of the alkane feed to the cracker grows. In any high temperature pyrolysis elemental carbon is also produced in quantities which increase with the severity of the reactions conditions.

(c) *Combustion*

When an alkane reacts with oxygen, the end products are carbon dioxide, water and a large quantity of energy, most of which is released as heat. Thus the reaction between methane and oxygen is fully represented by the equation:

$$CH_4 + 2O_2 \rightarrow CO_2 + 2H_2O \qquad \Delta H = -890 \text{ kJ}$$

The exothermic reaction of hydrocarbons with oxygen is made use of in a wide range of gaseous and liquid fuels derived from petroleum. The heat produced may be used directly, or it may be converted to a power stroke as in the internal combustion engine.

(d) *Controlled oxidation*

Oxidation of alkanes need not be carried to completion. If the reaction is carried out under controlled conditions, a variety of important oxygen-containing compounds is obtained. As yet no extensive application has been made of reactions of this sort, for too often even so-called "controlled" oxidation yields product mixtures too complex to be useful.

ALKENES

Alkenes constitute an homologous series C_nH_{2n}. The early members are listed in Table 21.1. The outstanding structural feature common to all alkenes, the functional group of the series in fact, is the carbon–carbon double bond.

Alkenes have in high degree a chemical versatility which has helped to make them a corner stone of the modern chemical industry. The range of materials now produced in whole or in part from the simple alkenes up to the C_4 members is quite remarkable.

TABLE 21.1. ALKENES

Name	Structure
Ethylene	$CH_2{=}CH_2$
Propylene	$CH_2{=}CH{\cdot}CH_3$
1-Butene	$CH_2{=}CH{\cdot}CH_2{\cdot}CH_3$
2-Butene	$CH_3{\cdot}CH{=}CH{\cdot}CH_3$
Isobutene (2-methylpropene)	$CH_2{=}\underset{\underset{CH_3}{\vert}}{C}{\cdot}CH_3$
1-Pentene	$CH_2{=}CH{\cdot}CH_2{\cdot}CH_2{\cdot}CH_3$

21.3. Synthesis of alkenes

(a) *Cracking of alkanes*

The important lower alkenes are produced by alkane cracking (Section 21.2). Natural gas is the preferred feedstock, although the naphtha fraction from petroleum is also widely used, particularly in Europe.

Lower alkenes can be obtained as pure components by fractional distillation of cracked vapours. *Steam cracking* is widely employed in a number of processes where ethylene production is the main aim. In this technique, the introduction of superheated steam into the cracking chamber decreases losses due to elemental carbon formation, and increases ethylene yield by permitting operation at a higher temperature than would otherwise be possible—cracking temperatures of about 900°C have been reported.

(b) *Dehydration of alcohols*

The functional group of the alcohol family is the hydroxyl —OH group. If a molecule of water is removed from an alcohol, an alkene results.

With petroleum sources readily available at present alcohol dehydration on a large scale has diminished in importance as a source of alkenes. Where practised, the alcohol is vaporized and passed over a suitable dehydrating catalyst.

$$CH_3 \cdot CH_2OH \xrightarrow[\text{Alumina}]{300°C} CH_2{=}CH_2 + H_2O$$
Ethyl alcohol   Ethylene

21.4.   Reactions of alkenes

Alkenes, being unsaturated, are very reactive and characteristically undergo a wide range of *addition reactions*, several of which are of interest.

(a) *Hydrogenation*

e.g.

$$CH_3 \cdot \underset{\underset{CH_3}{|}}{\overset{\overset{CH_3}{|}}{C}} \cdot CH{=}\overset{\overset{CH_3}{|}}{C} \cdot CH_3 + H_2 \xrightarrow[\text{200°C/5–10 atm}]{\text{Ni catalyst}} CH_3 \cdot \underset{\underset{CH_3}{|}}{\overset{\overset{CH_3}{|}}{C}} \cdot CH_2 \cdot \overset{\overset{CH_3}{|}}{C}H \cdot CH_3$$
2,4,4-Trimethyl-2-pentene   2,2,4-Trimethylpentane

(b) *Halogen addition*

$$
\underset{\text{Propylene}}{CH_3 \cdot CH = CH_2} + Cl_2 \rightarrow \underset{\text{1, 2-Dichloropropane}}{CH_3 \cdot \overset{\displaystyle Cl}{\overset{|}{C}H} \cdot CH_2Cl}
$$

e.g.

(c) *Halogen acid addition*

e.g.
$$
\underset{\text{Ethylene}}{CH_2 = CH_2} + HBr \rightarrow \underset{\text{Ethyl bromide}}{CH_3 \cdot CH_2Br}
$$

This reaction has limitations. If the alkene has a non-symmetrical structure, isomers are produced:

$$
\underset{\text{1-Butene}}{CH_3 \cdot CH_2 \cdot CH = CH_2} + HCl
\begin{cases}
\rightarrow \underset{\text{sec-Butyl chloride}}{CH_3 \cdot CH_2 \cdot \overset{\displaystyle Cl}{\overset{|}{C}H} \cdot CH_3} \\[2em]
\rightarrow \underset{\text{n-Butyl chloride}}{CH_3 \cdot CH_2 \cdot CH_2 \cdot CH_2Cl}
\end{cases}
$$

In such a case, the major reaction product can be predicted by applying an empirical rule which states that the *hydrogen of the acid adds to the carbon atom already the richer in hydrogen.* In the above example, the chief product would therefore be sec-butyl chloride.

(d) *Sulphuric acid addition: alkene hydration*

The reactions of sulphuric acid, H_2SO_4, in organic chemistry are better understood in the light of a more explicit structural formula, $HO \cdot SO_2 \cdot OH$. It will add across a double bond as the acid HX, X being the group $-O \cdot SO_2 \cdot OH$,

e.g.
$$
\underset{\text{Ethylene}}{CH_2 = CH_2} + HO \cdot SO_2 \cdot OH \rightarrow \underset{\text{Ethyl hydrogen sulphate}}{CH_3 \cdot CH_2 \cdot O \cdot SO_2 \cdot OH}
$$

Subsequent hydrolysis of the addition compound yields an alcohol and regenerates sulphuric acid:

$$CH_3 \cdot CH_2 \cdot O \cdot SO_2 \cdot OH + H \cdot OH \rightarrow CH_3 \cdot CH_2 \cdot OH + H_2SO_4$$
$$\text{Ethyl alcohol}$$

This is a very important method of synthesizing alcohols. We may consider the reaction to be the addition of water (hydration) to a double bond under the influence of acid catalyst. Whether it be water, $H \cdot OH$, which adds directly, or sulphuric acid, $HO \cdot SO_2 \cdot OH$, which first reacts, the direction of addition is similar to that of the halogen acids. Hence propylene yields isopropyl alcohol, not *n*-propyl alcohol,

$$CH_3 \cdot CH{=}CH_2 + H \cdot OH \xrightarrow{H^+} CH_3\overset{\displaystyle OH}{\underset{\displaystyle |}{C}}H \cdot CH_3$$
$$\text{Propylene} \qquad\qquad\qquad \text{Isopropyl alcohol}$$

(e) *Alkylation*

Alkylation is the term used to describe the addition of an alkane to an alkene. The general reaction may be represented by

$$>C{=}C< + RH \rightarrow >CH\underset{\displaystyle R}{\overset{\displaystyle |}{C}}<$$

Alkylation is a key step in a number of petroleum-based processes. Alkanes used in gasoline production can be formed by this method:

e.g.

| Isobutene | Isobutane | Iso-octane (2,2,4-Trimethylpentane) |

If in place of the alkane aromatic hydrocarbons are used important aromatic derivatives can be obtained.

(f) *Polymerization*

Polymerization is the process whereby a large number of small molecules are linked together chemically to form a final product of very high molecular weight. The starting material is called *monomer*, the end product *polymer*. Polymers are very important, and a later chapter (Chapter 28) is devoted to them.

Alkenes can be induced to undergo polymerization. We may picture the initial reaction to be one between some highly reactive chemical fragment B— and an alkene molecule CH_2=CHX:

$$B{-}+CH_2{=}\underset{X}{CH} \rightarrow B{-}CH_2{\cdot}\underset{X}{CH}{-}$$

Initiation is followed by successive addition of monomer molecules to the reactive site, until a polymer chain containing perhaps thousands of monomer units is obtained:

$$B{\cdot}CH_2{\cdot}\underset{X}{CH}{-}+n{\cdot}CH_2{=}\underset{X}{CH} \rightarrow B{\cdot}H_2C{\cdot}\underset{X}{CH}{\cdot}\left(\underset{X}{CH_2{\cdot}CH}\right)_n$$

The reaction is terminated by one of a number of possible side reactions.

21.5. Dienes

Dienes are compounds containing two carbon–carbon double bonds. Much of their chemistry is similar to that of simple alkenes, i.e. typically they take part in addition reactions involving the unsaturated bonds.

Three important members of this group are shown below:

$$CH_2{=}CH{\cdot}CH{=}CH_2 \qquad CH_2{=}\overset{\overset{\textstyle CH_3}{|}}{C}{\cdot}CH{=}CH_2$$

Butadiene   Isoprene

$$CH_2{=}\overset{\overset{\textstyle Cl}{|}}{C}{\cdot}CH{=}CH_2$$

Chloroprene

These three dienes form the basis of the modern synthetic rubber industry (Chapter 28). Butadiene and isoprene can be produced by cracking of suitable C_4 and C_5 petroleum fractions; other methods are also available and in use. Chloroprene may be synthesized from acetylene (Section 21.7).

ALKYNES

Alkynes, or acetylenes, constitute the homologous series C_nH_{2n-2}. The recurring functional group in the alkynes is a carbon–carbon triple bond. The series is typified by the parent member acetylene $HC\equiv CH$, the only alkyne of any industrial significance.

21.6.   Synthesis of acetylene

(a) *The carbide process*

When carbon is fused with calcium oxide in an electric furnace at about 2500–3000°C, calcium carbide is formed. This in turn decomposes in water to form acetylene.

$$CaO + 3C \rightarrow \underset{\substack{\text{Calcium} \\ \text{carbide}}}{CaC_2} + CO$$

$$CaC_2 + 2H_2O \rightarrow C_2H_2 + Ca(OH)_2$$
$$\text{Acetylene}$$

Coke is used as the carbon source, limestone as the calcium oxide source. The ready availability of these materials and the comparative simplicity of the process has made this the premier method of industrial acetylene production in the past.

(b) *Petroleum sources*

Like alkenes, acetylene can be obtained from alkane feedstock. Methane is cracked at temperatures of about 1500°C to give acetylene and hydrogen:

$$2CH_4 \rightarrow C_2H_2 + 3H_2$$

The high temperatures required can be achieved in a furnace, but electric arc cracking is preferred in some processes.

Methane cracking is superseding the carbide route in acetylene manufacture, particularly in areas where cheap supplies of natural gas are available. Higher hydrocarbon fractions such as naphtha can be cracked to acetylene, but they are less suitable than methane.

21.7. Reactions of acetylene

Being highly unsaturated acetylene undergoes a wide range of addition reactions, and much of its chemistry bears comparison with that of the alkenes. The triple bond in acetylene increases its potentiality in industrial synthesis, so much so that acetylene may be classified as a raw material in its own right, distinct from both coal and petroleum. Some of the more important reactions are summarized below:

(a) *Halogenation*

e.g.　　$CH \equiv CH + 2Cl_2 \rightarrow ClCH = CHCl + Cl_2 \rightarrow Cl_2CH \cdot CHCl_2$
　　　　　　　　　　　　　Dichloroethylene　　　　　Tetrachloroethane

(b) *Addition of HX compounds*

e.g.　　　　　　$CH \equiv CH + HCl \rightarrow CH_2 = CH \cdot Cl$
　　　　　　　　　　　　　　　Vinyl chloride

$CH \equiv CH + HCN \rightarrow CH_2 = CH \cdot CN$
　　　　　　　　　　　Acrylonitrile

$CH \equiv CH + HO_2C \cdot CH_3 \rightarrow CH_2 = CH \cdot O_2C \cdot CH_3$
　　　　　　Acetic acid　　　　　　Vinyl acetate

Each of these reactions yields a vinyl compound capable of polymerizing to important products.

As with alkenes, the hydration of acetylene may be considered as an "HX-type" addition. The initial product, however, is unstable, but the

21

spontaneous rearrangement which occurs provides a route (of decreasing importance) to acetaldehyde:

$$CH\!\equiv\!CH + H_2O \rightarrow [CH_2\!=\!CH\cdot OH] \rightarrow CH_3\cdot CHO$$

"Vinyl alcohol"   Acetaldehyde

(c) *Dimerization*

Polymerization was defined as the linking together of many molecules. *Dimerization* is the linking together of two molecules, and applied to acetylene gives a route to the diene chloroprene.

$$2CH\!\equiv\!CH \rightarrow CH_2\!=\!CHC\!\equiv\!CH \xrightarrow{\text{HCl}} CH_2\!=\!CH\cdot\overset{\overset{\textstyle Cl}{\textstyle |}}{C}\!=\!CH_2$$

Vinylacetylene   Chloroprene

(d) *Combustion*

If acetylene is burned in oxygen, a flame temperature of about 2300°C can be obtained, a temperature which is substantially higher than that given by any other common gaseous fuel. This feature is made use of in the oxyacetylene burner commonly used in welding and metal-cutting operations.

Exercises

21.1. Write out structural formulae for the following compounds:

isobutane;
2,2,3-trichlorohexane;
3-chloro-1-butene;
1,3-pentadiene;
propyne.

21.2. Write out all of the products which might arise from the chlorination of ethane.

21.3. Complete the following reactions:

(i) $\overset{\displaystyle OH}{\underset{\displaystyle |}{CH_3CH_2CHCH_3}}$ $\xrightarrow[\text{Alumina}]{300°C}$

(ii) $CH_3CH_2CH{=}CH_2 + Cl_2 \rightarrow$

(iii) $CH_3CH_2CH{=}CH_2 + HCl \rightarrow$

(iv) $CH_3CH_2CH{=}CH_2 + H_2SO_4 \rightarrow \xrightarrow{H_2O}$

(v) $CH_3CH{=}CH_2 + \underset{\displaystyle \underset{\displaystyle CH_3}{|}}{CH_3CHCH_3} \xrightarrow{\text{Alkylation}}$

21.4. Suggest how (i) acetylene and (ii) isoprene might react with excess hydrochloric acid.

CHAPTER 22

Hydrocarbons: II, Aromatic Compounds

EARLY in the development of organic chemistry there emerged a group of compounds clearly related to one another, and distinctly different from the aliphatics. They generally had pleasant odours, so were called *aromatic*. Michael Faraday is credited with the first isolation in 1825 of the parent compound of the aromatic class, benzene.

22.1. The structure of benzene

Benzene is a hydrocarbon; its molecular formula is C_6H_6. On the basis of the hydrogen:carbon ratio we would expect it to display a high degree of unsaturation:

		H : C *ratio*	*Characteristics*
Methane	CH_4	4 : 1	saturated
Ethylene	C_2H_4	2 : 1	unsaturated
Acetylene	C_2H_2	1 : 1	highly unsaturated
Benzene	C_6H_6	1 : 1	?

Benzene can under certain conditions undergo addition reactions of the type associated with multiple carbon–carbon bonds of the alkene or alkyne class. This behaviour, however, is not typical of benzene and its derivatives. Aromatic compounds more characteristically undergo substitution reactions. The behaviour of benzene, then, is that of a saturated hydrocarbon despite its hydrogen:carbon ratio.

The apparent anomalies in benzene chemistry pose problems in structure formulation which have been reconciled by postulating the ring structure shown below:

$$H-C\underset{\underset{H}{\overset{C}{\|}}}{\overset{\overset{H}{\|}}{C}}C-H \quad \text{or}$$

Benzene

The lack of alkene double-bond character is due to delocalization of the *p*-orbital electrons which form the second bond of any carbon–carbon double bond (Section 1.8). The structure of benzene, i.e. a planar arrangement of carbon and hydrogen atoms with symmetrical delocalized electron clouds above and below the plane of the ring, is best represented by the symbol:

Benzene: a symmetrical structure

Molecular orbital theory has led to a definition of aromaticity, or aromatic character, much wider than once was the case. For our purposes, however, we shall simply define an aromatic compound as being a compound containing at least one benzene-type ring. Compounds so defined are sometimes called *benzenoid compounds*. Although the chemical behaviour exhibited by the ring in any one benzenoid compound is modified in some way by substituted side groups, the reactions characteristic of benzene are characteristic of the aromatic ring in any compound of this class.

22.2. Synthesis of aromatic hydrocarbons

(a) *Coal sources*

When coal is carbonized it decomposes into four primary fractions: coal gas, ammoniacal liquor, coke and coal tar. This last fraction, which generally accounts for less than 10% of the coal charge, is rich in essential aromatic compounds which can be isolated by distillation and further refining. The more important hydrocarbon constituents obtained from coal tar sources are listed in Table 22.1. Their detailed structures we shall consider later.

TABLE 22.1. ESSENTIAL
HYDROCARBONS FROM
COAL TAR

Product	Molecular formula
Benzene	C_6H_6
Toluene	C_7H_8
Xylene	C_8H_{10}
Naphthalene	$C_{10}H_8$
Anthracene	$C_{10}H_{14}$
Phenanthrene	$C_{14}H_{10}$

The first fraction collected from coal tar is called *benzol*, essentially a benzene–toluene mixture containing in addition a little xylene. Motor benzol finds good use as a high-octane gasoline fuel. (Crude benzol actually separates in the course of the primary coal carbonization, and is often considered to be a fraction in its own right.)

(b) *Petroleum sources*

A number of routes are used to obtain aromatic hydrocarbons from petroleum. Some crude oils contain a minor but significant percentage of aromatics which can be extracted directly. This is not common, and synthetic techniques usually have to be adopted.

Aromatization of other hydrocarbon types can be applied most obviously to cycloalkanes, the cycloalkane structure defining the aromatic compound obtained:

$$
\underset{\text{Cyclohexane}}{
\begin{array}{c}
H_2 \\
C \\
H_2C \diagup \quad \diagdown CH_2 \\
| \qquad\qquad | \\
H_2C \diagdown \quad \diagup CH_2 \\
C \\
H_2
\end{array}
}
\quad\longrightarrow\quad
\underset{\text{Benzene}}{\bigcirc} \;+\; 3H_2
$$

Methylcyclohexane   Toluene

Conversion to aromatic structures is not confined to cycloalkanes, however. Alkanes can also undergo cyclization and aromatization, the alkane structure again determining the aromatic compound or mixture obtained:

n-Hexane   Benzene

n-Octane   Xylenes   Ethylbenzene

The above operation, whether applied to cycloalkanes or alkanes, is called hydroforming, or platforming if a platinum-containing catalyst is used. The preferred feedstock is a light petroleum fraction rich in suitable cycloalkanes.

A second and distinctly different technique has been applied to alkane-rich fractions whereby cracking is used first to produce alkenes and dienes which then interact to form aromatics:

Butadiene   Ethylene   Cyclohexene   Benzene

Replacement of ethylene by propylene in the above reaction would yield toluene.

Aromatization procedures lead to the formation of complex mixtures which are separated into fractions of required specification by a number of further refining processes.

22.3.   Homologues of benzene

Some of the more important aromatic hydrocarbons together with their molecular formulae have been listed in Table 22.1. There is no obvious pattern within these formulae that we would readily recognize as representing an homologous series similar to those of the aliphatic hydrocarbons. How then does homology occur in aromatic compounds? The answer to this question is that benzene has the distinction of being the parent member of more than one hydrocarbon family. The compounds indicated in Table 22.1 belong to one or other of the two major aromatic hydrocarbon series, the first of which results from alkyl radical substitution in the benzene ring:

Benzene, C_6H_6 Toluene, C_7H_8 Xylene, C_8H_{10} Ethylbenzene, C_8H_{10}

The above series, to which is given the name *arenes*, is a mixed hydrocarbon family in that each member is part aromatic, part aliphatic; the presence of the benzenoid ring, however, has such a dominating effect on the chemistry of the arenes that they are usually regarded as aromatic.

Aromatic hydrocarbon structures can also be obtained by the fusing together of two or more benzene rings, as in naphthalene, $C_{10}H_8$:

Naphthalene

In naphthalene all carbon atoms are no longer identical; nor indeed are the eight hydrogen atoms, either in position or in relative reactivity. Naphthalene is the second member of the series of *fused polynuclear hydrocarbons*. Two isomers, anthracene and phenanthrene, constitute the next member of this series, $C_{14}H_{10}$,

Anthracene   Phenanthrene

and multiple ring systems are derived by permutations on this type of ring extension. A number of the larger structures have received much attention as possible cancer-inducing compounds (carcinogens). They arise in several high temperature pyrolytic or combustion systems. Tobacco smoke is currently the most notorious source of hydrocarbon carcinogens; active constituents have also been reported in, for example, coal tar and soot.

22.4. Isomerism and nomenclature in benzene derivatives

If the benzene ring is monosubstituted by some group X, the resulting structure, viz.

     or C_6H_5X or   Ph·X

is quite unambiguous. No isomerism is possible, for in the completely symmetrical benzene ring all the positions are identical. Some important monosubstituted benzenes are given in Table 22.2.

Monosubstitution in benzene, however, results in a non-symmetrical molecule in which *three* different positions may be discerned. If we

TABLE 22.2. MONOSUBSTITUTED
DERIVATIVES OF BENZENE

Name	Formula
Nitrobenzene	$C_6H_5 \cdot NO_2$
Aniline	$C_6H_5 \cdot NH_2$
Chlorobenzene	$C_6H_5 \cdot Cl$
Phenol	$C_6H_5 \cdot OH$
Benzene sulphonic acid	$C_6H_5 \cdot SO_3H$
Benzoic acid	$C_6H_5 \cdot CO_2H$

number round the ring from the principal functional group,

Numbering in benzene rings

then positions 2 and 6, called the *ortho positions*, will be seen to be equivalent; positions 3 and 5, called the *meta positions*, will be seen to be equivalent; and the remaining position 4, called the *para position*, to be unique in itself. On this basis, disubstitution should yield three isomers. This argument is well proven, and xylene provides us with a convenient example:

Ortho- or *o*-Xylene   Meta- or *m*-Xylene   Para- or *p*-Xylene

If more than two groups are substituted in the ring, the number of possible isomers may or may not increase; thus there can only be one hexamethylbenzene, while there are three trimethylbenzenes.

22.5. Reactions of aromatic hydrocarbons

While reactions in a particular aromatic nucleus are affected by groups substituted in the nucleus, there is a clear pattern of behaviour which permits us to consider the reactions of benzene to be as typical of those of any aromatic compound as, say, the reactions of ethylene are of any alkene. The difference between benzene and its many relations can be regarded for our purposes as one of degree rather than of type. If the chemical reactivity of benzene is taken as the norm for aromatic compounds, other aromatic rings are found by comparison to exhibit either enhanced or suppressed reactivity, and reaction conditions have to be adjusted accordingly; but the same type of chemical change can usually be brought about.

Substitution is the normal mode of reaction in aromatic nuclei. Two addition reactions in benzene which are important are described below; these, however, must be regarded as abnormal.

(a) *Substitution reactions*

Substitution in the aromatic nucleus is an extensively used reaction in industry. Some more important substitution reactions are shown below:

(i) *Alkylation*

$$\text{Benzene} + CH_3 \cdot CH{=}CH_2 \xrightarrow{H_2SO_4} \text{Cumene}$$

Benzene   Propylene   Cumene

$$\text{Benzene} + CH_3Cl \xrightarrow{AlCl_3} \text{Toluene} + HCl$$

Methyl chloride   Toluene

Alkylating reagents other than alkenes or alkyl halides can be used.

(ii) *Nitration*. The nitro-NO_2 group is substituted in the benzene ring by reaction with concentrated "mixed acid", a mixture of nitric and sulphuric acids. Unless the temperature is held at a moderate level, disubstitution may occur:

Benzene   Nitro-   *m*-Dinitrobenzene
                benzene

(iii) *Sulphonation*

Benzene   Benzene-
               sulphonic
               acid

Sulphonation is much used in dyestuff, drug and other industries to confer water solubility on organic products.

(iv) *Halogenation*

Benzene   Chloro-
             benzene

(b) *Addition reactions*

Benzene and chlorine undergo an addition reaction in the presence of ultraviolet light, forming benzene hexachloride, $C_6H_6Cl_6$, one form of which is a potent insecticide, Gammexane.

Benzene and hydrogen also take part in an addition reaction if a suitable catalyst is present, reversing the hydroforming reaction:

Benzene   Cyclohexane

22.6.   Polysubstitution in the benzene ring

We have seen that the introduction of a second group into the benzene ring can yield three isomers. An empirical rule permits us to predict the *orientation* of a second substituent very easily. It states that if a compound C_6H_5X is to be disubstituted, *the orientation of the second group is dependent solely on the nature of the group X; if X is a saturated group, a mixture of o- and p-isomers results, if unsaturated the m-isomer results.*

There are exceptions to this rule, but they are comparatively rare. The structures of the principal substituent groups and their directing influences are indicated in Table 22.3. Substitution beyond the second stage will be affected by the directing influence of each group already in the ring, i.e. the groups X and Y. There is no simple rule to predict the net effect.

TABLE 22.3. PRINCIPAL SUBSTITUENT GROUPS
IN AROMATIC RINGS

o- and *p*-directing	*m*-directing
—R (alkyl benzenes)	—N$\diagdown$$\diagup$$^O_O$ (nitro compounds)
—OH (phenols)	$\overset{O}{\underset{O}{\overset{\|}{\underset{\|}{—S}}}}$—OH (sulphonic acids)
—NH₂ (amines)	$\overset{}{\underset{O}{\overset{}{\underset{\|}{—C}}}}$—OH (carboxylic acids)
—Halogen (halogen compounds)	$\overset{}{\underset{O}{\overset{}{\underset{\|}{—C}}}}$—R (aldehydes and ketones)

Exercises

22.1. What aromatic structures can be ascribed to the following molecular formulae (write out all possible isomers)?

$$C_7H_8 \qquad C_8H_{10} \qquad C_7H_7Cl$$

22.2. Write out structural formulae for the following compounds:

naphthalene;
p-nitrobenzoic acid;
m-chloroaniline;
2,4-dibromophenol;
2,4,6-trinitrotoluene.

22.3. Complete the following reactions:

(i)   Benzene $+ CH_2{=}CH_2 \xrightarrow{\text{AlCl}_3}$

(ii)   Toluene $+ CH_3Cl \xrightarrow{\text{AlCl}_3}$

(iii)   *m*-Xylene $\xrightarrow[40°C]{\text{HNO}_3}$

(iv)   *p*-Toluic acid $+ Cl_2 \xrightarrow{\text{FeCl}_3}$

CHAPTER 23

Hydroxy Compounds

COMPOUNDS in which the hydroxyl—OH group is the principal functional group may be divided into two classes, *alcohols* and *phenols*. In the former class the hydroxyl function is bonded to an aliphatic carbon atom, in the latter to an aromatic carbon atom:

$$C_{al}\text{—OH} \qquad C_{ar}\text{—OH}$$

Alcohol   Phenol

There are similarities between the behaviour of alcohols and that of phenols but there are also striking differences, and we shall consider the two classes separately.

ALCOHOLS

23.1. Structure and nomenclature

Alcohols are the key intermediates in many syntheses, and by virtue of their physical properties form a valuable group of industrial and commercial solvents. In part their usefulness as solvents stems from the freely water-soluble properties displayed by the lower members. This affinity for aqueous systems is not surprising in the light of the structural similarities between water, $H \cdot OH$, and alcohols, $R \cdot OH$. As the organic part of the molecule increases in size, so the water solubility decreases. The comparison of alcohols with water can be extended; e.g. both exhibit hydrogen bonding (Section 1.9).

Some common alcohols are listed in Table 23.1 where water solubilities are also given. Use of systematic names (methanol, etc.) for the lower alcohols is widespread.

TABLE 23.1. ALCOHOLS

Name	Structure	Solubility in water (g/100 g)
Methyl alcohol (methanol)	$CH_3 \cdot OH$	Soluble in all proportions
Ethyl alcohol (ethanol)	$CH_3 \cdot CH_2 \cdot OH$	Soluble in all proportions
n-Propyl alcohol	$CH_3 \cdot CH_2 \cdot CH_2 \cdot OH$	Soluble in all proportions
Isopropyl alcohol	$CH_3 \cdot \underset{\underset{OH}{\mid}}{CH} \cdot CH_3$	Soluble in all proportions
n-Butyl alcohol	$CH_3 \cdot (CH_2)_3 \cdot OH$	7·9
Isobutyl alcohol	$CH_3 \cdot \underset{\underset{CH_3}{\mid}}{CH} \cdot CH_2OH$	10·0
sec-Butyl alcohol	$CH_3 \cdot \underset{\underset{OH}{\mid}}{CH} \cdot CH_2 \cdot CH_3$	12·5
tert-Butyl alcohol	$(CH_3)_3 \cdot C \cdot OH$	Soluble in all proportions
Benzyl alcohol	$C_6H_5 \cdot CH_2OH$	4

23.2. Synthesis of alcohols

On the industrial scale alcohols are usually obtained from alkenes, either by the hydration reaction (Section 21.4) or by the Oxo process. In the latter process carbon monoxide–hydrogen mixtures (known generally as *synthesis gas*) are reacted with alkenes to give mixtures of aldehydes which can then be catalytically hydrogenated to alcohols:

$$R \cdot CH{=\!=}CH_2 + CO + H_2$$

$$R \cdot CH_2 \cdot CH_2 \cdot CHO \xrightarrow{\text{H}_2/\text{Ni}} R \cdot CH_2 \cdot CH_2 \cdot CH_2OH$$
Aldehyde   Alcohol

$$R \cdot \underset{\underset{CHO}{\mid}}{CH} \cdot CH_3 \xrightarrow{\text{H}_2/\text{Ni}} R \cdot \underset{\underset{CH_2OH}{\mid}}{CH} \cdot CH_3$$
Aldehyde   Alcohol

The reaction is often referred to as *hydroformylation* to indicate the addition of a hydrogen atom H and a formyl group —CHO across the double bond.

Fermentation can also be used to obtain alcohols. Fermentation is the degradation of naturally occurring substances into mixtures of simpler units brought about by the action of highly specific biological catalysts called enzymes. Alcohols are among the products of carbohydrate fermentation. Early man was quick to recognize this, and fermentation processes as practised in the brewing and distilling industries have a long history. Prior to the emergence in the last 40 years or so of comprehensive petroleum-based methods, however, fermentation was also used to produce a number of aliphatic organic chemicals for general industrial use. These included alcohols, ethyl alcohol usually being the major component, while others up to the C_5 members were produced in commercially useful quantities. Despite competition from other sources, fermentation remains an important route to industrial alcohols. It has been suggested that when petroleum eventually runs out, fermented alcohol will become both a primary fuel for internal combustion engines and the major source of ethylene for the organic chemical industry.

Fermentation does not result in the production of methyl alcohol, which is highly toxic and must never be confused with its near neighbour ethyl alcohol. Methyl alcohol is also unobtainable from the other alkene routes. It is made from synthesis gas of appropriate composition in a catalysed high pressure process:

$$CO + 2H_2 \rightarrow CH_3OH$$

Methyl
alcohol

For many purposes, alcohol mixtures are often quite suitable. Ethyl alcohol solvent is usually produced as an azeotropic mixture of about 96% alcohol and 4% water. *Methylated spirits* is a name applied to a number of mixtures, usually rich in ethyl alcohol and containing a little methyl alcohol and other constituents. Essentially pure ethyl alcohol is referred to as *absolute alcohol*.

23.3. Reactions of alcohols

Alcohols can readily take part in a wide variety of reactions. They are ideal starting materials for the synthesis of most of the more important aliphatic compounds, as indicated below:

(a) *Dehydration*

Alkenes can be prepared by the dehydration of an alcohol (Section 21.3). This reaction is less useful industrially than the reverse hydration and is likely to remain so for as long as alkenes from petroleum remain cheaply available.

(b) *Alkyl halide formation*

The hydroxyl group in alcohols can be replaced by the halogen atom of a hydrogen halide.

$$R \cdot OH + HX \rightarrow R \cdot X + H_2O \qquad X = halogen$$

(c) *Ether formation*

Ethers form a family of compounds of general structure $R \cdot O \cdot R'$. Their principal use is as solvents. The best known, diethyl ether, $CH_3 \cdot CH_2 \cdot O \cdot CH_2 \cdot CH_3$, is also used as an anaesthetic.

Alcohols can form ethers according to the general reaction

$$R \cdot OH + HO \cdot R \rightarrow R \cdot OR + H_2O$$

This dehydration may be achieved on the large scale by a vapour phase catalysed reaction, or by reaction with concentrated sulphuric acid.

(d) *Ester formation*

Esters, whose general structural formula is

$$R'CO_2R \quad \text{or} \quad R' \cdot \overset{\displaystyle}{\underset{\displaystyle O}{C}} - OR$$

are formed by reaction of alcohols with *carboxylic acids*:

$$R' \cdot \overset{\overset{\displaystyle O}{\|}}{C}\text{—OH} + \text{HO} \cdot R \rightleftharpoons R' \cdot \overset{\overset{\displaystyle O}{\|}}{C}\text{—OR} + H_2O$$

Carboxylic   Alcohol   Ester
acid

The two-part name of an ester indicates first the alcohol fragment, secondly the acid. For example, ethyl acetate is formed by reaction of *ethyl* alcohol with *acetic* acid:

$$CH_3 \cdot \overset{\overset{\displaystyle O}{\|}}{C}\text{—OH} + \text{HO} \cdot CH_2 \cdot CH_3 \rightleftharpoons CH_3 \cdot \overset{\overset{\displaystyle O}{\|}}{C}\text{—O} \cdot CH_2 \cdot CH_3 + H_2O$$

Acetic acid   Ethyl alcohol   Ethyl acetate

These reactions are reversible, ester formation commonly being catalysed by acid while the reverse *ester hydrolysis* reaction is base catalysed by, for example, sodium or potassium hydroxide. Ester hydrolysis results not in the formation of the free acid but of the appropriate salt. Thus the hydrolysis of ethyl acetate in sodium hydroxide is represented by the equation:

$$CH_3 \cdot \overset{\overset{\displaystyle O}{\|}}{C}\text{—O} \cdot CH_2 \cdot CH_3 + NaOH \rightarrow CH_3 \cdot \overset{\overset{\displaystyle O}{\|}}{C}\text{—ONa} + \text{HO} \cdot CH_2 \cdot CH_3$$

Ethyl acetate   Sodium acetate   Ethyl alcohol

(e) *Oxidation*

The products of alcohol oxidation depend on the structure of the alcohol, i.e. on whether it is primary, secondary, or tertiary. Mild oxidations proceed readily in accordance with the following general schemes:

$$R \cdot CH_2OH \xrightarrow{\text{[O]}} R \cdot CHO \xrightarrow{\text{[O]}} R \cdot CO_2H$$

Primary   Aldehyde   Carboxylic
alcohol   acid

$$\underset{\substack{\text{Secondary}\\\text{alcohol}}}{\overset{R}{\underset{R'}{>}}CH \cdot OH} \xrightarrow{[O]} \underset{\substack{\text{Ketone}}}{\overset{R}{\underset{R'}{>}}C=O} \longrightarrow \text{further oxidation unusual}$$

$$\underset{\substack{\text{Tertiary}\\\text{alcohol}}}{R'-\overset{\overset{\displaystyle R}{|}}{\underset{\underset{\displaystyle R''}{|}}{C}}\cdot OH} \longrightarrow \text{oxidation unusual}$$

Initial oxidation of both primary and secondary alcohols is formally a dehydrogenation, one molecule of hydrogen being removed. The carbon skeleton of the alcohol is unaltered in the course of these mild oxidations. Aldehydes readily oxidize further, but ketones, like tertiary alcohols, will only oxidize under more severe conditions and in such a manner that the carbon skeleton is broken up into smaller fragments. The industrial oxidation of alcohols is carried out by passing vaporized alcohol and air over silver.

Aldehydes and *ketones* are closely related chemical types, as we might anticipate from their structures:

$$\underset{\substack{\text{Aldehyde}}}{\overset{R}{\underset{H}{>}}C=O \quad \text{or} \quad R \cdot CHO} \qquad \underset{\substack{\text{Ketone}}}{\overset{R}{\underset{R'}{>}}C=O \quad \text{or} \quad R \cdot CO \cdot R'}$$

The chemical behaviour of these compounds is dominated by the carbonyl $C=O$ group. Carbonyl compounds figure in a number of industrial syntheses; two of the ketone series, acetone, $CH_3 \cdot CO \cdot CH_3$, and methyl ethyl ketone, $CH_3 \cdot CO \cdot CH_2 \cdot CH_3$, are also important solvents. Alcohol oxidation is not the sole industrial means of obtaining carbonyls in bulk; direct catalytic oxidation of alkenes is now in use, ethylene giving acetaldehyde, propylene giving acetone, and so on. Acetone is also produced by the cumene process (Section 23.5).

The above reactions illustrate the central position which alcohols hold in aliphatic chemistry. It is useful at this point to summarize the relationship which the more important aliphatic compounds have to one another:

$$
\begin{array}{ccc}
\overset{\displaystyle X}{\underset{\displaystyle >CH \cdot C <}{|}} & & >CH \cdot \overset{|}{\underset{|}{C}} \cdot O \cdot \overset{|}{\underset{|}{C}} \cdot CH < \\
\text{Alkyl halides} & & \text{Ethers}
\end{array}
$$

$$
>CH \cdot CH < \;\rightarrow\; >C{=}C < \;\rightleftharpoons\; \boxed{>CH \cdot \overset{\displaystyle OH}{\underset{\displaystyle |}{C}} <} \;\rightleftharpoons\; >CH \cdot \overset{\displaystyle O \cdot OC \cdot R}{\underset{\displaystyle |}{C}} <
$$

$$
\begin{array}{cccc}
\text{Alkanes} & \text{Alkenes} & \text{Alcohols} & \text{Esters}
\end{array}
$$

$$
>CH \cdot CO_2H \;\leftarrow\; >CH \cdot \overset{\displaystyle O}{\overset{\displaystyle \|}{C}}{-}
$$

$$
\begin{array}{cc}
\text{Carboxylic} & \text{Aldehydes} \\
\text{acids} & \text{and} \\
 & \text{ketones}
\end{array}
$$

PHENOLS

23.4. Structure and nomenclature

Phenols have the general formula Ar·OH. When a hydroxyl group is attached directly to an aromatic nucleus, the hydrogen atom becomes distinctly acidic. This is perhaps the outstanding difference between phenols and alcohols, although of course the two classes reveal the normal range of differences which exist between aliphatic and aromatic compounds. The presence of the hydroxyl function confers on many phenols a limited degree of water solubility.

Carbolic acid was the name once given to the parent member of the series, phenol, reflecting the nature of the group as a whole. Phenol itself is quite the most important member of the class to which it lends its name. It and a few of its homologues are shown below.

Phenol   o-Cresol   3,5-Xylenol   2-Naphthol

23.5. Synthesis of phenols

Coal tar (Section 22.2) includes a phenolic fraction from which are extracted phenol, the cresols, and the xylenols. This traditional source of phenol itself has for long been supplemented by a number of synthetic routes which, in the main, use benzene as a starting material. A petroleum-based process, the cumene process, is one of the more important synthetic routes to phenol. In this process, benzene is alkylated to cumene (isopropyl benzene) which is then readily oxidized to a mixture of phenol and acetone:

Benzene   Propylene   Cumene   Phenol   Acetone

23.6. Reactions of phenols

Parallels can be drawn between the reactions of alcohols and those of phenols. The latter, for example, are capable of forming *aryl esters* and *aryl ethers* in reactions analogous to those of the alcohols. The comparison need not be carried very far, however, before it becomes clear that we are dealing with two quite distinct classes of compound. Two general reactions illustrate this.

(a) *Acidic behaviour*

Phenols react with strong alkalis in a classical acid–base fashion to form salts and water. Thus phenol reacts with sodium hydroxide according to the following equation:

Phenol   Sodium phenate

These salts exhibit typical behaviour, dissociating to ionic species in water to give conducting solutions.

From the above equation we see that any simple phenol is a mono-basic acid, and that its dissociation in water can be represented by the equation:

$$Ar \cdot OH + H_2O \rightleftharpoons ArO^- + H_3O^+$$

and thus its dissociation constant K_a is given by:

$$K_a = \frac{[Ar \cdot O^-][H_3O^+]}{[Ar \cdot OH]}$$

The K_a values for a range of phenols clearly indicate that they are very weak acids when compared to the traditional range of inorganic acids; phenol itself has a K_a value of $1 \cdot 2 \times 10^{-10}$ at 25°C.

(b) *Nuclear reactions*

Phenols undergo typical aromatic substitution reactions. The aromatic ring is highly activated by the presence of a phenolic group, and *o-*, *p-*substitution takes place very readily. Paradoxically, the supremely important example here is provided not by any of the better known aromatic substitutions, but by the reaction of phenol with formaldehyde.

Phenol   Formal-dehyde   *o-*, *p-*Hydroxybenzyl alcohol

The reaction is the first step in the production of a wide range of phenol–formaldehyde resins.

Exercises

23.1. Classify the following into primary, secondary and tertiary alcohols:

> ethyl alcohol;
> isobutyl alcohol;
> 2,4-dimethyl-2-pentanol;
> *n*-hexyl alcohol;
> benzyl alcohol.

23.2. Ethyl alcohol boils at 78°C, whereas ethyl chloride boils at 12·5°C. Explain this difference. (*Hint*: water has a much higher boiling point than hydrogen chloride.)

23.3. Suggest how the following could be synthesized from *n*-propyl alcohol:

> propyl butyrate;
> 2-chloropropane;
> 1,2-dibromopropane;
> propionic acid ($CH_3CH_2CO_2H$).

23.4. What products would you expect from the mononitration of

> (i) phenol;
> (ii) *m*-cresol;
> (iii) 2-naphthol?

Are conditions for mononitration liable to be more or less severe than those for the mononitration of benzene?

CHAPTER 24

Organic Acids and Bases

CARBOXYLIC AND SULPHONIC ACIDS

24.1. Carboxylic acids

The compounds represented by the general structural formula

$$R-C\begin{subarray}{l}\diagup O \\ \diagdown OH\end{subarray} \quad \text{or} \quad Ar-C\begin{subarray}{l}\diagup O \\ \diagdown OH\end{subarray}$$

are the most important of the organic acids. These are the carboxylic acids. They are capable of donating a proton from the hydroxyl group to suitable acceptors, and are therefore acids in the classical sense.

Like the alcohols, the lower carboxylic acids are freely water soluble because of structural similarities to water, but this solubility decreases with increasing size of hydrocarbon radical. They dissociate in water according to the equation:

$$R \cdot CO_2H + H_2O \rightleftharpoons R \cdot CO_2^- + H_3O^+$$

whence

$$K_a = \frac{[R \cdot CO_2^-][H_3O^+]}{[R \cdot CO_2H]}$$

The K_a values quoted in Table 24.1, which lists some common carboxylic acids, indicate that these are weak acids compared to the inorganic acids, but much stronger acids than phenols. It will be seen that substituent groups can affect the acid strength of both aliphatic and aromatic acids

337

quite markedly. In the examples shown, chloracetic acid and *p*-nitro-benzoic acid each contain a substituent group which exercises a pronounced electron attracting effect, the result of which is to make the carboxylic acid function more positive than in the unsubstituted structure, and therefore more able to split off a proton.

TABLE 24.1. CARBOXYLIC ACIDS

Name	Structure	K_a (25°C)
Formic acid	$H \cdot CO_2H$	$1 \cdot 77 \times 10^{-4}$
Acetic acid	$CH_3 \cdot CO_2H$	$1 \cdot 75 \times 10^{-5}$
Chloracetic acid	$ClCH_2 \cdot CO_2H$	$1 \cdot 55 \times 10^{-3}$
Propionic acid	$CH_3 \cdot CH_2 \cdot CO_2H$	$1 \cdot 4 \times 10^{-5}$
Butyric acid	$CH_3 \cdot CH_2 \cdot CH_2 \cdot CO_2H$	$1 \cdot 5 \times 10^{-5}$
Benzoic acid	$C_6H_5 \cdot CO_2H$	$6 \cdot 4 \times 10^{-5}$
p-Nitrobenzoic acid	$p\text{-}NO_2 \cdot C_6H_4 \cdot CO_2H$	$4 \cdot 0 \times 10^{-4}$

24.2. Synthesis of carboxylic acids

The oxidation of aldehydes (Section 23.3) is an easy route to carboxylic acids; acetaldehyde oxidation is the most important industrial example of this method.

$$CH_3 \cdot CHO \xrightarrow{[O]} CH_3 \cdot CO_2H$$

Acetaldehyde　　　　Acetic acid

Vinegar, which is about a 5% aqueous solution of acetic acid, is made by enzyme-induced air oxidation of ethyl alcohol. This reaction can be the ruination of wines and beers, producing as it does a sour acidic flavour in the liquid.

Glyceride hydrolysis provides a suitable route to a group of carboxylic acids which would be difficult to obtain by other methods. Glycerides are naturally occurring esters in which the acid fragments contain long carbon chains. On hydrolysis they yield long chain acids, the most important of which are listed in Table 24.2.

TABLE 24.2. CARBOXYLIC
ACIDS FROM GLYCERIDES

Name	Formula
Palmitic acid	$C_{15}H_{31} \cdot CO_2H$
Stearic acid	$C_{17}H_{35} \cdot CO_2H$
Oleic acid	$C_{17}H_{33} \cdot CO_2H$
Linoleic acid	$C_{17}H_{31} \cdot CO_2H$
Linolenic acid	$C_{17}H_{29} \cdot CO_2H$

Alkyl side chains of any length on aromatic nuclei are readily oxidized to carboxyl functions, this being the most useful method of obtaining aromatic acids.

Toluene   Benzoic
acid

24.3. Reactions of carboxylic acids

The dominating feature of the carboxylic acids is their orthodox acidic behaviour, typified by the formation of ionizable, water-soluble salts on reaction with bases.

e.g.

Benzoic   Ammonium
acid   benzoate

Other reactions important from the industrial standpoint are ester formation (Section 23.3) and reduction to alcohols. This latter reaction is used in industry for the production of long chain alcohols employed chiefly in synthetic detergent manufacture. The feedstock is a long chain fatty acid mixture obtained from glycerides. The hydrogenation, which is catalysed, is carried out at high pressures and temperatures.

$$R \cdot CO_2H \xrightarrow{H_2} R \cdot CH_2OH$$

24.4. Sulphonic acids

Only the aromatic sulphonic acids are of any consequence; these are prepared by direct sulphonation of the aromatic nucleus (Section 22.5). They have the general formula

$$Ar \cdot SO_3H \quad \text{or} \quad Ar—\overset{\displaystyle O}{\underset{\displaystyle O}{\overset{\|}{\underset{\|}{S}}}}—OH$$

and readily donate a proton from the hydroxyl group.

Dissociation of sulphonic acids in water is virtually complete:

$$Ar \cdot SO_3H + H_2O \rightleftharpoons Ar \cdot SO_3^- + H_3O^+$$

They are therefore by far the strongest of organic acids, and are more comparable in acid behaviour to sulphuric acid than to the carboxylic acids. Nuclear reactions can occur; the ring is deactivated by the unsaturated sulphonic acid group and *m*-orientation is the principal result of disubstitution.

Sulphonation is much used to render dyestuffs and drugs water soluble, for sulphonic acids dissolve easily in water and other polar solvents. The major outlet of sulphonic acid derivatives is in the synthetic detergent industry (Chapter 29).

ORGANIC BASES

24.5. Amines

As carboxylic acids are the principal acids of organic chemistry, so amines are the principal bases. Amines are related to ammonia, and their position in the relationship is used as a method of sub-classification:

$$NH_3 \qquad R\!-\!NH_2 \qquad \overset{\overset{\displaystyle R'}{\displaystyle |}}{R\!-\!NH} \qquad \overset{\overset{\displaystyle R'}{\displaystyle |}}{\underset{\underset{\displaystyle R''}{\displaystyle |}}{R\!-\!N}}$$

Ammonia	Primary amine	Secondary amine	Tertiary amine

The feature common to all amines is a nitrogen atom which behaves as a basic function for reasons which we shall presently examine. Three typical amine structures are shown below:

$$CH_3 \cdot NH_2 \qquad (CH_3 \cdot CH_2)_2 \cdot NH$$

Methylamine   Diethylamine   Aniline

24.6. Synthesis of amines

Aromatic primary amines are obtained by reduction of the corresponding nitro compound,

Nitro-   Aniline
benzene

Secondary and tertiary amines are readily prepared from the primary amine.

Aliphatic amines may be obtained from alcohols, which react in consecutive stages with ammonia to yield eventually tertiary amines. The presence of a large excess of ammonia ensures that the primary amine is the principal product:

e.g. $3CH_3 \cdot CH_2OH + NH_3 \rightarrow CH_3 \cdot CH_2 \cdot NH_2 \rightarrow$
        Ethyl alcohol   Ethylamine

$$(CH_3 \cdot CH_2)_2 \cdot NH \rightarrow (CH_3 \cdot CH_2)_3 \cdot N$$
Diethylamine   Triethylamine

Higher aliphatic amines are better prepared from the corresponding alkyl halides by reaction with ammonia.

24.7.   Reactions of amines

Amines are basic. This can best be illustrated if we consider the dissociation of a water-soluble amine. This system equilibrates according to the equation:

$$R \cdot NH_2 + H_2O \rightleftharpoons RNH_3^+ + OH^-$$

in which the amine is seen as a classical base, acting as a proton acceptor. The equilibrium constant K_b for the dissociation is given by

$$K_b = \frac{[RNH_3^+][OH^-]}{[R \cdot NH_2]}$$

As with K_a values for organic acids, K_b values vary with amine structure. Some are listed in Table 24.3, from which it will be seen that the aliphatic amines are more basic than ammonia, the aromatic amines less so.

TABLE 24.3. AMINES

Name	Formula	K_b (25°C)
Methylamine	$CH_3 \cdot NH_2$	$4 \cdot 4 \times 10^{-4}$
Dimethylamine	$(CH_3)_2 \cdot NH$	$5 \cdot 1 \times 10^{-4}$
Trimethylamine	$(CH_3)_3 \cdot N$	$0 \cdot 6 \times 10^{-4}$
Ethylamine	$CH_3 \cdot CH_2 \cdot NH_2$	$4 \cdot 7 \times 10^{-4}$
Ammonia	NH_3	$1 \cdot 8 \times 10^{-5}$
Aniline	$C_6H_5 \cdot NH_2$	$4 \cdot 2 \times 10^{-10}$
p-Toluidine	$p\text{-}CH_3 \cdot C_6H_4 \cdot NH_2$	$1 \cdot 2 \times 10^{-9}$

We should expect amines to undergo normal acid–base reactions, and this is the case. Thus water-soluble ionic salts are formed with acids:

Aniline + HCl ⟶ Aniline hydrochloride

cf.

$$H \cdot NH_2 + HCl \rightarrow H \cdot NH_3^+ Cl^-$$

Ammonia   Ammonium chloride

Salts can also be formed with carboxylic acids; these in turn can be converted to substituted amides. This reaction sequence is made use of in nylon production.

$$R \cdot CO_2H + H_2N \cdot R' \rightarrow R \cdot CO_2^- + H_3^+N \cdot R' \rightarrow R \cdot CO \cdot HN \cdot R' + H_2O$$

Acid   Base   Salt   Substituted amide

Amines take part in a wide range of reactions over and above those in which they behave as orthodox bases. One example which is of interest within the context of synthetic detergents (Chapter 29) is the sequence of reactions which a primary amine undergoes with an alkyl halide to produce eventually a *quaternary ammonium salt*:

$$R \cdot NH_2 \xrightarrow{RX} R_2 \cdot NH \xrightarrow{RX} R_3N \xrightarrow{RX} R_4N^+X^-$$

Quaternary ammonium salt

Exercises

24.1. Place the following in ascending order of K_a value:

     benzene sulphonic acid;
     benzoic acid;
     p-nitrophenol;
     dichloracetic acid.

24.2. Suggest how acetic acid might be synthesized starting from

   (i) ethylene;
   (ii) acetylene.

24.3. Starting from toluene indicate how you would synthesize

   (i) *p*-nitrotoluic acid;
   (ii) *m*-nitrotoluic acid.

24.4. Which of the following amines is likely to be the most basic, and which the least basic:

   o-toluidine;
   aniline;
   p-nitroaniline;
   p-aminobenzoic acid?

CHAPTER 25

Coal

THE fuel consumption of any society is directly related to its standard of living, and currently we are seeing an increase in world fuel demands so large as to be unimaginable even 40 years ago.

Any system which alters exothermically is a prospective fuel. Nuclear fuels supply heat not from orthodox chemical reactions, but from sub-atomic changes in the nucleus. These changes are dealt with in Chapter 2, where attention is drawn to the great promise that nuclear energy offers. We still, nevertheless, depend almost entirely on traditional fossil fuel sources, which release their potential energy by exothermic reaction with aerial oxygen.

Fuels must be available in very large quantities at as cheap a price as possible. This has always been a primary requirement, and it always will be. Men first used *wood* as an answer to their fuel needs, and defore-station on a massive scale occurred in many areas over long periods. *Peat* where available was also a convenient fuel. *Coal* became important with the very great increase in fuel demands which accompanied the in-dustrial revolutions. *Petroleum* (including natural gas), a relative new-comer to the world fuel scene, has become a dominant member of this group.

In this chapter we shall concern ourselves with coal and its near rela-tions. This fuel, like the other major fossil fuels, is also a valuable chem-ical raw material, and should always be so regarded.

25.1.　Chemical constitution of coal

"Coal" is a term applied to a range of mineral materials which the geologist would regard as organic rock types. Coals, then, vary one from another. Moreover, any given coal cannot be represented by a uniquely defined chemical structure; a single coal is itself a mixture of complex chemical structures.

Raw coal we can conveniently regard as being a three-component material:

$$\text{``Raw coal''} \longrightarrow \begin{cases} \rightarrow \text{``Pure'' coal} \\ \rightarrow \text{Inorganic material (mineral matter)} \\ \rightarrow \text{Moisture} \end{cases}$$

We are here using the term "pure" coal to cover the bulk of the material, which is organic in type and constitutes the essential fuel. The remaining two components have no fuel value as such, and may be regarded as impurities.

The organic material in coal, derived from the decay of woody tissue and other vegetable matter over very long periods of time of up to about 300 million years, contains carbon and hydrogen along with smaller quantities of oxygen, sulphur and nitrogen. The manner in which these elements are ordered in coal structures poses a fascinating problem, and the whole question of coal constitution is still a very open one.

Coal is a polymeric material, i.e. it consists of many high molecular weight molecules, all similar in chemical structure but not necessarily identical. This very definition of polymeric material, whether applied to coal or any other natural or synthetic polymer, implies that we cannot speak of a specific coal structure but only of the structural features which predominate. Coal, containing largely carbon and hydrogen, is chiefly a polymeric hydrocarbon. Much of this hydrocarbon material is aromatic, as indicated by carbon:hydrogen ratios in coal and by the aromatic nature of, for example, coal carbonization and coal oxidation products; but some fully saturated alicyclic rings are also thought to be present. These part aromatic, part alicyclic ring systems exist as condensed clusters (hydroaromatic clusters) which in turn are linked together by short aliphatic bridges, probably methylene —CH_2— bridges mainly.

How do other elements fit into this type of structural unit? Oxygen certainly appears as a phenolic function, as a carbonyl function, and in geologically young coals as a carboxylic acid function. Nitrogen and even more so sulphur are less well defined, but they are likely to occur in heterocyclic structures forming part of the main ring clusters. Partial evidence for this latter view is the appearance of heterocyclic compounds among the products of coal carbonization.

D. W. van Krevelen, an outstanding contributor to the field of coal constitution, has emphasized that coal structure cannot be tied to a visual model but should be expressed as a series of structural parameters. Nevertheless, a visible model helps all of us to get a clearer picture of what a segment of a coal molecule might look like. One such model is

FIG. 25.1. Representation of a coal "molecule" (P. H. Given, *Nature*, 184, 980 (1959)).

reproduced in Fig. 25.1. It attempts to represent a segment of a coal of specific analysis, and does not therefore include every structural feature mentioned above; thus no sulphur appears, nor any carboxylic acid function.

The aromatic nature of coal helps to explain why it is a solid. The ring clusters tend to be planar, and packing of these clusters or lamellae is efficient enough to produce a solid mass with a high softening temperature. This contrasts with petroleum, where the large number of flexible aliphatic structures leads to a low packing efficiency and a liquid phase.

25.2. Analysis of coal

The complex nature of coal rules out, for many purposes, analytical techniques which measure precisely some clearly defined chemical or physico-chemical property. Some of the more important tests are empirical; they do not measure absolute functions, and if meaningful results are to be obtained such empirical tests must be carried out according to standard specified conditions. In the United Kingdom the British Standards Institution is responsible for the standardization of coal and coke testing, the various procedures being embodied in BS 1016 and 1017. The equivalent body in the United States is the American Society for Testing Materials.

(a) *Proximate analysis*

Perhaps the most important coal test, the proximate analysis, records three parameters: *moisture, volatile matter* and *ash*, the residue being reported as *fixed carbon*.

Coal is a hygroscopic material and moisture is always present to a small extent. This hygroscopic moisture is usually referred to as *inherent moisture*, being the equilibrium moisture content of a coal sample air-dried in the laboratory prior to proximate analysis. Coal may also contain varying quantities of *free moisture*, which can readily be determined, but must not be confused with the proximate analysis figure. Moisture in coal contributes nothing to the fuel potential of the material; but damp coal burns more evenly than dry coal on a fuel bed, and for this reason a coal feed to a furnace may be sprayed with water to a predetermined moisture content before firing.

Volatile matter is an important coal parameter, and goes a long way towards classifying a given coal. It is defined as the percentage weight loss from a coal sample heated for 7 minutes at 900°C±5°C (precise conditions vary a little in different countries).

Ash in coal is determined by burning off the organic material and weighing the residue. This ash residue cannot be directly equated with original mineral matter of the coal. During the process of combustion, carbonates will decompose to oxides, which together with other like transformations will yield an ash figure perhaps quite different from the true mineral matter figure. The latter can be determined indirectly, but laboriously, and if the ash figure is less than 5% this conversion is usually avoided. Ash in coal is, like moisture, of no fuel value but it does have the virtue of protecting grate bars from direct contact with incandescent carbon, a contact which would certainly lead to totally unacceptable corrosion rates. Ash can be troublesome if it has too low an *ash fusion temperature*, for ready softening at high temperatures leads to clinker formation on cooling. Clinker on grates interferes with combustion air supply, is difficult to discharge, and is very undesirable.

If proximate analyses for different coals are to be compared, this is best done by reducing the analytical figures to a dry, ash-free (d.a.f.) basis, i.e. by eliminating the "impurities" from the analyses, and simply stating volatile matter against fixed carbon. Otherwise the presence of different levels of ash and moisture in two coals may obscure very real differences or similarities which exist between the materials. In very accurate work, a coal analysis may ultimately be quoted on a d.mm.f. (dry, mineral-matter free) basis.

(b) *Ultimate analysis*

The ultimate analysis of coal is not an empirical test, but a precise determination of the elemental constituents using standard analytical techniques. The test, as defined by the British Standard, determines carbon, hydrogen, nitrogen and sulphur; oxygen may be determined by difference. Although not considered part of the ultimate analysis, methods are also specified for determining trace elements, in particular chlorine, phosphorus and arsenic.

Fɪɢ. 25.2. Gray–King standard coke types. (BS 1016, part 12, 1959. Reproduced by permission of the British Standard Institution, London, W.1. The figures are not to scale, and are unsuitable for laboratory use.)

(c) *Gray–King assay*

Each year large quantities of coal are carbonized in process of making metallurgical coke, coal gas, and other products. In the United Kingdom the behaviour of a coal during carbonization is assessed by means of a Gray–King assay. The standard low temperature assay is carried out by carbonizing a sample of coal at a maximum temperature of 600°C, the various decomposition products being collected as required.

The real importance of the Gray–King assay is its use as a British Standard test for classifying cokes. By comparison with a series of photographs of standard cokes, shown in Fig. 25.2, the coke residue from the assay is accorded a classification which runs from A to G, and thence from G_1 to G_9 type cokes. Thus an effort is made to indicate the two chief properties of a coke, viz. the degree to which it has swollen, and the degree to which individual particles have agglutinated to form one mass. The significance of key points on the scale is indicated below:

Coke type:	A . . . . G . . . . G_9
Description:	pulverent hard highly
	highly swollen no swelling or shrinking swollen
	(standard coke)

(d) *Calorific value*

The calorific value, or heat content, of any fuel is of self-evident importance. For solid and liquid fuels, it is determined by burning a sample in a steel calorimeter under 25 atm of oxygen, and measuring the heat output in a surrounding calorimeter. For coals, calorific values range from about 9000 to 16,000 Btu per pound.

25.3. Classification of coal

We need only remind ourselves of the highly complex chemical structures which are believed to occur in coals to realize that rigorous classification of these materials is likely to be very difficult.

(a) *Coal rank: the Seyler classification*

The *rank* of a coal denotes its geological age, i.e. its position on the scale describing the change of vegetable materials through various fairly well-defined stages to the final product of the coalification process, graphite:

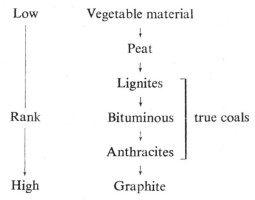

Change in rank is reflected by a steady change in other parameters, notably carbon content and volatile matter, but also hydrogen content, oxygen content, calorific value, and inherent moisture content. The progression outlined above is not in itself comprehensive enough to serve as a classification system, but most of the more detailed schemes in use are based on rank changes, and are commonly formed by plotting or tabulating two or more parameters which are themselves functions of rank, and so obtaining a correlation over a wide range of coals. We can see this clearly in the most detailed of the classification schemes devised for British coals, that due to Seyler. Here carbon and hydrogen contents for a series of coals are plotted to obtain a *coal band*, which is subdivided into a number of groups. The primary divisions in the Seyler scheme are shown in Table 25.1.

Having constructed the basic coal band, other relationships can be represented within the Seyler scheme (as indeed in other classification schemes). Thus lines of constant volatile matter and constant calorific value are readily superimposed on the initial chart. Much information

TABLE 25.1. DIVISION IN THE SEYLER
CLASSIFICATION SCHEME

Coal type	% C	% H
Peat (average)	56·5	5·5
Lignite	65–76	5·25–5·1
Ortho-lignitious	76–80	5·1–5·05
Meta-lignitious	80–84	5·05–5·0
Para-bituminous	84–87	5·0–4·95
Ortho-bituminous	87–89	4·95–4·8
Meta-bituminous	89–91	4·8–4·25
Carbonaceous	91–93	4·25–3·3
Anthracite	93–94·8	3·3–2·5

beyond this is given in the fullest form of the Seyler chart, which is presented and discussed in some detail elsewhere.[†] A restricted version is reproduced in Fig. 25.3.

Coals in the main are derived from woody tissues, but small quantities of fossilized material from other vegetable sources also occur. If too much of this latter material is present, an abnormal coal which does not fall within the coal band of the Seyler chart may result. For such abnormal or *dull* coals corrections have to be made in applying the Seyler chart to them. A majority of British coals are normal, or *bright*, coals.

(b) *National Coal Board classification*

The complexity of the Seyler classification makes it unwieldy for many practical purposes. The National Coal Board now uses a scheme in which two readily determined and very practical parameters, viz. volatile matter and Gray–King coke type, form the basis of a division of British coals into nine primary groups, numbered 100 to 900. (These must not be confused with the division of coals into five groups for the domestic and commercial market.) The groups are in turn subdivided into classes,

[†] Ed. H. M. Spiers, *Technical Data of Fuel*, World Power Conference, London, 1961.

Fig. 25.3. The Seyler chart.

101 and 102, 201, and so on, which may be split yet again into, for example classes 201a and 201b. Separation of different coals within the National Coal Board scheme is not always satisfactory, but it does suffice for most purposes. The scheme is summarized in Table 25.2(a) and (b).

TABLE 25.2. NATIONAL COAL BOARD CLASSIFICATION OF BRITISH COALS

(a) Anthracites, Low-volatile Steam Coals and Medium-volatile Coals

Group, VM$_{d.mm.f.}$ and general description	Class	Sub-class	VM$_{d.mm.f.}$	Gray-King coke type	Specific description
100 Under 9·1 Anthracites	101 102		Under 6·1 6·1–9·0	A A	Anthracites
200 9·1–19·5	201	201a 201b	9·1–13·5 9·1–11·5 11·6–13·5	A–C A–B B–C	Dry steam coals
Low-volatile steam coals	202 203 204		13·6–15·0 15·1–17·0 17·1–19·5	B–G E–G$_4$ G$_1$–G$_8$	Coking steam coals
	206		9·1–19·5	A–B (VM up to 15·0) A–D (VM above 15·0)	Heat-altered low-volatile steam coals
300 19·6–32·0	301	301a 301b	19·6–32·0 19·6–27·5 27·6–32·0	G$_4$ and over	Prime coking coals
Medium-volatile coals	305 306		19·6–32·0 19·6–32·0	G–G$_3$ A–F	(Mainly) heat-altered medium-volatile coals

TABLE 25.2 *(cont.)*

(b) *High-volatile Coals* ($VM_{d.mm.t.}$ above 32.0%)

Group	$VM_{d.mm.t.}$		Gray-King coke type	Description
	32·1–36·0	Over 36·0		
	Class			
400	401	402	G_9 and over	Very strongly caking
500	501	502	G_5–G_8	Strongly caking
600	601	602	G_1–G_4	Medium caking
700	701	702	E–G	Weakly caking
800	801	802	C–D	Very weakly caking
900	901	902	A–B	Non-caking

Several other classification schemes have been proposed and are similar to those already described in as far as they relate general properties within the framework of a two- or three-parameter plot. The discussion of this subject set out by Francis[†] includes coverage of American and European systems.

25.4.   Coal combustion

Coal may be burned in industrial furnaces on a number of different types of grate. The reactions which occur are essentially independent of the type of grate in use; indeed many of the phenomena occurring in a large industrial installation differ from those in a domestic grate only in magnitude.

When coal is fed to an incandescent fuel bed it decomposes. Combustible volatile matter is evolved, and a coke residue is left on the bed itself. We must therefore consider the combustion process to be two-stage, the first stage occurring by interaction of primary air with coke in the fuel bed, the second occurring by interaction of secondary air with volatile combustibles in the vapour space above the bed. We shall deal with the more important features of primary and secondary combustion in turn.

Primary combustion is best illustrated by considering a typical system in which fresh fuel is overfed to the top of the bed. Decomposition occurs

[†] W. Francis, *Coal: Its Formation and Composition*, Arnold, London, 1961.

rapidly, and the coke residue then moves down through the bed, reacting as it travels with primary air rising up through the grate bars. This counter-current system is a heterogeneous one, being the reaction of a gas at a solid surface. In the ideal case reaction will be complete before solid combustible material reaches the grate bars, at which point the residue should consist solely of inorganic ash. In practice this is very difficult to achieve, and a little unburned carbon is usually rejected with the ash.

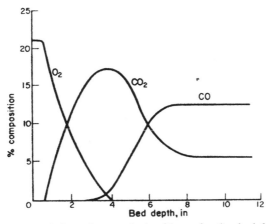

FIG. 25.4. Variation of gas analysis with combustion bed depth.

If gas samples are taken through a fuel bed of the type described and subsequently analysed, a profile is obtained in which distinct zones emerge. This is shown in Fig. 25.4, where changes in gas content against bed depth are plotted, and Fig. 25.5, which interprets these gas changes diagrammatically.

Air rising through the inert *ash zone* first reacts with carbon in an *oxidation zone* to produce mainly carbon dioxide. (Formation of the dioxide is known to be a two-step process, the first step being a hetero-geneous reaction of oxygen at a carbon surface to yield carbon monoxide. This product subsequently diffuses from the carbon surface and reacts very rapidly with further oxygen to yield carbon dioxide.) Reaction is virtually complete in the oxidation zone, all oxygen being consumed. Carbon dioxide then rises through the upper reaches of the bed, and in so doing

FIG. 25.5. Reaction zones in a combustion bed.

reacts with further quantities of carbon to form carbon monoxide—hence the defining of a *reduction zone*. This latter zone is essentially the region in which the important gas–solid system carbon–carbon monoxide–carbon dioxide attempts to reach equilibrium. The degree to which equilibrium is approached is determined chiefly by the residence time in the bed; carbon monoxide is generally the principal product. The top of the fuel bed is called the *distillation zone* wherein occurs the evolution of volatile matter from fresh coal decomposition.

We can now see that from the upper surface of the fuel bed there emerges not only volatile matter, but also large quantities of carbon monoxide and some carbon dioxide, products of the primary combustion process. The escape of unburned carbon monoxide from a furnace cannot be tolerated because of the high potential heat loss which would result.

Thus the process of secondary combustion, in which the gaseous products of the primary procedures occurring in the fuel bed should be completely oxidized, is in a sense the critical stage in determining the overall combustion efficiency of a furnace.

Secondary combustion is a homogeneous gas-phase reaction in which the end-products should be carbon dioxide and water vapour. If the burning gases are ejected too quickly from the furnace into cooler regions, combustion may be interrupted. Injection of preheated secondary air into the combustion chamber is therefore so designed that a high degree of turbulence and consequently rapid mixing of the gases results. If the gases are accidentally cooled below their ignition temperature, elemental carbon will appear as black sooty smoke at the chimney head. It is difficult to avoid soot emission when starting a furnace from cold, hence in the United Kingdom the Clean Air Act permits industrial furnaces to make smoke for 15 minutes after preliminary firing. Properly operated, a hot furnace which nowadays will most likely be automatically stoked should not encounter soot-emission troubles.

Smoke emission of a different sort can occur if the secondary air supply is insufficient. In this case unburned gases will escape, and this will be signalled by brown smoke from volatile hydrocarbon material appearing at the chimney head. In such instances, however, the big loss in potential heat is due not to the visible hydrocarbon loss, but to its invisible companion, carbon monoxide.

Air supplies to a furnace have to be carefully controlled if maximum efficiency is to be attained. Both primary and secondary air streams must be adequate enough to ensure complete combustion at their respective sites. Complete combustion can never be achieved with theoretical quantities of air, and an excess must always be present. We must take account of another heat loss, however, when dealing with combustion air supply. Waste gases will be ejected at the chimney head at temperatures higher than ambient temperature, and this effluent stream therefore constitutes a considerable sensible heat loss. Most of this heat loss is due to nitrogen, accounting as it does for 79% of the air supply. Any increase in excess air therefore increases substantially the volume of waste gases, and with it the associated sensible heat loss.

A deficiency in air supply means a loss in potential heat due to in-complete combustion. An unnecessary amount of excess air means a loss in sensible heat in the waste gases. These opposing factors must be optim-ized in any furnace system, as indicated in Fig. 25.6. The optimum excess air figure for a coal-fired furnace may be of the order of 50%. Liquid and gaseous fuels lend themselves to a much higher degree of automated control, and with these fuels greatly reduced excess air figures can be obtained.

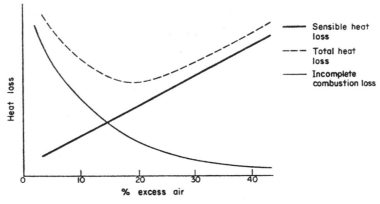

FIG. 25.6. Relations between excess air and heat losses
in a furnace.

Much of the sensible heat in a waste gas stream is recovered by the use of heat exchangers, in which combustion air streams are preheated. In suitable cases fuels may also be preheated. We might reasonably postulate that the more heat that can be extracted in a heat exchanger, i.e. the lower the ejection temperature of the waste gases, the more efficient will be the overall process. Here again, however, a compromise has to be made. Sulphur in the course of fuel combustion oxidizes to a mixture of sulphur dioxide and sulphur trioxide. If a waste gas stream is cooled below its dew point, such that water condensation occurs, the sulphur gases partially dissolve to form highly corrosive acidic solutions. It is essential, therefore, to discharge the effluent gas at a temperature higher than its dew point in order to avoid acid attack on the plant.

(The discharge of sulphur gases to atmosphere is itself by no means desirable. Such discharges are controlled by statutory limitation of the sulphur content of the gases.)

25.5. Pulverized fuel

If a solid fuel is ground to a very finely divided state it can be transported in pressurized air streams and generally handled as if it were a fluid. Fluidized particle systems are now important in a number of fields, nowhere more so than in coal-fired furnaces.

Firing a furnace with pulverized coal offers several important advantages. Like any truly fluid system, automation and very fine control become possible. Much less excess air is required compared to a furnace firing on lump coal. Poor-quality coals which would otherwise be unsuitable can be burned. High combustion intensities and high efficiencies can be achieved. These and other advantages have made pulverized fuel firing the normal mode in large installations, particularly coal-fired power stations, where the high cost of grinding can be absorbed by very big unit operations. Medium-sized pulverized fuel units are also quite common, however.

Pulverized fuel furnaces differ in some ways from orthodox coal-fired furnaces. There is no grate, the fuel particles (perhaps up to 40 million per cubic inch) being carried on an air stream into a stationary three-dimensional flame through which they travel as they combust. Secondary air is injected into the flame zone at a carefully selected point. The mechanism of combustion, however, parallels that of lump coal in that two distinct stages have been observed. Each particle first devolatilizes, the volatile matter burning round about the minute coke residue. A heterogeneous reaction then occurs between coke particle and oxygen. Thus the first stage is analogous to secondary combustion in an orthodox furnace, the second stage to the processes occurring in primary combustion in a fuel bed.

The high temperatures which are reached in these systems necessitate furnace walls which are water cooled by embedded tubes. If the ash fusion temperature is low enough, molten ash may result, and this can

24

be tapped off (a "wet-bottom" furnace). Otherwise ash will be ejected as fly ash in the hot waste gases, and will be separated from them in ancillary units.

Recently there has been much interest in burning coal in fluidized beds. In this technique finely divided coal is kept in constant agitation by the passage of combustion air, but unlike pulverized fuel firing, the coal mass is retained within a fixed reaction zone. The technique offers many of the advantages attached to pulverized firing but requires a smaller and therefore less expensive combustion chamber for the same heat production. Moreover, by the addition of limestone (calcium oxide) to the fluidized bed much of the sulphur in coal can be fixed in the involatile ash residue rather than being discharged to the atmosphere as a highly polluting mixture of sulphur oxides. Fluidized bed reactions may become very important in coal technology in the future, not only in combustion, but also as a part means of converting coal to chemical raw materials.

25.6. Coal carbonization

Coal decomposes at high temperatures to yield complex gaseous, liquid and solid products in the distillate, and a coke residue. Carbonization of coal may be carried out at temperature maxima of between 700°C and 1350°C, high temperature processes being the more common.

Coal gas is the name given to the truly gaseous fraction from which has been removed a variety of undesirable impurities, principally hydrogen sulphide, hydrogen cyanide, ammonia, and benzole vapour. A coal gas from high temperature carbonization contains about 50% hydrogen, 25% methane and 10% carbon monoxide, and will have a calorific value of approximately 500 Btu per cubic foot. This gas has been the foundation of towns' gas supplies in the past in areas rich in coal. Coal carbonization is no longer economic, however, as a route to towns' gas when compared with oil-based processes, or with readily available natural gas. If in the future gas is again made in large quantities from coal, it is likely to be via a fluidized-bed process of the sort discussed above.

Coal tar contains in essence the liquid and volatile solid products from degradation of the main coal substance. It has already been described as a rich source of aromatic compounds (Section 22.2). A further outlet provides a range of *coal-tar fuels,* these being coal-tar fractions, classified according to viscosity, which may be used as liquid fuels in furnace combustion in analogous manner to petroleum fuel oils. Coal-tar fuels have lower sulphur contents than corresponding petroleum products, an attractive feature in metallurgical operations where sulphur can have a detrimental effect on the end product.

Coke is, in any carbonization process, the principal product on a weight basis and consists largely of carbon along with the ash residue of the coal and a little moisture. Its properties vary markedly with the type of coal carbonized. *Metallurgical coke,* which in the blast-furnace process for pig-iron production acts both as fuel and reducing agent, must be hard and dense enough to carry a heavy burden without breaking into fines, and must also be low in sulphur and phosphorus content. Good coking coals are now in short supply, and one technique adopted by the iron and steel industries to reduce coke usage has been the introduction of oil injection into blast furnaces. *Gas coke* from the towns' gas industries is softer and more reactive than metallurgical coke, usually having been derived from a high volatile matter bituminous coal. This byproduct has been used by the gas industry and others to manufacture water gas and producer gas; some gas cokes are also suitable for use as domestic solid smokeless fuels. These latter materials are technically defined as *semi-cokes,* i.e. carbonization residues which have been stripped of smoke-producing hydrocarbons, but which still retain sufficient volatile matter (mainly hydrogen) to maintain combustion in an open grate. The most reactive semi-cokes are made by low temperature carbonization processes and are marketed under a variety of different brand names (e.g. Coalite and Rexco in the United Kingdom).

25.7. Coal: a chemical raw material

The importance of coal in the future may lie not in its potential as a solid fuel but as a precursor of other more highly refined chemical materials. We can regard secondary fuels derived from coal as belonging to this class, for such fuels are materials which have been processed to some extent. The distinguishing feature of the fuel-processing industries is that the required end product is not the chemical itself, but its heat or power content.

Coal carbonization was the earliest technique developed to convert the raw material into other products. Consider the wide range of secondary fuels which may be derived by the thermal decomposition of coal: coke, semi-coke, motor benzole, coal-tar fuels, coal gas. Producer gas and water gas (Section 12.5) are coal-based products. Carbonization we have noted, moreover, as a route to coal chemicals: coal-tar products are obvious, other products perhaps less so; for example, we would hardly think of Neoprene rubber as being a coal chemical until we realize that Neoprene is made from chloroprene, which is made from acetylene, much of which has been made from calcium carbide, which is made from coke, which is made from coal!

For the production of liquid and gaseous fuels and of chemicals, however, carbonization is falling out of favour. It is not a very flexible process, and therefore not a very efficient one. Newer, more sophisticated processes for the conversion of coal into a range of useful products are available, the more important of which we shall note.

Hydrogenation of coal degrades the main structure and yields a series of liquid products which can replace petroleum fuels. The *Bergius process*, in which a mixture of coal and coal tar is hydrogenated at high pressure and temperature in the presence of a catalyst, was the forerunner of a number of hydrogenation processes which were used during the Second World War to produce fuel oils, diesel oils and gasolines. Coal hydrogenation oils are not at present competitive with petroleum oils.

Synthesis gas manufacture from coal leads to a number of further uses. Synthesis gas, i.e. a mixture of carbon monoxide and hydrogen, can be made as an equimolar mixture by the water–gas reaction:

$$C + H_2O \rightleftharpoons CO + H_2$$
$$\text{Synthesis}$$
$$\text{gas}$$

These gases may subsequently be converted into hydrocarbon mixtures and a wide range of oxygenated compounds (see, for example, Section

23.2) by catalytic processes the best known of which is the *Fischer–Tropsch process*. This latter process can be made astonishingly flexible by altering the reaction variables, especially the molar ratio of carbon monoxide to hydrogen and the catalyst.

The Lurgi process is also used to produce synthesis gas. Here coal is completely gasified by blowing it with a mixture of steam and oxygen at pressures of the order of 25 atm. Lurgi gas is enriched with hydrocarbon gases and distributed as towns' gas. In South Africa, a Lurgi plant is in operation to make synthesis gas for ultimate conversion to liquid fuels. The Lurgi process has become uneconomic either for towns' gas manufacture or liquid fuels synthesis compared to petroleum-based processes, unless special factors such as cheap labour or cheap coal weigh in its favour.

25.8. Future prospects

In this century coal has steadily lost ground to petroleum at almost every point. Since the OPEC crisis of October 1973, however, petroleum has risen dramatically in price. That, combined with a very much heightened awareness that petroleum products are likely to be in short supply before the end of the century, and may not be available at all in any significant quantities by the middle of the twenty-first century, has opened up new prospects for coal.

Coal is abundantly available. The lifetime of known coal reserves can be measured in centuries; those of petroleum can be measured in decades. Coal suffers technical disadvantages in lacking the fluidity of liquid or gaseous systems, and in presenting a considerable ash disposal problem. However, these disadvantages can be partly or wholly overcome by employing the pulverizing and fluidizing techniques referred to above. Coal can also be gasified underground, *in situ*, by treatment with mixtures of air and steam or oxygen and steam. Underground gasification is very attractive, but poses technical problems which have yet to be solved. Nevertheless, there is now much activity in investigating better ways of winning and using coal. The past belonged to coal: it may be that coal will also claim the future.

Exercises

25.1. A coal has the following proximate analysis on an air-dried basis:

> moisture, 3·1%;
> volatile matter, 27·2%;
> ash, 4·7%;
> residue, fixed carbon.

What is the proximate analysis on a dry, ash-free basis?

25.2. On a dry, mineral-matter free basis a coal is analysed to give C, 87·0%; H, 5·0%. From the Seyler chart, compile:

  (i) the coal type;
 (ii) its gross calorific value in Btu lb^{-1};
(iii) its volatile matter;
(iv) its British Standard swelling number.

25.3. List the advantages and disadvantages of burning coal

  (i) on a fixed grate;
 (ii) in a pulverized fuel-fired furnace;
(iii) in a fluidized-bed system.

CHAPTER 26

Petroleum-based Industries

THROUGH the last 70 years or so petroleum products have acquired a dominating central role in the industrialized nations in two areas, viz. the chemical industry and transportation. The weakness of such heavy reliance on one natural resource was revealed in the severe economic problems which followed the large cost increases in petroleum products after 1973, but petroleum-based activities will continue to rely on continuing supplies of crude oil at least until the turn of the century.

The longer-term problem with petroleum is not cost but supply. In recent years, for the first time, the rate of discovery of new deposits has fallen behind the growth rate of petroleum usage. When new discoveries are made, such as those of the North Sea or of western Canada, they are of great benefit to the local national community. Such discoveries, however, make only small impacts on the world scene. The North Sea deposits, for example, represent about 2% of total known reserves; upward of 60% of known reserves are still those accounted for in the Middle East, and there is no good reason to believe that massive finds comparable to the Middle East are likely. The present expectation is that petroleum and its products will be in serious short supply by about 1990, and are likely to double in real price terms by then; and that supplies will run out by about the middle of the twenty-first century.

26.1. Chemical nature and classification of petroleum

Petroleum, like coal, is a generic name. Any one oil is a highly complex mixture of (mainly) hydrocarbons, and one oil may be very different from another in its detailed composition. Unlike coal, however, individual

367

compounds can be distinguished in crude oil. Among these we have already noted the very simplest of hydrocarbons, low molecular weight materials like methane and its immediate homologues. Although relatively high molecular weight material does occur, petroleum is not polymeric in the same sense as coal.

Crude petroleum, which has probably arisen from the decomposition of marine organisms, consists of carbon and hydrogen and minor quantities of oxygen, nitrogen and sulphur. The hydrocarbon groups which constitute the bulk of any petroleum oil are *alkanes, cycloalkanes (naphthenes)* and *aromatics*, the latter usually being of minor importance. Alkenes may also occur in very small amounts.

The minor elements in petroleum are difficult to characterize. Oxygen occurs as a carboxylic acid function in a series of complex structures referred to as the *naphthenic acids*, and is also present in *heterocyclic ether structures*. Nitrogen is present in a range of *heterocyclic nitrogen base* compounds and in neutral forms. Sulphur, which is particularly undesirable in a fuel, occurs in petroleum in the form of *mercaptans* or *thioalcohols*, RSH, as *thioether* ($-C-S-C-$) and *disulphide* ($C-S-S-C-$) linkages, as *heterocyclic sulphur* in both aromatic and saturated ring structures, and as elemental sulphur.

Petroleum crudes do not display the same type of family relationship as coals. The ultimate analyses of a wide range of oils are distributed around mean values of 84·5% carbon, 12·5% hydrogen, and residue minor elements, and variations from these mean values are only of the order of a few per cent. Primary classification is simply based on the nature of the residue obtained on distillation of an oil. Three classes exist:

   (i) *paraffin base oils*, in which the residue is largely paraffin wax;
   (ii) *asphalt base oils*, in which the residue is largely asphalt (or bitumen), i.e. an aromatic residue;
   (iii) *mixed base oils*, in which the residue is a mixture of the above two classes.

While the primary nature of an oil determines the manner in which it is refined, individual fractions are in essence defined and classified by physical properties such as boiling range, viscosity and so on. We are here saying, as with coal, that the history of a given fraction or its detailed chemical composition is not of first consequence. If the fraction performs its given task satisfactorily, that is sufficient. We shall soon see, however, that satisfactory performance of, for example, a gasoline can better be achieved by taking account of chemical structure as well as gross physical properties.

26.2. Analysis of petroleum oils

The methods of analysing petroleum oils are too numerous to consider in detail, but we shall note some of the more important tests. The authoritative manuals in this field are, in the United Kingdom, the Institute of Petroleum handbook *Standard Methods for Testing Petroleum and its Products*; and in the United States the ASTM handbook *Standards on Petroleum Products and Lubricants*. Inspection of standard methods reveals that while some of the tests measure specific physical parameters, many of them are empirical. Hence, as with solid fuels, it is important to carry out a determination according to the prescribed method.

(a) *Specific gravity*

This value is determined orthodoxly. It gives some indication of the oil composition, being inversely proportional to the alkane (or paraffin) content of an oil. In the United States a reciprocal gravity figure, the API (American Petroleum Institute) gravity, is often quoted.

(b) *Viscosity*

Oil viscosity is very important in a number of fields. *Kinematic viscosities* are measured in a U-tube viscometer. Often used in the United Kingdom is the *Redwood viscosity*, which is simply the time of flow of an oil sample through the standard orifice of a Redwood viscometer. The equivalent American unit is the *Saybolt viscosity*. For lubricating oils, the variation of viscosity within a normal range of working temperatures must also be determined, and is expressed by means of a *viscosity index* (Section 29.2).

(c) *Vapour pressure*

Vapour pressure is particularly relevant to motor fuels, indicating as it does the ease of cold-weather ignition, and the danger of vapour lock occurring. The value is determined by the *Reid method*.

(d) *Distillation range*

Again within the field of motor fuels, the distillation range is a primary test. Specific values (e.g. the initial boiling point, the temperature of 10, 50, and 90% distillate, and the final boiling point) are laid down for gasolines, diesel oils and related fuels.

(e) *Aniline point*

This very simple test in which the solubility in aniline of diesel oils is measured indicates empirically the fraction of alkanes present in the sample. The aniline point is often combined with the API gravity and expressed as a *diesel index* (Section 26.12).

(f) *Flash point*

The flash point of an oil (not to be confused with the spontaneous ignition temperature) indicates the temperature of ignition from an open flame. Flash points are important in assessing the fire hazard associated with the bulk storage of liquid fuels.

(g) *Calorific value*

As with solid fuels, a bomb calorimeter is used to measure calorific values of liquid fuels.

(h) *Carbon residue*

The carbon residue of an oil is the coke residue left after all volatiles have been driven off in non-oxidizing conditions. Carbon deposits can be very troublesome in a number of fuel-injection systems, in oil burners, and in lubricating oils; on the other hand, carbon residues greater than about 6% are desirable in some oil-based gas-making processes. Carbon residues are measured by either the *Conradson* or the *Ramsbottom* test.

26.3. Petroleum refining

Petroleum crudes are never used as such. It is the task of a petroleum refinery to divide its raw material into fractions, each tailored to a particular job. A number of refining operations are commonly used, the most important of which is physical separation by distillation. Distillation procedures are fully discussed in Chapter 17.

Initial distillation in the oil industry is carried out as a continuous process operating at atmospheric pressure. Secondary distillation of the heavier ends may be required, depending on refinery demand and the nature of the oil. Where needed, fractionation of high boiling material takes place at reduced pressures in order to lower working temperatures and avoid thermal cracking.

A schematic representation of crude petroleum fractionation is given in Fig. 17.4, which simplifies the whole process and indicates an ease of separation difficult to obtain in practice.

Simple distillation is seldom a self-sufficient process, and a variety of other refining techniques (dealt with in Chapters 17 and 18) are used in developing petroleum end products. These include steam and azeotropic distillation, absorption, solvent extraction, and other methods of separation which are applied in a variety of ways.

In addition to physical methods of separation and refining, the oil industry also employs processes designed to change the chemistry of the products. Into this category would fall cracking procedures of various kinds. Chemical processes are particularly important in petrochemicals production and in gasoline technology.

Of the multiplicity of fractions obtained from crude petroleum refining, we shall consider in detail in the following sections only the more important liquid products, and in particular those which are liable to be of interest to the engineer.

26.4. Gasoline: production

Gasoline, or petrol, is the fuel used in the spark-ignition engine. In mild climates a fraction distilling in the range 30–150°C is required. Primary distillation of a crude petroleum will never yield more than about 30% gasoline ("straight-run gasoline"), and sometimes considerably less than this. Market demands are now so severe that straight-run gasoline supplies have to be heavily supplemented by cracking higher boiling fractions.

Thermal cracking of suitable fractions was first used to increase gasoline yields, but cracking in the presence of a catalyst is now favoured. The milder conditions which can be used in *catalytic cracking* give higher yields of a better-quality gasoline and less elemental carbon.

In catalytic cracking, the elemental carbon produced is deposited at the surface of the catalyst, duly rendering it inactive. A regeneration cycle, during which this carbon is burned off, is therefore necessary. Interrupting a production line is expensive, and continuous processes have been developed in catalytic cracking whereby catalyst regeneration can be carried out constantly without interfering with gasoline production. The best way of doing this is to use a *fluidized-bed* system identical in principle to that discussed for coal combustion in Section 25.5. Fluidized bed catalytic cracking is represented in Fig. 26.1. The catalyst bed is "fluidized" by the upward passage of oil-feed vapour. Spent catalyst is drawn continuously from the bottom of the cracking chamber, transported on an air stream to the regeneration chamber in which the carbon deposit is burned off, and returned in due course to the incoming feed stream and thence back to the cracking chamber. Cracked gases are taken from the top of the cracking chamber directly to a fractionating column. Operating temperatures are typically about 550°C at pressures a little above atmospheric pressure, conditions much milder than those used in the earlier thermal processes.

The "cat cracker" is a prominent part of the modern refinery. Processes of the type described above can be operated for periods sometimes measured in years before shut-down is necessary, with throughput running up to one million gallons per day.

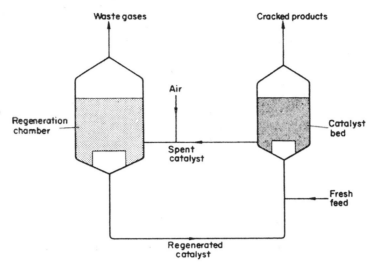

FIG. 26.1. Schematic representation of a fluidized-bed cracking plant.

26.5. Gasoline: quality

Fuel is consumed in a spark-ignition engine by igniting a mixture of gasoline and air inside the cylinder head fractionally before the piston reaches the top of its travel. The ensuing combustion should take place rapidly, but smoothly, at flame speeds of about 15 metres per second.

The efficiency of power production in an internal combustion engine is related to the *compression ratio*, i.e. the ratio of the cylinder volume when the piston is at the bottom of its stroke to the cylinder volume when the piston is at the top of its stroke. The variation of E, the efficiency, with r, the compression ratio, is shown in Fig. 26.2. Compression ratios have steadily increased with the development of more efficient and powerful engines. Before the Second World War, a family saloon car might have a compression ratio of 6·5 : 1; now values of 8–10 : 1 are common.

Increased compression ratios, however, have brought problems as well as benefits. A fuel–air mixture can ignite spontaneously on compression, and should this happen in a spark-ignition engine a very rapid

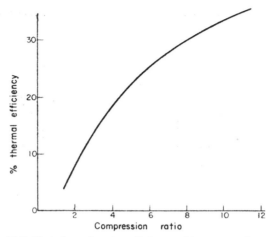

Fɪɢ. 26.2. Variation of engine efficiency with compression ratio.

combustion occurs, producing shock waves of much greater velocity than normal. These waves, reverberating throughout the cylinder head, produce a characteristic sound which gives this pre-ignition phenomenon the name *knocking* or *pinking*. Engine knock results in a drop in efficiency, and can also cause mechanical damage. As we might expect, the higher the compression ratio of an engine, the greater the tendency for knocking to occur.

Gasolines vary in their resistance to knocking. The increase in compression ratios has therefore been accompanied by an increase in gasoline quality with respect to knock resistance. This latter property of a gasoline cannot be defined in absolute terms but is indicated on an arbitrary scale called the *octane rating*.

Octane numbers are simply defined. The hydrocarbon *n*-heptane, C_7H_{16}, which has very poor resistance to knocking, is given an octane number of zero; the hydrocarbon iso-octane (2,2,4-trimethylpentane, C_8H_{18}), which has a high resistance to knocking, is given an octane number of 100. *The octane number of a gasoline is the percentage of iso-octane in a mixture of iso-octane and n-heptane having the same knock resistance as the gasoline under test.* Testing is carried out in a standard engine of variable compression ratio, the ratio being increased gradually until the onset of knocking for the fuel under test. A number of standard engines have been specified, and each need not necessarily give the same result. Octane ratings for motor gasolines may be reported as the octane number (motor method) or the octane number (research method). The latter rating, usually a few points higher for high octane fuels, is most commonly quoted.

26.6. Octane rating and chemical structure

There is a clear relationship between the octane number of a hydrocarbon and its structure. In Table 26.1 we can compare the octane numbers of a series of C_7 hydrocarbons. The progression is such that the octane number rises through the series n-alkanes → branched-chain alkanes → cycloalkanes → alkenes → aromatics. Multiple branches in an alkane, particularly if the branches are methyl groups, give rise to very good octane ratings, e.g. 2,2,3-trimethylbutane.

Compression ratios are such today that gasolines of octane number above 90 are required if engine knock is to be avoided. Straight-run gasoline from crude petroleum is likely to have an octane number no greater than about 60. Clearly some upgrading procedure is necessary. This can be achieved by blending low rating gasolines with, for example, aromatic hydrocarbons. Much more important, however, are a number of high temperature conversion processes.

26.7. Production of high octane gasoline

We have seen (Section 26.4) that the heavy demand for greater quantities of gasoline has been met by the use of thermal and catalytic cracking processes. The demand for a better quality of gasoline has been met in the same way.

If a gasoline fraction of low octane number is exposed to high temperatures for a short period of time (perhaps 1 to 10 sec) many of the molecules isomerize rather than break down into smaller fragments. Straight-chain alkanes isomerize to branched-chain structures, and to cycloalkanes which in turn partially dehydrogenate to aromatics. Simple

TABLE 26.1. OCTANE NUMBER (RESEARCH) OF SOME C_7
HYDROCARBONS

Compound	Structure	Octane number
n-Heptane	$CH_3 \cdot (CH_2)_5 \cdot CH_3$	0
3-Methylhexane	$\overset{\displaystyle CH_3}{\overset{\displaystyle \vert}{CH_3 \cdot CH_2 \cdot CH \cdot CH_2 \cdot CH_2 \cdot CH_3}}$	52·0
3-Ethylpentane	$\overset{\displaystyle CH_2 . CH_3}{\overset{\displaystyle \vert}{CH_3 \cdot CH_2 \cdot CH \cdot CH_2 \cdot CH_3}}$	65·0
Methylcylohexane		74·8
2-Methyl-1-hexene	$\overset{\displaystyle CH_3}{\overset{\displaystyle \vert}{CH_2{=}C \cdot CH_2 \cdot CH_2 \cdot CH_2 \cdot CH_3}}$	90·4
2,3-Dimethylpentane	$\overset{\displaystyle CH_3\,CH_3}{\overset{\displaystyle \vert\quad\vert}{CH_3 \cdot CH \cdot CH \cdot CH_2 \cdot CH_3}}$	91·1
2,4-Dimethyl-1-pentene	$\overset{\displaystyle CH_3 \qquad CH_3}{\overset{\displaystyle \vert\qquad\quad\vert}{CH_2{=}C \cdot CH_2 \cdot CH \cdot CH_3}}$	99·2
2,2,3-Trimethylbutane		100 + 1·8[†]
Toluene		100 + 5·8[†]

[†] See Section 26.8.

alkenes are also produced. The information quoted in Table 26.1 indicates that these changes must result in an increase in the octane number of a gasoline so treated.

Thermal and thermo-catalytic processes for upgrading gasoline are called *reforming* processes; a number of these are in widespread use. *Hydroforming* employs a molybdenum catalyst at an elevated pressure of about 250 pounds per square inch and temperature of 500°C. Hydrogen gas is present to saturate alkenes produced by cracking and so to decrease coke formation at the catalyst surface. Cyclization and dehydrogenation of cycloalkanes to aromatics still takes place, however. The closely related and more recent *platforming* process, like hydroforming, is widely applied to gasoline reforming. Both procedures are also used for the synthesis of primary aromatic hydrocarbons (Section 22.2).

Low temperature processes are also used to obtain structures of high octane rating. Catalytic isomerization of *n*-butane to iso-butane followed by cracking of the product to isobutene is used to produce large quantities of the branched C_4 alkene. This fraction can subsequently undergo alkylation (Section 21.4) to produce highly branched high octane rating hydrocarbons. The product, called *alkylate*, has an octane rating of greater than 90 and is blended with other suitable streams.

Gasoline fuel need not be hydrocarbon material. Alcohols have good octane ratings and methyl and ethyl alcohol are sometimes used as components of a final gasoline blend.

26.8. Additives

Commercial gasolines contain a number of additives, the most important of which is tetraethyl lead, $Pb(CH_2CH_3)_4$.

A number of organometallic compounds help to combat knocking tendencies in a gasoline and therefore effectively raise the octane rating. Tetraethyl lead is not the best of these compounds but it is to date the cheapest. Up to about 4 cm³ per gallon may be added, the effect of the additive varying in different gasolines. Use is made of tetraethyl lead in classifying fuels of higher octane rating than the standard iso-octane. Thus toluene, which is said to have an octane rating of "100+5·8",

25

gives identical performance under test conditions to iso-octane containing 5·8 cm³ tetraethyl lead per gallon.

There has been growing concern that the amount of lead emitted from car exhausts in urban areas with a high car population has reached levels at which the health of the community is at risk. Some lead-free gasolines are now marketed, the appropriate octane rating being achieved by blending in a higher percentage of high octane fuel. This trend, along with improved engine design to reduce all exhaust pollutants, is likely to continue.

Other additives in a typical gasoline will include anti-oxidants (often amines) to prevent gum formation, rust inhibitors, and carburettor de-icing compounds.

26.9.   Sulphur in gasoline

The presence of sulphur is as undesirable in gasoline as it is in any fuel. Apart from producing acid gases on combustion, sulphur compounds interact with tetraethyl lead, rendering it ineffective and so reducing the octane number of the fuel. The fuel is said to have a depressed *lead susceptibility*. Moreover, sulphur present as mercaptans produces really foul smells which do nothing for the market image of the product.

Sulphur compounds can be removed by sulphuric acid treatment, a refining process which can also be used for the removal of alkenes, the precursors of polymeric gums in gasoline. Solvent extraction with, for example, strong alkali is also used to reduce the sulphur content. Improving a gasoline by the removal of mercaptans is termed "sweetening"; this may entail mercaptan extraction by a process of the type mentioned above, or it may be a process in which the sulphur content is not reduced, but the mercaptans are merely converted to a more innocuous form of sulphur. Hence the *Doctor process* is used to convert mercaptans to disulphides by lead plumbite treatment, carried out, paradoxically, in the presence of a carefully calculated amount of free sulphur:

$$Na_2PbO_2 + 2RSH + S \rightarrow RSSR + PbS + 2NaOH$$

A similar result can be achieved using copper chloride.

26.10. "What petrol should I use?"

Gasoline technology is complex. The procedures we have been considering are well established, and it is reasonable to assume that a customer will find little, if any, difference between different brands of the same grade of petrol. The only real variable wherein the consumer can exercise choice is that of octane rating. In the United Kingdom a British Standard defining different grades of gasoline according to octane rating was introduced in 1967, and this went a long way towards rationalizing what had been a rather confusing situation. The standard divides gasolines into four groups, each of which is given a star grading. The scheme is summarized in Table 26.2, and discussed in an independent report.[†]

TABLE 26.2. BRITISH STANDARD
GASOLINE GRADINGS
(BS 4040, part 11, 1967)

Star rating	Research octane number
5 star	100 or greater
4 star	97 or greater
3 star	94 or greater
2 star	90 or greater

An earlier report from the same source[‡] emphasized that nothing is to be gained by using a quality of gasoline better than is necessary to prevent knocking in an engine. Indeed, car manufacturers will not recommend five-star fuel for medium compression engines, where it can cause damage. The advice of the aforementioned reports should be followed, viz. use the cheapest grade of gasoline which does not cause knocking in your car.

[†] *Which?*, April 1967, Consumer Association, London.
[‡] *Which?*, January 1964, Consumer Association, London.

26.11.   Diesel fuels

The diesel engine is a compression-ignition engine which uses petroleum fuels distilling between about 180°C and 360°C. In the cycle of operations, air is passed into the cylinder head during an upstroke. Compression results in a sharp rise of pressure and temperature, and fuel, sprayed in towards the end of the compression stroke, begins to ignite spontaneously. Fuel feed and ignition continue during the downward power stroke.

The process of injecting oil into a combustion chamber such that it breaks up into very small droplets is called *atomization*. Efficient atomization is necessary in a compression-ignition engine (or for that matter in an oil fired furnace) if combustion is to be complete. For diesel fuels, which are usually selected from the gas oil range of the "petroleum spectrum", viscosity is an important factor in determining ease of atomization. Diesel oil viscosities range from about 30 Redwood seconds at 100°F up to 100 times that figure, heavier low-speed engines being able to accommodate more viscous oils.

Carbonaceous deposits are undesirable in diesel engines, as they interfere with the fuel-injection system and contaminate the lubricating oil. Here again specifications become less exacting as the engine speed decreases. A high speed engine may require a light gas oil with a Conradson carbon residue of 0·05% or less, whereas a very low speed engine may accept an oil of carbon residue in excess of 10%.

26.12.   Cetane numbers

If we contrast the mode of operation of a gasoline engine, where pre-ignition by compression must be avoided, to that of a diesel engine, where compression ignition is central to the whole power cycle, we might well expect logical differences in the fuels most suitable for each type of internal combustion engine.

The *cetane rating* for diesel fuels is equivalent to the octane rating for gasolines, and is a measure of the ease with which a given fuel will undergo compression ignition. Two hydrocarbons are specified as the low and high points on the cetane scale, viz.

CH₃

$$CH_3(CH_2)_{14}CH_3$$

1-Methylnaphthalene
Cetane no. 0

n-Cetane
Cetane no. 100

the cetane number of a fuel being the percentage of cetane in a mixture of the two standards giving equivalent performance to the fuel under test. Thus the structural requirements for a diesel fuel are precisely those which have to be avoided in gasolines; straight-chain alkanes like *n*-cetane ignite readily on compression, aromatics like 1-methylnaphthalene are too stable to be good diesel fuels. Cetane numbers in practice range from about 25 to 50, increasing with increasing engine speed.

To avoid the comparatively complex cetane number determination, a related quantity called the *diesel index* is often used as a test of ignition quality, being defined as

$$\text{Diesel index} = \frac{\text{Aniline point (°F)} \times \text{API gravity}}{100}$$

The diesel index figure is often, but not always, close to and a little higher than the cetane number.

Additives can be used to raise the cetane number of a diesel fuel. Alkyl nitrates such as ethyl nitrate, $CH_3 \cdot CH_2 \cdot O \cdot NO_2$, are among the compounds suitable for this purpose.

26.13. Kerosine

Kerosine, or "paraffin", once important because of its illuminating flame, decreased in value until the appearance of the aviation jet engine. Kerosine, or kerosine blended with gasoline, is used as fuel for turbine-driven aircraft, and this fraction is consequently now rated an important product of the petroleum industry.

26.14. Fuel oils

While the light petroleum fractions we have been considering are generally used as sources of heat and power, the term "fuel oil" is reserved for viscous oils of the gas oil and distillate residue type which are used in furnace combustion as alternatives to coal or gaseous fuels. Liquid fuels lend themselves to automatic control to a greater extent than coal and they have a higher thermal storage capacity than the solid fuel. (Fuel oils have a gross calorific value of about 19,000 Btu per pound, a good-quality coal about 15,000 Btu per pound.)

The most important feature of an oil fired furnace is the burner or atomizer through which the fuel passes to the furnace. The ultimate atomizing device is a simple wick from whose surface oil vaporizes and passes into the combustion chamber. Wick burners are very limited in throughput and are normally suitable only for domestic heaters and small load boilers serving domestic central heating units. In industrial furnaces a number of atomizing burners are in use. They may depend on imparting a strong centrifugal force to the oil feed which sprays out into the combustion chamber, breaking up into a mist of very fine droplets; alternatively, a pressurized air or steam jet impinging on the oil feed can be used to bring about atomization, the mixture then being passed into the combustion chamber. The viscosity of a fuel is important in determining its ease of atomization and also the ease and cost of pumping operations. Fuel oils are classified according to viscosity, and commonly the heavy oils must be heated to a predetermined temperature before they can be satisfactorily handled.

The mechanism of combustion in an oil fired furnace parallels that of pulverized fuel firing. Each oil droplet entering the flame envelope first loses much volatile matter which burns off in a homogeneous gas phase reaction, leaving as residue a carbon particle (called a cenosphere) which undergoes reaction with oxygen in precisely the same way as a coke residue derived from coal. Both oil droplets and pulverized fuel particles follow a burning time law approximately represented by the equation

$$t_c = F_c \, K d_o^2$$

where t_c is the burning time of the volatile matter of a droplet or particle of initial diameter d_o, F_c is a factor dependent on the amount of excess air present, and K a complex burning constant dependent on a number of system variables. The d_o^2 term emphasizes the importance of the atomization process in fuel oil combustion.

Considerations of the sort which have been discussed in Section 25.4, in particular those dealing with air supply and smoke emission, are equally applicable to an oil fired system. Oil requires less excess air than coal for complete combustion and therefore has the advantage of a higher thermal efficiency.

26.15. Other petroleum products

From the heavy residues of crude petroleum a number of valuable products are extracted. Most important of these, perhaps, are the *lubricating oils* (dealt with separately in Chapter 29). A valuable range of *petroleum waxes* and *greases* is obtained by extensive refining of the back-end fractions. *Bitumen* finds an outlet chiefly as a road-making material, but has many other applications. The ultimate solid residue of petroleum distillation is *petroleum coke*. Its very low ash content makes it a suitable source of electrode carbon, and like coal coke it may also be used as a solid smokeless fuel, although the lack of ash leads to a high rate of corrosion of grate bars unless precautions are taken.

If we include in "other petroleum products" the full range of petrochemicals, our list becomes almost unending. Many times in earlier chapters petrochemical processes were mentioned. Most organic materials can now be made from petroleum, but the industry is also the source of rather more obscure products. Thus *carbon black*, used mainly as a rubber additive to increase the wear resistance of motor tyres, is made from a variety of petroleum fractions, depending on the process being used. *Ammonia*, very much an inorganic product, is now manufactured from hydrogen obtained by steam reforming of light petroleum fractions, the hydrogen subsequently being reacted with nitrogen (Chapter 14).

Exercises

26.1. Crude petroleum includes alkanes, naphthenes, aromatics, naphthenic acids and mercaptans. Write out the structural formula for a typical simple member of each of these classes.

26.2. What are likely products from the thermal cracking of ethane, isobutane, and ethyl benzene?

26.3. List the following compounds in likely descending order of octane rating: *n*-octane; 2,3-dimethylhexane; benzene; 2-heptene.

26.4. What are the advantages of using heavy fuel oil in place of coal in furnace combustion?

CHAPTER 27

Chemical Explosives

LIKE traditional fuels, traditional explosives release their energy by chemical reaction. In this chapter we shall concern ourselves with explosives of this class, omitting those which depend on nuclear reactions and which are dealt with separately in Chapter 2.

Fuels and explosives have much in common. In both cases strongly exothermic oxidation reactions serve as energy donors. The essential difference is the *speed* at which oxidation occurs; fuels react at a controlled rate while explosives react almost instantaneously, and violently. (Precisely the same dividing line can be drawn between nuclear fuels and nuclear explosives.)

27.1. Physical characteristics

Two phenomena are particularly associated with explosive reactions, viz.

(i) the release in a very short period of time of a large quantity of energy, mainly as heat;

(ii) a great increase in the volume of products (principally gases) compared to the volume of reactants; e.g. nitroglycerine combustion gives a 10^4 volume increase at 3000°C.

Allowed to take place in a confined space, a reaction with these characteristics will produce gases at high temperatures and pressures, and the subsequent shock waves associated with this high energy system will cause the disintegration of surrounding material.

Explosives vary in their detailed behaviour on detonation, and a number of specification tests have been established. Of these, two are especially useful in assessing different preparations:

(i) *Explosive strength* is a measure of the energy usefully made available by a detonation. It is quoted in calories per gram ("weight strength") or calories per cm^3 ("bulk strength") and is commonly measured against a standard charge of blasting gelatin. This can be done as in the Trauzl test, in which standard charge and explosive under test are each detonated in a small cavity in a lead block, and the volume increases of the cavities compared.

(ii) *Velocity of detonation* (v.o.d.) is a measure of the velocity of the detonation wave passing through the explosive. It can be determined by optical and electrical techniques, or by comparative testing against a standard detonating fuse. Values are commonly quoted in metres or kilometres per second.

While other tests are significant, these two will tell the experienced operator a great deal about an explosive. The job in hand determines the explosive used. For example, shattering a hard material will require a charge with a high explosive strength and a high v.o.d. Avoiding breakage (e.g. in mining lump coal) will entail choosing a charge with just the opposite criteria.

27.2. The chemical structure of explosives

We have noted that explosive systems must generate both a large amount of heat and a large volume of gas. It is possible to compound chemical mixtures which have these characteristics although the individual components may be innocuous in themselves. The first known explosive, gunpowder or blackpowder, belongs to this category, being a mixture of sodium or potassium nitrate, sulphur and carbon which react together violently to give a complex mixture of products including carbon monoxide, carbon dioxide and nitrogen. The typical modern explosive, however, is a single substance containing the necessary structural features within the one constitution to provide a high speed, exothermic, gas-forming reaction.

What, in practice, are the necessary structural features? Two requirements may be specified and used to classify explosives on a chemical basis:

(i) There must be present in the molecule at least one chemical bond which can be easily broken. Such bonds may be quite stable in so far as their resistance to chemical decay is concerned. (Indeed, chemical stability is a desirable feature of industrial explosives, and decay inhibitors are often added to improve this property.) These bonds, however, must break readily if energy (usually mechanical, electrical or thermal) is applied; they are bonds which have a low *energy of dissociation*, and their dissociation leads to a very fast disintegration of the molecule as a whole. Comparatively few chemical groups are satisfactory in this respect; inspection shows that most explosives contain nitrogen–nitrogen, nitrogen–oxygen, nitrogen–chlorine, or chlorine–oxygen bonds. There are a number of groups containing such bonds, and some of the more important ones are listed in Table 27.1. Most explosives contain more than one of these groups per molecule, although a single group can in some cases confer explosive properties on a molecule.

(ii) There should be present in an explosive molecule a good proportion of oxygen, if possible sufficient to bring about combustion without recourse to aerial oxygen. The *oxygen balance* of a molecule is used to express the quantity of oxygen internally available in the substance, and is calculated on the basis of carbon being only partially oxidized to carbon monoxide. Nitroglycerine, for example, has a positive oxygen balance and is therefore suitable for mixing with a combustible material deficient in oxygen,

$$C_3H_5N_3O_9 \rightarrow 3CO + 2{\cdot}5H_2O + 1{\cdot}5N_2 + 1{\cdot}75O_2$$

Nitroglycerine   Excess oxygen

while trinitrotoluene has to be treated in just the opposite way because of a negative oxygen balance.

$$C_7H_5N_3O_6 \rightarrow 6CO + 1{\cdot}5N_2 + (C + 5H)$$

Trinitro-toluene   Combustible balance

TABLE 27.1. BOND TYPES WITH POTENTIAL EXPLOSIVE PROPERTIES

Bond type	Group structure	Group name
Nitrogen–oxygen	$-N{=}O$	Nitroso
	$-N{\nearrow}^O_{\searrow O}$	Nitro
	$-O-N{\nearrow}^O_{\searrow O}$	Nitrate
	$-O-N{=}C$	Fulminate
Nitrogen–nitrogen	${=}N-N{=}$	—
	$-N{=}N-$	Diazo
	$-N{=}\overset{+}{N}{\equiv}\overset{-}{N}$	Azide
Nitrogen–chlorine	NCl_3	Nitrogen trichloride (and related compounds)
Oxygen–chlorine	$-O-\overset{\uparrow O}{\underset{\downarrow O}{Cl}}\to O$	Perchlorate

27.3.   Classification of explosives

Explosives may be divided into primary, secondary and tertiary classes according to the ease with which they can be detonated. Sensitivity to detonation varies greatly through a range of substances, but the main classes are well recognized.

(a) *Primary explosives* are those capable of very ready and powerful detonation, perhaps in extreme cases by the lightest of mechanical shocks. Substances in this class are much too dangerous to be handled in bulk. They are used principally as fillers for detonators, their ignition being used to set off explosion of the main mass of material. Mercury fulminate, $Hg(ONC)_2$, is the classical example of a primary or initiating explosive. More recently lead azide, $Pb(N_3)_2$, has come into widespread use, and one or two other members of this highly sensitive group are produced commercially.

(b) *Secondary explosives* form the real meat of commercial preparations. These comprise the group of compounds known as high explosives. While not so sensitive to detonation as primary explosives, some high explosives are nevertheless very readily detonated. The methods by which they can be made stable for normal handling are outlined in Section 27.4. The majority of modern secondary explosives in widespread use contain either nitrate ester groups, or nitro groups.

Nitrate ester explosives are formed by reaction of a suitable alcohol with mixed acid (nitric acid and sulphuric acid).

$$\text{—OH} + \text{HO·NO}_2 \xrightarrow{\text{H}_2\text{SO}_4} \text{—O·NO}_2 + \text{H}_2\text{O}$$

The most important industrial explosive, nitroglycerine, is made by this route, as are pentaerythritol tetranitrate (PETN) and the polymeric nitrocellulose.

$$
\begin{array}{ll}
\text{CH}_2\text{·O·NO}_2 & \quad\quad \text{CH}_2\text{·O·NO}_2 \\
| & \quad\quad\quad\quad | \\
\text{CH·O·NO}_2 & \text{O}_2\text{N·O·CH}_2\text{—C—CH}_2\text{·O·NO}_2 \\
| & \quad\quad\quad\quad | \\
\text{CH}_2\text{·O·NO}_2 & \quad\quad \text{CH}_2\text{·O·NO}_2 \\
\text{Nitroglycerine} & \quad\quad \text{Pentaerythritol} \\
& \quad\quad\quad \text{tetranitrate}
\end{array}
$$

Nitroglycerine is a particularly potent material, having a high explosive strength and a high velocity of detonation. It is a liquid with a high sensitivity to detonation, and has to be treated with the utmost caution. The more expensive PETN has an even higher velocity of detonation than nitroglycerine, but cost considerations restrict its use to a few special cases. Nitrocellulose is widely used as the solid combustible phase in commercial mixtures. Other nitrate explosives are in limited use.

$$
\begin{array}{c}
\text{CH}_2\text{·O·NO}_2 \\
| \\
\text{O——CH} \\
\diagup \quad\quad \diagdown \\
\sim \text{O—CH} \quad\quad \text{CH} \sim \\
\diagdown \quad\quad \diagup \\
\text{CH—CH} \\
| \quad\quad\quad | \\
\text{O}_2\text{N·O} \quad\quad \text{O·NO}_2 \\
\text{Nitrocellulose}
\end{array}
$$

Nitro explosives are usually aromatic compounds, formed by the nitration with mixed acid of suitable aromatic rings. The outstanding example in this class is trinitrotoluene (TNT), a solid explosive of medium explosive strength. Another interesting nitro explosive is cyclotrimethylenetrinitramine (RDX) in which the nitro groups are attached to heterocyclic nitrogen atoms. It is a powerful explosive of high velocity of detonation, but is costly.

Trinitrotoluene   Cyclotrimethylenetrinitramine

(c) *Tertiary explosives* are less widely recognized as forming a class in their own right. They are materials which of themselves are not generally regarded as explosives but which in contact with other combustible material can form very dangerous mixtures. The really important member of this class is ammonium nitrate, which decomposes exothermically to give a positive oxygen balance and a high gas content in the products:

$$NH_4 \cdot NO_3 \rightarrow 2H_2O + N_2 + \tfrac{1}{2}O_2$$

The explosion of ammonium nitrate stores has been the cause of a number of disasters. The characteristics indicated by the above equation make it a valuable constituent of explosive mixtures.

27.4.   Commercial preparations

Commercial explosive preparations are, typically, multicomponent mixtures containing a high explosive mixed with other material which in itself may or may not be combustible. The use of mixtures permits control not only of explosive strength and velocity of detonation, but also of cost and safety factors. The classification of commercial explosives is most conveniently made according to the high explosive used.

Nitroglycerine-based explosives were first properly developed by Alfred Nobel, a Swedish engineer. Nobel discovered that sensitivity to detonation of the highly dangerous oily liquid could be greatly reduced by absorption on an inert porous material, kieselguhr, which can absorb up to three times its own weight of nitroglycerine. *Dynamite* is such a mixture, although combustible absorbents such as sawdust, and sodium or potassium nitrate, are now commonly added in place of kieselguhr. *Ammonia dynamites* are mixtures in which nitroglycerine has been partially replaced by ammonium nitrate. In the range of nitroglycerine gelatins, such as *blasting gelatin* (Nobel's term) or *gelignite*, the absorbent is nitrocellulose. A nitroglycerine gelatin may be totally organic, and therefore have a high resistance to water, making these very powerful explosives suitable for wet working conditions.

Trinitrotoluene-based powders are widely used in engineering practice. They generally consist of TNT and ammonium nitrate; Amatol is one such mixture. A special class, the Ammonals, contain in addition aluminium powder.

Ammonium nitrate mixed with fuel-oil forms a widely used explosive. Neither ingredient is of itself a high explosive and each can therefore be transported in bulk to the place of usage with less stringent safety measures than usual and mixed on the spot. Preprepared commercial mixtures are also available, and are considered to be safer in handling characteristics than other explosives (although ammonium nitrate in bulk should always be treated with caution).

Exercises

27.1. Distinguish between *explosive strength* and *velocity of detonation*.

27.2. How would you explain the difference between a fuel and an explosive?

CHAPTER 28

High Polymers

A POLYMER is a substance composed of a number of repeating units, each repeating unit being derived from a starting material called the *monomer*. The linking together of monomer units is called *polymerization*, and if this reaction proceeds to an extensive degree the high molecular weight product is called a *high polymer*.

Polymeric materials occur widely in nature. Thus protein polymers in one form or another are one of the main constructional materials of animal life; carbohydrate polymers, particularly cellulose, serve the same purpose in plant life. Inorganic polymers also exist naturally; many minerals are complex silicate polymers, while diamond and graphite are high molecular weight aggregates of carbon. These are some of the more obvious examples of polymers in nature. There are many others.

Despite the abundance of first-hand examples, it was not until well into the twentieth century that chemists began to understand the true nature of polymers. This understanding has led to a situation where something like a half of the organic chemical industry is now involved in synthetic polymer production. It is difficult to imagine our surroundings without the range of polymeric materials now available, although these materials have emerged in the main only in the last 40 years or so. For most of us this is much more the Polymer Age than the Atomic Age.

28.1. Polymerization

Consider the reaction whereby an ester is formed from an alcohol and a carboxylic acid:

$$R \cdot OH + HO \cdot OC \cdot R' \rightarrow R \cdot O \cdot OC \cdot R' + H_2O$$

Alcohol   Carboxylic   Ester
            acid

In the above example, both alcohol and acid have only one reactive functional group, i.e. for each component the functionality number f is 1, and together they are said to form an $f(1,1)$ system. The ester product in the above reaction has no free functional groups and the value of f is zero after a one-step reaction.

If we start with an $f(2,1)$ system, i.e. dialcohol and monoacid or diacid and monoalcohol, two steps are required before a diester with zero functionality is obtained:

$$HO \cdot R \cdot OH + HO \cdot OC \cdot R' \rightarrow HO \cdot R \cdot O \cdot OC \cdot R' + H_2O$$

Dialcohol   Monoacid   Half ester
$f = 2$ $f = 1$ $f = 1$

$$HO \cdot R \cdot O \cdot OC \cdot R' + HO \cdot OC \cdot R' \rightarrow R' \cdot CO \cdot O \cdot R \cdot O \cdot OC \cdot R' + H_2O$$

                                          Diester
$f = 1$ $f = 1$ $f = 0$

The feature of these reactions is a decrease in total functionality with each step. If, however, we commence with an $f(2,2)$ system, i.e. diacid and dialcohol, each ester-forming step results in a product which itself has $f = 2$. There is no loss in functionality:

$$HO \cdot R \cdot OH + HO \cdot OC \cdot R' \cdot CO \cdot OH \rightarrow HO \cdot R \cdot O \cdot OC \cdot R' \cdot CO \cdot OH + H_2O$$

$f = 2$ $f = 2$ $f = 2$

$$HO \cdot R \cdot O \cdot OC \cdot R' \cdot CO \cdot OH + HO \cdot R \cdot OH \rightarrow$$

$f = 2$ $f = 2$

$$HO \cdot R \cdot O \cdot OC \cdot R' \cdot CO \cdot O \cdot R \cdot OH + H_2O$$

                    $f = 2$

The reaction, then, can go on indefinitely to form high polymeric material.

26

This reaction illustrates a general principle, viz. *only* $f(2,2)$ *systems yield polymeric material*. Reaction in systems of lower functionality ceases at an early stage.

The above example of a polymerization sequence involves condensation reactions in which a water molecule is split out each time alcohol and acid functions react. *Condensation polymers* are those formed by reactions in which some small molecule (e.g. water, ammonia, hydrogen chloride) splits out at each step. The other major class of polymers, *addition polymers* (Section 21.4), are formed by the linking together of initially unsaturated monomer units, usually alkenes or substituted alkenes. No byproduct splits out. In addition polymerization the mechanism of the reaction differs from condensation polymerization, but the end result is the same—a large increase in molecular weight occurs. The functionality concept is still valid, for each carbon–carbon double bond, if broken, provides two reactive sites, i.e. an alkene is potentially difunctional:

$$CH_2{=}CH_2 \xrightarrow[\text{or catalyst}]{\text{heat}} {-}CH_2{\cdot}CH_2{-}$$

Ethylene $\qquad\qquad\qquad$ Difunctional fragment

$$CH_2{=}\underset{\underset{\text{X}}{|}}{CH} \xrightarrow{\text{polymerize}} \sim CH_2{\cdot}\underset{\underset{\text{X}}{|}}{CH}{\cdot}CH_2{\cdot}\underset{\underset{\text{X}}{|}}{CH} \sim$$

Substituted $\qquad\qquad\qquad$ Polyalkene
alkene

28.2. Thermoplastic and thermosetting polymers

The division of polymers into addition and condensation types is useful, but limited. A more practical classification based on the behaviour of the finished product divides a large number of polymers into *thermoplastic* and *thermosetting* materials.

Thermoplastic polymers can be heated to softening without any chemical degradation occurring. Thermosetting polymers, on the other hand, once moulded, cannot be melted down without chemical degradation occurring; old material, therefore, cannot be used in new moulding processes.

This difference in behaviour is simply explained in terms of chemical structure. We have seen that an $f(2,2)$ system is the lowest order from which polymer may be obtained. With only two reactive sites available in the growing polymer, such a system can only form long chain-like structures and no chemical linkages will occur between chains. An $f(2,2)$ system is said to form *linear* polymer, and Fig. 28.1 represents an idealized picture of a mass of amorphous linear polymer in which M signifies the repeating unit in the chain. In melting a linear polymer, i.e. in rendering

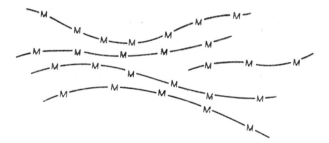

FIG. 28.1. Linear polymer.

the system fluid such that any one polymer chain can move freely with respect to its neighbour, the only forces which have to be overcome are relatively weak physical forces of attraction between chains. Much chain entanglement will occur, and polar attractions will also be present, but these are easily overcome. More particularly, no chemical bonds need be broken to free one chain from another, and melting is not accompanied by thermal degradation. Linear polymers, then, behave as thermoplastics. The low heats of fusion values of these materials, which are usually between 5 and 50 kJ mol^{-1}, reflect the relative ease of the melting process.

If now we consider a system of higher functionality, e.g. an $f(2,3)$ system such as a diacid reacting with a trialcohol, a new factor will be seen to emerge. Let us imagine that initial polymerization yields linear polymer containing the still unreacted third alcohol function.

$$\underset{\substack{\text{Trialcohol} \\ f = 3}}{\underset{\overset{|}{\text{OH}}}{\text{HO·R·OH}}} + \underset{\substack{\text{Diacid} \\ f = 2}}{\text{HO·OC·R'·CO·OH}} \longrightarrow \underset{\substack{\text{Polymer with free} \\ \text{alcohol function} \\ f = 1}}{\underset{\overset{|}{\text{OH}}}{\sim \text{O·R·O·OC·R'·CO} \sim}}$$

In this case it is now formally possible that one chain can be linked to another by chemical bonding through further reaction of the free alcohol functions with a diacid molecule.

$$\begin{array}{ccc} \sim \text{O·R·O·OC·R'} \sim & & \sim \text{O·R·O·OC·R'} \sim \\ \overset{|}{\text{OH}} & & \overset{|}{\text{O}} \\ \overset{|}{\text{CO·OH}} & & \overset{|}{\text{CO}} \\ \overset{|}{\text{R'}} & \longrightarrow & \overset{|}{\text{R'}} \\ \overset{|}{\text{CO·OH}} & & \overset{|}{\text{CO}} \\ \overset{|}{\text{OH}} & & \overset{|}{\text{O}} \\ \sim \text{O·R·O·OC·R'} \sim & & \sim \text{O·R·O·OC·R'} \sim \end{array}$$

A *cross-linking reaction* of this kind leads to the formation of a three-dimensional rigid polymer network in which the principal chains are firmly linked to one another. Any effort to melt down such material means the breaking of chemical bonds, a high energy process compared to that of melting thermoplastic polymer, requiring about 400 kJ mol^{-1}. The rupture of chemical bonds on heating would mean the substance being irreversibly degraded and destroyed. Cross-linked systems are therefore described as thermosetting.

Any polymer-forming system which has a functionality of $f(2, 3)$ or higher can cross-link. In manufacture, the degree of cross-linking can be varied continuously by varying the amount of trifunctional material present. Hence a gradation in physical properties from a highly thermoplastic material to a highly cross-linked thermosetting material can be achieved.

28.3. Molecular weight distribution in polymers

Organic monomers, often gases or low boiling liquids, can be polymerized to useful constructional materials. The great increase in molecular weight which the reaction produces is the key to the usefulness of the process. The experimentally determined molecular weight of a polymer sample is an average value, for polymerization yields a product which is a mixture of large molecules, each of essentially the same chemical structure, but of varying size and molecular weight. In a typical thermoplastic with an average molecular weight of 100,000 there may well be individual chains with molecular weights as low as 5000 or as high as 2,000,000. The physical properties of a polymer vary with its average molecular weight, which must therefore be controlled. For example, polyacrylonitrile, used as a synthetic fibre, does not exhibit fibre-forming characteristics until a molecular weight of between 10,000 and 20,000 is achieved. Again, the glass transition temperature of thermoplastics, discussed in the following section, rises rapidly with increasing molecular weight before approaching a final steady value.

28.4. Physical properties of polymers

Polymers differ markedly in physical behaviour from other more traditional materials of construction. They are not so susceptible to certain modes of weathering and corrosion, and so are used for making, for example, pipes and drains, gear wheels, perhaps in the future car bodies. They are light and easily moulded, and so permit the possibility of large-scale prefabrication in the building industry. Being essentially covalent, they have low thermal and electrical conductivities, and this fact, combined with their low water-retention characteristics, makes them excellent insulating materials. They can be made with a lack of rigidity extremely useful in kitchen utensils and toys. Examples of the wide use of polymers are not difficult to find. The major disadvantage of organic polymers is that of carbon compounds in general: they do not respond well to high temperatures, at which they will undergo either combustion or thermal degradation. Even this drawback has been partly overcome by the development of new organic and inorganic polymers.

Linear and partially cross-linked materials can exist in a number of different physical states which in turn affect their bulk properties. At sufficiently low temperatures many plastics exist in the solid phase as amorphous inflexible aggregates called *glasses*. If a polymeric glass is heated a fairly well-defined temperature called the *glass transition temperature*, T_g, is observed at which the polymer chains become more flexible and the material is converted to a *rubber*, i.e. a material capable of undergoing considerable reversible elastic deformation. As a polymer passes through the glass transition temperature bulk properties such as the specific heat or the modulus of elasticity undergo change and it is by following the change in such a property or in its temperature coefficient that the glass transition temperature is measured. The glass transition temperature, which is not an absolute value, varies widely through a range of polymers, and essentially reflects the ease with which a polymer chain can be mobile by rotation about the carbon–carbon bonds in the backbone. Polyethylene, for example, has a reported T_g of $-122°C$, polyvinyl acetate a value of 28°C and polystyrene a value of 102°C. Through this series the side group becomes steadily larger, and therefore hinders the ease with which rotation, and therefore elastic behaviour, can occur. If the side group is polar, then the polar attractions between chains must also be overcome before a glass transition is observed. Thus polyvinyl chloride, which has a comparatively small but distinctly polar side group, has a T_g of 70°C. Glass transition values can be altered by, for example, altering the method of polymerization or by adding *plasticizers* to the polymer. Plasticizers, which are bulky molecules like dibutyl phthalate, act as internal lubricants in a polymer, inserting themselves between the polymer chains and so weakening the cohesive forces which hold them together in the solid state. The result is that the glass transi-

$$CO\cdot O\cdot CH_2\cdot CH_2\cdot CH_2\cdot CH_3$$

$$-CO\cdot O\cdot CH_2\cdot CH_2\cdot CH_2\cdot CH_3$$

Dibutil phthalate

tion temperature is lowered and the material becomes more flexible. Loss of plasticizer can occur by accidental solvent extraction, or perhaps by evaporation in continually warm conditions, and it is often made evident by an obvious increase in brittleness in the bulk material.

In the glass and rubber states polymers are amorphous, but they can also exist in a crystalline state. Crystallinity can be equated with order, and in a polymer can vary from zero in a totally amorphous system to almost 100% in a highly ordered system. The appearance of both crystalline and noncrystalline regions in a polymer mass is represented in Fig. 28.2. We shall see (Section 28.6) that the degree of crystallinity in a

C crystalline region

FIG. 28.2. Crystalline and non-crystalline regions in a polymer mass.

polymer can vary with the method of initiating the reaction. It can also be encouraged by process techniques such as annealing or cold drawing or, as in the case of polyamides (Section 28.8), by having present in the polymer chemical structures which have an inherent tendency to arrange themselves in ordered patterns. Highly crystalline polymers exhibit a well-defined *crystalline melting point*, T_m. With increasing crystallinity in a polymer there is a corresponding decrease in elasticity, and polymer crystallinity is therefore an important parameter used in controlling the mechanical properties of the finished material.

The interrelations between the glassy state, the rubber state and the crystalline state in a polymer cease to apply in a highly cross-linked system. Progressive cross-linking leads to progressive rigidity, and a fully thermosetting rigid material is likely to have poorer elastic deformation characteristics than, for example, steel.

28.5. Cellulose plastics

Cellulose is a naturally occurring polymer of complex structure, forming the major part of vegetable cell-wall structures. The raw material, which is extracted from sources such as cotton, wood pulp and straw, can be processed to give a range of useful plastics. Inspection of a segment of the cellulose chain shows that each ring in the chain contains three

Partial structure of cellulose

alcohol hydroxyl groups. Cellulose usually undergoes chemical modification to improve its characteristics by carrying out one of a number of reactions at these hydroxyl functions.

Cellulose may be nitrated by reaction with mixed acid. The extent of nitration can be varied, so varying the properties of the product. *Cellulose*

$$-OH + HNO_3/H_2SO_4 \longrightarrow -O \cdot NO_2 + H_2O/H_2SO_4$$

nitrate, mixed with camphor as plasticizer, is marketed as a moulded plastic under such names as Celluloid or Xylonite. It tends to discolour in time, and is highly flammable. Indeed, fully nitrated cotton cellulose, guncotton, is a principal ingredient of many industrial explosives.

Reaction of cellulose with acetic acid yields the ester *cellulose acetate*,

$$-OH + HO_2C \cdot CH_3 \longrightarrow -O \cdot OC \cdot CH_3 + H_2O$$
Acetic acid

less flammable than the nitrate and suitable for moulding or film castings; partially acetylated cellulose is used as the base for photographic film. Fully acetylated cellulose is suitable for fibre production and appears under such names as Tricel.

Regenerated cellulose is the name given to a cellulose which has been extracted, chemically treated, and then reconverted to be used as cellulose rather than as one of its derivatives. Regeneration processes usually cause a drop in the molecular weight of the product. The proprietary transparent sheeting Cellophane is a regenerated cellulose; so too is most rayon, or artificial silk.

28.6. Vinyl and related thermoplastics

Thermoplastics formed by addition polymerization account for a very large and important fraction of polymer production. Within this field there exists the sub-group of *vinyl polymers*, which are obtained from monomers of general formula $CH_2{=}CH{\cdot}X$ whose structures include the vinyl $CH_2{=}CH{-}$ radical. Most of the principal thermoplastics are vinyl polymers, and with their monomers these are listed in Table 28.1.

TABLE 28.1. VINYL MONOMERS AND POLYMERS

Monomer	Structure	Polymer
Ethylene	$CH_2{=}CH_2$	Polythene
Propylene	$CH_2{=}CH$ $\qquad$ CH_3	Polypropylene
Vinyl chloride	$CH_2{=}CH$ $\qquad$ Cl	Polyvinyl chloride (PVC)
Vinyl acetate	$CH_2{=}CH$ $\qquad$ $O{\cdot}CO{\cdot}CH_3$	Polyvinyl acetate (PVA)
Styrene	$CH_2{=}CH$	Polystyrene
Acrylonitrile	$CH_2{=}CH$ $\qquad$ CN	Polyacrylonitrile

Some thermoplastics are also derived from unsaturated monomers other than vinyl compounds, and three of the more significant examples are listed in Table 28.2.

TABLE 28.2. NON-VINYL MONOMERS AND POLYMERS

Monomer	Structure	Polymer
Vinylidene chloride	$CH_2{=}CCl_2$	Polyvinylidene chloride
Methyl methacrylate	$CH_2{=}C \cdot CO \cdot O \cdot CH_3$ $\quad\quad\;\; \mid$ $\quad\quad\;\; CH_3$	Polymethyl methacrylate
Tetrafluorethylene	$CF_2{=}CF_2$	Polytetrafluorethylene (PTFE)

There are a number of ways in which unsaturated monomers can be induced to polymerize. Generally some sort of initiator is used. Alkene polymerization has already been described briefly (Section 21.4) and the three-step process of initiation, propagation and termination is summar-

Initiation

$$B{-} + CH_2{=}\underset{\displaystyle X}{\overset{\displaystyle |}{CH}} \longrightarrow B{-}CH_2 \cdot \underset{\displaystyle X}{\overset{\displaystyle |}{CH}}{-}$$

Initiation   Monomer
fragment

Propagation

$$B{-}CH_2 \cdot \underset{\displaystyle X}{\overset{\displaystyle |}{CH}}{-} + n \cdot CH_2{=}\underset{\displaystyle X}{\overset{\displaystyle |}{CH}} \longrightarrow B{-}CH_2 \cdot \underset{\displaystyle X}{\overset{\displaystyle |}{CH}} \cdot (CH_2\underset{\displaystyle X}{\overset{\displaystyle |}{CH}})_n$$

Termination

$$B{-}CH_2\underset{\displaystyle X}{\overset{\displaystyle |}{CH}} \cdot (CH_2 \cdot \underset{\displaystyle X}{\overset{\displaystyle |}{CH}})_n + R{-} \longrightarrow B(CH_2 \cdot \underset{\displaystyle X}{\overset{\displaystyle |}{CH}}) \cdot R_{n+1}$$

Growing chain   Reactive   Dead polymer
fragment

ized above. The final termination step (which in part can be used to control the final molecular weight) is envisaged as a growing chain reacting with another fragment R— which itself may be a second growing chain. Other termination mechanisms are possible. From beginning to end a very short time—perhaps one-thousandth of a second—is required for any one chain to be formed.

The initiating system used can affect quite profoundly the properties of the final polymer. Many vinyl polymers have bulky side groups attached to the backbone, and these tend to interfere with the chain packing efficiency. If polymerization is initiated by generating free radicals[†] in the presence of monomer, an initiating technique commonly used, the side groups of the product are arranged quite haphazardly along each chain. We can see that this must lead to interference of one group with another at many points, and therefore poor packing results. This state of affairs is represented in Fig. 28.3, which it should be noted is a gross distortion of the true shape of the main chain in which each carbon atom has four bonds tetrahedrally distributed about it.

If, instead of using a free radical system, polymerization is initiated by any one of a variety of partially inorganic ionic initiators, e.g. a mixture

FIG. 28.3. Interfering side-groups in free-radical initiated polymer.

[†] When a chemical bond is broken symmetrically the atomic nuclei concerned each contain a single unpaired electron in the outer shell. Such particles, called *free radicals*, are highly reactive and play important roles in a number of industrial processes; for example, the cracking of ethane can yield two methyl free radicals:

$$H_3C\!-\!CH_3 \rightarrow 2H_3C\cdot$$

of aluminium triethyl, $Al(CH_2 \cdot CH_3)_3$, and titanium tetrachloride, $TiCl_4$, chain growth takes place in a controlled manner such that a regular arrangement of side groups about the polymer backbone results.

Figure 28.4 represents the two regular arrangements, known as *syndiotactic* and *isotactic* forms, which can arise for a simple vinyl monomer. Regularity at molecular level encourages regularity, or crystallinity, in the mass. Tactic polymer chains pack together efficiently, and often yield

Syndiotactic   arrangement

Isotactic   arrangement

FIG. 28.4. Regularity in polymer chains.

highly crystalline material. The chains actually pack in helical arrangements, reflecting the helical conformations found in naturally occurring polymers such as the genetic code material deoxyribonucleic acid (DNA).

Polymers which are highly tactic are called *stereoregular polymers*; their tendency to higher degrees of crystallinity manifests itself in other properties. Thus a highly tactic polymer will have a higher density than its free radical initiated counterpart. In the extreme case of propylene, the free radical initiated material is a gum with no useful properties, so great is the interference effect of the bulky methyl side groups; stereoregular polypropylene, a crystalline solid, is of considerable commercial significance. The Italian chemist Natta and the German Ziegler, the two men most closely connected with developments in stereoregular polymerization, were awarded jointly the Nobel prize in chemistry in 1963 for their work, so important is it judged.

Copolymerization is another useful and widely applied method of controlling polymer properties. Here two or more unsaturated monomers are polymerized together to give a mixed polymer whose properties depend on the initial monomer ratios used. Thus styrene and acrylonitrile can be co-polymerized:

Styrene   Acrylonitrile   A–S Copolymer

A short summary of some of the more important thermoplastics is given below. Much more comprehensive information about their properties will be found elsewhere.[†]

Polyethylene

Polyethylene, or polythene, was one of the earliest vinyl polymers to be marketed (in 1937 in the United Kingdom). The early ICI process used very high pressures and elevated temperatures to polymerize the gaseous monomer, but newer processes employing Ziegler–Natta type initiators have greatly reduced the intensity of reaction conditions, and almost eliminated the formation of chain branching which can arise due to side reactions. Low density polythene (sp.gr. 0·91–0·93) produced by the earlier processes begins to soften at about 85°C; high density polythene (sp.gr. *ca.* 0·935–0·96) shows a crystalline melting point at about 110°C, and sometimes considerably higher. The high density material, therefore, can be used for carrying hot water where the former may fail.

[†] For example, *British Plastics Yearbook*, Iliffe, London.

Polypropylene

Stereoregular polypropylene has qualities which make it suitable for synthetic fibre production; with, in addition, low water-retention figures, polypropylene is being used for rope production as well as being an alternative to polythene.

Polyvinyl chloride

Vinyl chloride polymerizes to a material which is less flammable than the related hydrocarbon polymers. PVC is now widely used for clothing. As a substitute for shellac, it has made the production of long-playing records possible. It is often produced in modified form by copolymerization with, for example, vinyl acetate or vinylidene chloride.

Polystyrene

Styrene is produced from petroleum sources by the synthesis and subsequent cracking of ethyl benzene:

This monomer reacts to give one of the very important vinyl polymers. The inflexibility of the benzene rings imparts a brittle quality to the product, which is transparent, and polystyrenes do not exhibit a glass transition until temperatures of about 102°C are reached. Both electrical and thermal resistances are high; expanded polystyrene is now much used as a thermal insulator in the form of tile or granule. Copolymerized with acrylonitrile and other monomers, it forms rubber-like products (Section 28.9).

Polymethylmethacrylate

This polymer, which first appeared in 1933, is marketed as, for example, Perspex (UK) and Lucite (USA). The monomer, methyl methacrylate, is a derivative of acrylic acid, the parent structure to which all acrylic polymers are related:

$$CH_2{=}CH{\cdot}CO_2H \qquad CH_2{=}CH{\cdot}CO{\cdot}O{\cdot}CH_3$$

Acrylic acid   Methyl acrylate

$$\overset{\displaystyle CH_3}{\underset{\displaystyle CH_2{=}C{\cdot}CO{\cdot}O{\cdot}CH_3}{|}}$$

Methyl methacrylate

Methyl methacrylate monomer is obtained by a multistep synthesis from acetone. The polymer has outstanding optical properties, and is widely used as a glass substitute.

Polytetrafluoroethylene

Fluorine-containing polymers are expensive, and PTFE, the simplest and best known of the group, is no exception. Sold under the trade names of Teflon and Fluon, this high density material (sp.gr. *ca.* 2·2) has some quite outstanding properties: a softening temperature of almost 400°C, negligible water-retention figures, a very low coefficient of friction which gives it special lubricating qualities, and a high degree of chemical stability.

28.7. Thermosetting resins

The principal thermosetting resins are condensation polymers. In the United Kingdom currently the output of thermosetting resins is running at about half that of thermoplastics. Finished products in this field often contain a large percentage of inert filler, e.g. sawdust or asbestos fibre in a moulded article, paper or cloth sheets in a laminate.

Phenol-formaldehyde resins

The reaction between phenol and formaldehyde (Section 23.6) introduces one or more methylol —CH_2OH groups into the benzene ring:

| Phenol | Formal-dehyde | Methylolated phenol |

These highly reactive groups can undergo further condensation reactions with one another to produce ether bridges and methylene —CH_2— bridges between the rings:

$$—CH_2OH + HOCH_2— \longrightarrow —CH_2 \cdot O \cdot CH_2— + H_2O$$

Ether bridge

$$—CH_2 \cdot O \cdot CH_2— \xrightarrow{heat} —CH_2— + CH_2O$$

Methylene bridge   Formal-dehyde

Each benzene ring has a potential functionality of three, and so extensive cross-linking can be achieved. If curing were taken to completion, a three-dimensional network, containing only methylene bridges in the ideal case, would be obtained.

By ensuring a deficiency of formaldehyde in the first stage of manufacture, i.e. by decreasing the average number of methylol groups per ring such that f is two or less in many rings, linear thermoplastic products,

Idealized representation of cured phenol-formaldehyde resin

called novolaks, can be obtained. Later addition of a source of formaldehyde followed by curing will give the fully cross-linked resin. (These and other resins are often supplied as Tube A, the thermoplastic intermediate, and Tube B, the cross-linking reagent. On mixing, the final fully cured product is obtained.)

Phenol-formaldehyde resins were the first totally synthetic polymers to be produced on any scale, appearing in the patent literature in the few years before the First World War. Their development was due to Henri Baekeland, whose name continues to be closely associated with these products in the proprietary name Bakelite.

Amine-formaldehyde resins

Curing reactions in the phenol-formaldehyde system involve a number of side reactions in addition to the cross-linking steps indicated above. One result of these minor reactions is to produce a dark-coloured product. To overcome this problem a variety of clear products based on amine-formaldehyde reactions have been developed.

The initial reaction between formaldehyde and an amine group is two-stage:

$$R \cdot NH_2 + 2CH_2O \longrightarrow R \cdot NH \cdot CH_2OH \longrightarrow R \cdot N(CH_2OH)_2$$

Primary amine	Formaldehyde		Monomethylol amine		Dimethylol amine

27

These methylolated amines can subsequently polymerize to give thermoplastic and thermosetting polymers. In part the condensation reactions are similar to the scheme set out for phenol-formaldehyde polycondensations:

$$\text{>N·CH}_2\text{OH} + \text{HOCH}_2\text{·N<} \longrightarrow \text{>N·CH}_2\text{·O·CH}_2\text{·N<} + \text{H}_2\text{O}$$

Ether bridge

$$\text{>N·CH}_2\text{·O·CH}_2\text{·N<} \longrightarrow \text{>N·CH}_2\text{·N<} + \text{CH}_2\text{O}$$

Methylene   Formal-
bridge   dehyde

$$\text{>N·CH}_2\text{OH} + \text{HN·CH}_2\text{OH} < \longrightarrow \text{>N·CH}_2\text{·N·CH}_2\text{OH} + \text{H}_2\text{O}$$

Methylolated methylene
bridge

Curing, however, must be much more complex than the above scheme would indicate, for some monomethylol amine derivatives, with an apparent functionality f of one, can give linear polymers, and some dimethylol derivatives ($f = 2$?), thermosetting resins.

The most important amine components used in industrial polycondensate production are urea and melamine, the latter heterocyclic compound offering the possibility of introducing six methylol groups per ring. Amine-based resins such as Formica and Melmex often contain both urea and melamine units, and other amine components can also be

H$_2$N\
      >CO
H$_2$N/

Urea

Melamine

used to modify the properties of the final resin. These clear resins can readily incorporate light pastel-coloured pigments, in either mould or laminate form. Their stability to light and heat and their resistance to dirt, solvents, and normal wear and tear are excellent, the more so when melamine is used in high proportion.

Polyester resins

Polyesters have a number of uses. The class includes a series of materials chiefly employed as protective coatings and adhesives.

Alkyd resins are polyester structures formed from dibasic carboxylic acids (e.g. phthalic acid) and the trifunctional alcohol glycerol.

$$CO_2H$$

Phthalic acid $(f = 2)$ + $HOCH_2 \cdot CH \cdot CH_2OH$ (Glycerol, $f = 3$) $\longrightarrow$ Alkyd resin

Alkyd resins are principal components of protective surface coatings of various kinds (e.g. paints, lacquers, enamels) and have good adhesive properties.

Epoxy resins are more difficult to classify chemically; a number are in part polyesters, cross-linked resins being formed by reacting phthalic or other dibasic acids with linear polymers containing free alcohol functions in addition to the characteristic epoxy $\diagup C——C \diagdown$ group.

$$CH_2—CH \cdot CH_2 \cdot O—\langle O \rangle—\underset{\underset{CH_3}{|}}{\overset{\overset{CH_3}{|}}{C}}—\langle O \rangle—O \cdot CH_2 \cdot \underset{\underset{OH}{|}}{CH} \cdot CH_2 \wedge +$$

Linear epoxy polymer

$+ HO_2C \cdot R \cdot CO_2H \longrightarrow$ Epoxy resin

Dibasic acid   Thermosetting polymer

27*

Their usefulness in part overlaps with the alkyd resins, epoxy resins having the advantage of even better adhesive properties and a flexibility which, in a coating, means high resistance to cracking. They are also used in laminates and mouldings, appearing in their various forms under trade names such as Araldite and Epon.

Polyurethanes are linear polymers which can be cross-linked by reaction with polyesters to give protective coatings of very high stability and wear resistance. They are expensive, but are now being used where a very high class of finish to an article is justified.

28.8.   Synthetic fibres

It will be appreciated that any one polymer may well be versatile, and may be used as a thermoplastic, or as thermosetting moulding material, or as a highly adhesive coating, and so on. We are merely emphasizing the uses to which particular polymers are best suited.

For a polymer to make a good synthetic fibre, it must crystallize to give a large proportion of chains lying in order parallel to one another, and having a high tensile strength in the longitudinal axis. Stereoregular polypropylene has the necessary characteristics to make good fibre; so also does polyacrylonitrile (e.g. Orlon fibre). The two outstanding synthetic fibres, however, are condensation polymers.

Nylons

Nylon is the name given to synthetic polyamides, in which the recurring amide —CO·NH— linkage is formed from a diamine and a dibasic carboxylic acid.

$$HO_2C \cdot R \cdot CO_2H + H_2N \cdot R' \cdot NH_2 \rightarrow \, \sim OC \cdot R \cdot CO \cdot NH \cdot R' \cdot NH \sim + H_2O$$

    Dibasic acid   Diamine   Polyamide or Nylon

When the American chemist W. H. Carothers chose to examine polyamides as potential fibre-forming substances he chose wisely, for he was reflecting the practice of nature (animal hair, for example, is a complex polyamide or protein). These structures form good fibres because there exists the possibility of hydrogen bonding between a strongly polarized oxygen atom of one chain and an amide group hydrogen of another chain. It will be seen from Fig. 28.5 that this effect, repeated many hundreds of times along any one chain length, must assist greatly in aligning the chains in order and producing fibrous, crystalline polymer.

Fig. 28.5. Hydrogen bonding in polyamides.

While the hydrogen bond effect is very important in polyamides, the way in which nylons and other synthetic fibres are spun and drawn also greatly affects the extent to which crystallinity occurs.

Nylons have a terminology all of their own. The most common nylon in the United Kingdom and the United States is that made from adipic acid and hexamethylene diamine. The product is called 6,6-Nylon, the

$$HO_2C \cdot (CH_2)_4 \cdot CO_2H + H_2N \cdot (CH_2)_6 \cdot NH_2 \rightarrow$$

Adipic acid   Hexamethylene
                  diamine

$$\sim OC \cdot (CH_2)_4 \cdot CO \cdot HN \cdot (CH_2)_6 \cdot NH \sim$$

6,6-Nylon

first digit indicating the number of carbon atoms in the amine segment, the second the number in the acid segment. Thus another nylon which has found favour in the United States, 6,10-Nylon, is formed from hexamethylene diamine and sebacic acid, $HO_2C\cdot(CH_2)_8\cdot CO_2H$. An interesting nylon now produced in both Europe and the United States is 6-Nylon, made by polymerization of a cyclic amide, caprolactam.

Caprolactam

$\sim NH\cdot CO\cdot(CH_2)_5 \sim$

6-Nylon

These three nylons dominate the market, the 6,6 version being the most important. The starting point for commercial production of 6,6-Nylon and 6-Nylon, rather surprisingly, is benzene (or phenol), from which may be made all the intermediates.

Nylons form excellent fibres, but their ability to retain physical dimensions after fabrication have also made them popular thermoplastic moulding materials.

Polyethylene terephthalate

Polyethylene terephthalate fibre is marketed as Terylene (UK) and Dacron (USA); it also appears as Melinex film and sheeting. It is a polyester, formed from terephthalic acid and ethylene glycol.

Terephthalic acid   Ethylene glycol

Polyethylene terephthalate

The polymer owes its fibre-forming qualities to the presence of recurring planar benzene rings. In a polymer chain the planar rings act as stiffeners along the chain length. This degree of rigidity in the longitudinal axis is the key structural feature of the polyester fibre. The importance of the linear structure can be emphasized by replacing terephthalic acid by phthalic acid, in which the two acid groups are *ortho-* to one another. With ethylene glycol, this yields a polymeric gum with no fibre-forming properties. The shapes of the chains are such that they cannot lie side by side, and the degree of crystallinity is reduced to negligible proportions.

28.9. Rubbers, natural and synthetic

A rubber is a substance with a low modulus of elasticity, i.e. it can be stretched considerably and still return to its original dimensions.

Natural rubber, which is obtained as a milk white latex from the rubber tree, is a polymer of the diene hydrocarbon isoprene (Section 21.5). In the main the isoprene units are linked "head-to-tail", a residual double bond occurring in the middle of each segment and within the main backbone. Other arrangements of the isoprene unit in which the double bond is external to the main chain make minor contributions to the overall structure of rubber latex.

$$\sim CH_2-\underset{\underset{CH_3}{|}}{C}=CH-CH_2-CH_2-\underset{\underset{CH_3}{|}}{C}=CH-CH_2 \sim$$

Segment of rubber chain showing "head-to-tail" arrangement

$$\sim CH_2-\underset{\underset{CH_3}{|}}{C}=CH-CH_2-CH_2-\underset{\underset{\underset{H_3C}{}}{\overset{}{C}}=CH_2}{CH} \sim$$

Segment of rubber chain with external double bond

Polymer of the first structure would be called 1,4-polyisoprene, indicating the "head-to-tail" linking of monomer units which involves carbon atoms one and four in each isoprene molecule. This name, however, is incomplete. When a non-symmetrical alkene double bond is present in a compound stereoisomers[†] called *cis*- and *trans*-isomers can occur, depending on the orientation of different groups about the double bond. In the polyisoprene molecule *cis*-isomerism occurs when all methyl groups lie on the same side of the chain, *trans*-isomerism when they alternate from one side to the other. These two quite distinct forms of the polymeric hydrocarbon, *cis*-1,4-polyisoprene and *trans*-1,4-polyisoprene, are represented below:

$$CH_3 \quad H \qquad\qquad CH_3 \quad H$$
$$C = C \qquad\qquad C = C$$
$$-CH_2 \qquad CH_2-CH_2 \qquad CH_2-$$

(a) *cis*-1,4-Polyisoprene

$$CH_3 \qquad CH_2-CH_2 \qquad H \quad CH_2-$$
$$C = C \qquad\qquad C = C$$
$$-CH_2 \quad H \qquad\qquad CH_3$$

(b) *trans*-1,4-Polyisoprene

In a rubber, individual polymer chains must have space enough to uncoil in response to external forces. The *cis* configuration in polyisoprene packs inefficiently and hence is equated with elastic behaviour. Natural rubber is about 95% *cis*-1,4-polyisoprene. The *trans* isomer, in which much more efficient packing can be achieved, also occurs naturally. It is called gutta percha, a hard white non-elastic substance used for the outside cover of golf balls. Both *cis*- and *trans*-polyisoprene are now produced synthetically from the monomer by the use of catalysts which ensure a stereoregular polymerization.

Natural rubber in its raw state is soft and sticky. It was of very limited use until Samuel Goodyear discovered the process of *vulcanization*. This process is now recognized as a cross-linking reaction between main

---

[†] When two compounds have the same molecular formula but differ from one another in the way in which their atoms are arranged in space, they are called *stereoisomers*.

chains which involves the residual double bonds in the polyisoprene molecule. The cross-links act as internal stiffeners in the polymer. Goodyear vulcanized rubber by heating it with sulphur. This produces sulphur

$$
\begin{array}{c}
\text{H}_3\text{C} \\
| \\
\sim \text{CH}_2\text{—C}=\text{CH—CH}_2 \sim \\
\\
\\
+\ \text{S} \longrightarrow \\
\\
\sim \text{CH}_2\text{—C}=\text{CH—CH}_2 \sim \\
| \\
\text{H}_3\text{C}
\end{array}
\qquad
\begin{array}{c}
\text{H}_3\text{C}\quad\text{S} \\
|\qquad| \\
\sim \text{CH}_2\text{—C—CH—CH}_2 \sim \\
| \\
\text{S} \\
| \\
\text{S} \\
| \\
\sim \text{CH}_2\text{—C—CH—CH}_2 \sim \\
|\qquad| \\
\text{H}_3\text{C}\quad\text{S}
\end{array}
$$

Rubber latex   Vulcanized rubber

bridges between the principal chains. Other cross-linking reagents are now used for vulcanization, but sulphur remains the important one.

Variation in the amount of sulphur added varies the stiffening effect, and rubber can be produced as a highly elastic material on the one hand, or as a highly cross-linked rigid material (ebonite or vulcanite) on the other.

Synthetic rubbers are increasingly taking the place of the natural product, which is stable in neither price nor continuity of supply. A number of monomers can be polymerized to elastic materials, among which not surprisingly isoprene and other dienes are prominent. A summary of some synthetic rubbers is given in Table 28.3, from which it will be seen that copolymers occupy an important place in the industry. This list is by no means exclusive, and a number of synthetic elastomers which include elements other than carbon in their backbone are of some significance, e.g. the sulphur-containing Thiokols, which have outstanding solvent resistance; polyurethane elastomers for rubber foam production; the inorganic silicone rubbers which withstand high temperatures to a far greater degree than the organic materials; and many others which are often related to the principal synthetics listed in Table 28.3.

TABLE 28.3. SYNTHETIC RUBBERS

Monomer(s)	Polymer	Characteristics
Isoprene	*cis*-Polyisoprene	Slightly inferior to the natural product
Butadiene	*cis*-Polybutadiene	Retains rubber state at low temperatures; low energy loss material
Butadiene + Styrene	Buna S, SBR, GR–S	Most important synthetic. High abrasion resistance; low energy losses; used for car tyres, etc.
Butadiene + Acrylonitrite	Buna N, NBR	High resistance to hydrocarbon solvents
Chloroprene	Neoprene	High resistance to solvents; to heat; and to oxidation
Isobutylene	Butyl rubber	Highly resistant to gaseous diffusion; resistant to oxidation
Ethylene + Propylene	Ethylene-propylene rubber	Low density rubber similar to SBR rubbers

The low energy losses noted as particularly characteristic of the butadiene and butadiene-styrene rubbers are worth some consideration. If a piece of rubber is stretched, and then allowed to contract again, a plot of the force applied against the deflection shows that stretching and contraction do not follow identical paths (Fig. 28.6) but form a hysteresis loop. The work done in each step is given by the area under each curve, and the work or energy loss over the whole cycle is therefore given by the area enclosed by the curves, the so-called hysteresis loss. This loss is easily demonstrated, for it appears as heat, and rubber continually expanded and contracted soon rises in temperature. In car tyres, a low hysteresis-loss rubber is essential, for energy loss here means not only a rise in tyre temperature but also an increase in petrol consumption. Butadiene-styrene rubbers are particularly suitable for tyre manufacture, having a high ratio of energy reabsorbed to energy expended ("high-resilience" rubbers) in the expansion–contraction cycle.

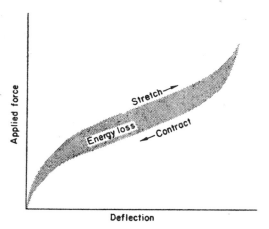

FIG. 28.6. Hysteresis loss in rubbers.

28.10.   High-temperature polymers

Any substance is limited in use by the strength of the chemical bonds existing throughout its mass. Inorganic polymers offer the promise of greater thermal and chemical stability than many of their organic counterparts, whose non-polar carbon–carbon bonds and weakly polar carbon–hydrogen bonds are too readily subject to degradation. *Silicone polymers*, in which the backbone is a chain of alternating silicon and oxygen atoms, have been extensively developed, and are available as liquid and solid lubricants, as rubbers, as resins, and so on. Each silicon atom carries organic side chains, commonly methyl groups, and this combined with the intrinsic strength (or weakness) of the backbone limits

$$-O-\underset{|}{\overset{|}{Si}}-O-\underset{|}{\overset{|}{Si}}-O-\underset{|}{\overset{|}{Si}}-O-$$

Backbone of silicone polymer

these polymers to a temperature ceiling of about 250°C, much better than most organic polymers, but rather less than many uses require. Other new polymers, both organic and inorganic, have been developed in recent years which can function in air at temperatures up to perhaps 400–500°C, and at even higher temperatures in inert atmospheres. They are expensive and are used where there is no alternative.

Exercises

28.1. Polystyrene is a glass at room temperature. What effect on structure would occur if styrene were polymerized with (a) butadiene; (b) *p*-divinylbenzene?

28.2. The fibre-forming properties of polyethylene terephthalate are destroyed if very small quantities of benzoic acid are present during polymerization. Explain why this is so.

28.3. Polyvinyl acetate can be hydrolysed in base solution to a water-soluble polymer. Write the equation for this reaction. Why is it not possible to synthesize the water-soluble polymer from the corresponding monomer? (*Hint*: see Section 21.7.)

28.4. Suggest a complete synthesis from simple starting materials of

(a) a rubber;
(b) a thermal insulating polymeric material;
(c) a fibre;
(d) a moulding resin.

CHAPTER 29

Lubricants and Detergents

MANY important phenomena are concerned with interactions at the boundaries of two different phases—at a liquid–liquid boundary or a liquid–solid boundary, for example. In this chapter we shall deal with two topics of particular interest to engineers where surface interactions are all-important, viz. lubrication and detergency.

LUBRICANTS

29.1. The nature and mechanism of lubrication

Any two surfaces in contact with one another have to overcome frictional forces if they are to move with respect to one another. The work done in overcoming these forces generates heat and absorbs a considerable amount of power. It is the task of a lubricant to lessen frictional forces and so lessen power losses and reduce heat output.

Lubricant action—the separation of two solid surfaces by a fluid in order to reduce shear stresses during motion—may differ from one situation to another. Three mechanisms are usually distinguished, and these we shall consider in turn.

Hydrodynamic lubrication, or fluid lubrication, occurs where there is a comparatively large separation of solid surfaces, i.e. of the order of 10^{-4} cm or greater. Given this condition, a continuous lubricant film can insert itself between the two surfaces. At the interface between solid and fluid the forces of attraction which exist cause an orientation of the fluid molecules, while within the main body of fluid the normal random distribution is maintained. This state of affairs is represented in Fig. 29.1.

Polar molecules, substances like carboxylic acids and esters, attach them-selves to solid surfaces more firmly than non-polar hydrocarbons and orientate more markedly, but their resistance to oxidation under hot working conditions is not favourable. A majority of hydrodynamic lubricants are petroleum-based hydrocarbons (containing, as we shall see, various additives).

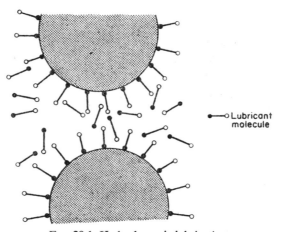

●─○ Lubricant
molecule

FIG. 29.1. Hydrodynamic lubrication.

Boundary lubrication is necessary when two surfaces are in close con-tact with one another. In practice, two apparently smooth surfaces cannot be in intimate contact. On the molecular scale, even a highly polished surface has many irregularities and contact will only be made at odd points (points A in Fig. 29.2). If a long chain compound, polar at one end, nonpolar at the other, is introduced between such surfaces, the polar end will again adhere strongly to the solid surface, while the rest of the molecule will orientate outward from the surface. Contact of the free non-polar end of one such molecule with a similar neighbouring molecule will produce low shearing stress conditions, i.e. will produce lubricating conditions. Thus carboxylic acids with long hydrocarbon chains, like stearic and oleic acid, have been used as boundary lubricants. Boundary lubrication (at points B in Fig. 29.2) will not prevent abrasion and wear occurring at contact points (A in Fig. 29.2).

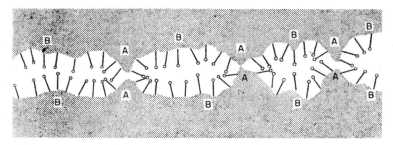

FIG. 29.2. Intermittent contact between two "close" surfaces.

Extreme pressure lubrication is a form of boundary lubrication designed to cope with high temperature operating conditions. In this case solid films, typically phosphide films, are formed at the surfaces in question. These films, when subject to stress, melt and shear much more readily than the metal itself, and so lubrication is achieved. The films are regenerated *in situ* at the high operating temperatures by the continual presence in the lubricant medium of a suitable reagent. Tricresyl phosphate, $(H_3C \cdot C_6H_4 \cdot O)_3 \cdot OP$, is widely used for this purpose.

29.2. The viscosity of lubricants

A liquid becomes "thinner" on heating. This change in viscosity with temperature is particularly important for lubricant oils, which often have to perform satisfactorily over considerable ranges of temperature.

Viscosity is a measure of the shear forces in a fluid. As these increase, i.e. as the viscosity of the oil increases, the work required to move a surface immersed in the oil will increase. (Every motorist is aware of starting difficulties in winter weather, difficulties mainly due to the extra load produced by the cold viscous engine oil.) However, an oil which becomes too thin at higher temperatures may fail to withstand the pressure applied to it, and may be ejected from the regions between solid surfaces in close contact. Ideally one wants a lubricant whose viscosity does not change with temperature.

The viscosity of an oil may be measured and reported in a number of ways. Popular among these is to measure the flow time through a standard orifice; in the United Kingdom the Redwood viscosity is commonly quoted, in the United States the Saybolt viscosity is used. In these and other methods a given viscosity value will apply only at a particular temperature which should be quoted, i.e. a viscosity value in itself does not give any information about the variation of viscosity with temperature. This important relationship is indicated by the *viscosity index* (VI) of an oil, which is arrived at by comparing the change in viscosity of the test oil over a range of temperatures to that of two standard oils over the same range. The standard oils are given arbitrary VI values of zero (in the case of the oil with poor viscosity–temperature characteristics) and 100 (in the case of the oil with good viscosity–temperature characteristics). A high VI therefore denotes an oil which tends to exhibit minimal change in viscosity with changing temperature.

As with the octane rating of gasolines, the VI of a lubricating oil reflects the chemical structures present in the oil. For example, naphthenes (cycloalkanes) have lower VI values than alkanes. Again like the octane rating, the VI of an oil can be improved by various additives, and a VI in excess of 100 is now possible. Synthetic viscous polymers are used for this purpose, leading to the production of a range of visco-static lubricants. The widespread use of such oils has greatly lengthened engine life in motor cars.

The car owner is familiar with a designation of crankcase oil, which may be described by a coding such as SAE 20W or SAE 30 or some similar grouping. These designations represent another method of indicating viscosity values. They refer to divisions in the viscosity classification scheme set out by the Society of Automotive Engineers. The letter W indicates a winter grade oil.

29.3.   Common lubricants

The structural requirements of a lubricant vary according to whether hydrodynamic, boundary or extreme pressure lubrication is necessary. In the simplest case, that of pure hydrodynamic lubrication, almost any fluid will do—e.g. compressed air can be used. Seldom, however, is a clear-cut case met in practice, and a more likely system is one in which all

three lubrication mechanisms are occurring at various points. Any lubricating medium in these circumstances must therefore meet a number of
separate requirements.

Petroleum-based lubricants comprise the biggest class of lubricants.
While a polar molecule adheres to a solid surface more firmly than a
non-polar hydrocarbon, the latter has a greater resistance to oxidation
and subsequent gum formation, and petroleum hydrocarbon fractions
isolated from the higher boiling end of the petroleum distillate form the
most important single source of lubricating materials. A "simple"
hydrocarbon lubricant in practice is likely to contain very small quantities
of polar compounds which are thought to be important in the overall
lubricating mechanism. (It is possible to refine too highly and lessen the
effectiveness of a lubricating oil.) Widely used are *compounded oils*,
which consist of a hydrocarbon fraction blended with polar oils derived
from animal or vegetable sources. Lubricant *greases* are similar, being
a blend of a hydrocarbon and a soap. We shall discuss these polar oils
and soaps when we look at the nature of detergent materials. Other
varieties of lubricants are the *cutting oils* used in production engineering;
these oils, often milky white in appearance, are emulsions of hydrocarbon
and polar oil in aqueous soap solution, the oil medium acting as a good
lubricant, the aqueous medium with its high specific heat and thermal
conductivity absorbing much of the heat generated in the process.

Synthetic lubricants are varied in type. Hydrocarbon polymers like
polyethylene and polybutylene are used either as direct substitutes for
petroleum lubricants, or as additives intended to raise the viscosity index
of the oil. Synthetic polymers containing polar groups are used where
special features such as low volatility or low temperature operation are
required. An interesting range of inorganic lubricants is provided by
silicone oils, which are expensive, but have outstanding properties in
terms of high viscosity indices and high thermal stabilities. *Fluorinated*
and *chlorinated hydrocarbons* are useful where a risk of flammability
exists or where a high degree of chemical stability is required.

Solid lubricants have long been in use. (The distinction between solid
and fluid lubricants may be more apparent than real. It has been demonstrated that fluids under very high pressure, as at an interface, ex-

hibit very great increases in effective viscosity and behave as solids.) *Graphite* is the best known example. Its lubricant properties were once solely attributed to its layered structure (Section 6.2) in which one graphite plane could slip easily with respect to another, but it has been demonstrated that gases adsorbed on the graphite surface are also necessary. Graphite is used as a component in oil and grease mixtures, sometimes being dispersed in the oil as a colloid. In this latter state it tends to cling to metal surfaces more readily than a purely fluid lubricant, and is resistant to oxidation. *Molybdenum disulphide*, which also has a layer type of structure, is used as a high temperature lubricant and as an additive to oils. The disulphide adheres to the solid surface, producing, it is thought, a low coefficient of friction because of the subsequent ease of shearing of molybdenum–sulphur bonds.

29.4. Additives

Lubricants contain a number of additives apart from those which themselves contribute to the lubrication mechanism. Engine oils normally have present a *detergent*, the purpose of which is to free the surfaces from dirt and carbonaceous deposits. *Antioxidant* additives are essential components of many organic preparations, including lubricating oils; amine compounds are commonly used to inhibit oxidation reactions. To minimize the effect of crystallization of heavier hydrocarbons from the fluid at low working temperatures a *pour point depressant* is added to lubricant oils. This does not necessarily prevent crystallization occurring, but it ensures that the crystals present are much smaller and less troublesome than would otherwise be the case. Lastly, *corrosion inhibitors* are added to protect metal bearings.

SOAPS AND DETERGENTS

29.5. Detergent action

The process of deterging is one of dirt and grease removal from a surface. The agents used to bring this about are called soaps and synthetic detergents ("syndets" in industrial jargon). Dirt is normally in itself partly or solely organic and therefore not readily soluble in or removed by water, and a detergent must be employed.

Detergent action is complex, but we can simplify it by dividing it into two distinct steps. The first of these is to bring the detergent solution into intimate contact with the surface to be cleaned, a process called *wetting*; the second is to remove the dirt from the surface region by *emulsification* in the aqueous phase.

Both wetting and emulsification mechanisms are dependent on the chemical structure of the detergent molecules. A typical structure consists of a long hydrocarbon chain terminating in a highly polar function. Thus the sodium salt of a long chain carboxylic acid like stearic acid, or the quaternary ammonium salt of a long chain amine like cetyltrimethylamine both have detergent properties.

$$C_{17}H_{35}{\cdot}CO_2^-Na^+ \qquad C_{16}H_{33}{\cdot}\overset{+}{N}(CH_3)_3Cl^-$$

Sodium stearate   Cetyltrimethylammonium chloride

The hydrocarbon part of these compounds, which is typically organic, covalent, and therefore non-polar, has no affinity for or tendency to dissolve in water and is termed the *hydrophobic* group. The polar part, conversely, will freely dissolve in water to give ionic species, and is termed the *hydrophilic* group. Every detergent contains in some form two such groups. In a two-phase system, a molecule will tend to orientate itself at the interface such that the hydrophobic group is in the organic phase, the hydrophilic end in the aqueous phase. This ordered arrangement at the interface lowers the surface tension in that region. It is this lowering of interfacial energy which constitutes the driving force in the wetting and emulsification processes.

Let us consider a surface covered with an oil film. Water will be repelled by the oil film in this system, and wetting will not occur. If a detergent is present, however, the hydrophobic group can penetrate the oil film, and become attached to the surface proper. Adsorption of the detergent at the surface displaces the oil, which rolls up to form a globule. This "rolling-off" phenomenon is beautifully illustrated in Fig. 29.3.

FIG 29.3. Detergent action at a fibre surface. (Reproduced by permission of Unilever Research.)

The oil globules, having formed, are removed by mechanical agitation and subsequent emulsification in the aqueous medium. In this step, the hydrophobic groups of the detergent molecules orientate towards the oil globules, while the ionized hydrophilic ends are directed outwards into the surrounding aqueous medium. Each oil globule therefore acquires

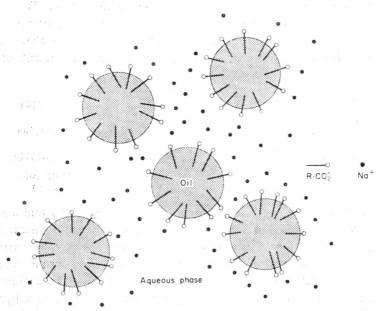

$R \cdot CO_2^-$ Na^+

Oil

Aqueous phase

FIG. 29.4. Mutual repulsion of emulsified globules.

a shield of charged particles. One globule will be repelled by another, coagulation will be prevented, and dispersion in the aqueous phase encouraged. This is represented in Fig. 29.4 in which the action of, for example, sodium stearate is seen to produce a negatively charged anionic species at the emulsified particle.

The picture outlined above, while simplified, contains the essential ingredients of detergent action. Instinctively we tend to associate a lather with good cleansing properties. A decrease in surface tension (such as occurs in a detergent solution) will always enhance the formation of bubbles, but these are not thought to be a necessary feature of a cleansing solution. Dirt may be entrained by and carried away in the foam, but this is a secondary effect.

29.6. Soaps

Soaps are detergents derived from naturally occurring oils, fats and waxes which in turn are available from vegetable sources. These oils are glycerides, i.e. they are esters of the trihydric alcohol glycerol and a number of long chain carboxylic acids. The hydrolysis of a glyceride, a reaction called *saponification*, yields glycerol and a mixture of salts of the respective acids:

$$
\begin{array}{lll}
C_2H \cdot O \cdot OC \cdot R & CH_2 \cdot OH & R \cdot CO \cdot ONa \\
| & | & \\
CH \cdot O \cdot OC \cdot R' \; + \; 3NaOH \; \longrightarrow & CH \cdot OH \; + & R' \cdot CO \cdot ONa \\
| & | & \\
CH_2 \cdot O \cdot OC \cdot R'' & CH_2 \cdot OH & R'' \cdot CO \cdot ONa \\
\text{Glyceride} & \text{Glycerol} & \text{Sodium salts} \\
& & \text{(Soap)}
\end{array}
$$

Some of the more common acid structures occurring in glycerides have been listed in Table 24.2, from which we can see that they consist of a long hydrocarbon chain (typically containing fifteen or seventeen carbon atoms, and perhaps containing one or more carbon–carbon double bonds) terminating in a polar carboxylic acid group. The sodium or potassium salts, with a polar hydrophilic group attached to a hydrophobic hydrocarbon group, have the necessary structural features required of a detergent, and are called *soaps*.

Glycerides, which vary from simple esters, in which the same acid fragment occurs at all three ester linkages, to complex esters, which yield mixed acids on saponification, are important not only in the soap industry but also in foodstuffs (animal and vegetable fats are glycerides essentially) and in the manufacture of paints, varnishes and linoleum. The alcohol component of saponification, glycerol, has several outlets which include its use as an anti-freeze, as a source of synthetic polyesters, and as the stock material for nitroglycerine manufacture; demand for glycerol is such that much has also to be synthesized from propylene.

29.7. Synthetic detergents

Soaps have a number of disadvantages. In hard water the carboxylic acid anion precipitates out as insoluble calcium and magnesium salts and becomes ineffective; soaps tend to be sensitive to acid conditions; moreover, the demand for soaps has outstripped the supply of raw materials. Synthetic detergents, which overcome these disadvantages, have become very important and now account for well over half of the detergent market.

The largest single class of synthetic detergents is that in which the hydrophilic group is the sodium salt of a sulphonic acid or sulphonated alcohol. The sulphonic acid type is prepared by alkylating the benzene ring with a group containing ten or twelve carbon atoms, and sulphonating the nucleus:

| Alkyl benzene | Alkyl benzene sulphonate | Alkyl benzene sulphonate (sodium salt) |

Alcohols for sulphonation are prepared by catalytic reduction of long chain acids derived from glycerides, or by synthesis from alkenes. If in these structures the group R is branched,

$$R \cdot CO_2H \rightarrow R \cdot H_2OH \xrightarrow[Na_2CO_3]{N_2SO_4} R \cdot CH_2 \cdot O \cdot SO_3Na$$

Alkenes

Long chain alcohol   Alcohol sulphate (sodium salt)

resistance to degradation by biochemical organisms is high. Such biologically "hard" detergents have in the past caused considerable problems of river pollution, and are now avoided.

Sulphates and sulphonates ionize in water to yield species in which the anion, $RO \cdot SO_3^-$ or $R \cdot SO_3^-$, is the active participant in the detergent process. *Anionic detergents* are by far the biggest class.

Cationic surface active agents are more often used as textile softeners and as germicides than as detergents. Quaternary ammonium salts (chlorides or bromides), whose structure may be written as $R_4N^+Cl^-$, typify this class.

Some surface active agents are *amphoteric*, containing both anionic and cationic species. They take the form of the sodium salt of an amino acid, i.e. $R \cdot NH \cdot (CH_2)_2 \cdot CO_2Na$.

Nonionic detergents are also produced, and indeed are much more important than cationic and amphoteric types. Here interfacial activity depends on the presence of a number of groups, polar in nature and therefore having an affinity for water, but not ionizable in water. Long chain ethers, made by reaction of alcohols or phenols with ethylene

$$R—\langle\bigcirc\rangle—OH + n \cdot CH_2—CH_2 \rightarrow R—\langle\bigcirc\rangle—(O \cdot CH_2 \cdot CH_2)_n \cdot OH$$
$$\underset{O}{\diagdown\diagup}$$

| Alkyl phenol | Ethylene oxide | Long chain polyether |

oxide, provide an important example. The interaction with water is thought to occur through hydrogen bonding at the ether oxygen.

$$—O— + H_2O \rightarrow —O—$$
$$H—O—H$$

| Ether link | Hydrogen bonded ether |

29.8. Detergent builders

The efficiency of detergents is improved by the presence of additives known as detergent builders. These are electrolytes, the most common of which at the present time are the sodium phosphates. One of the main functions of these substances, whose role in the detergent process is recognized as complex, is to encourage the emulsification of greases.

Exercises

29.1. Define and distinguish between hydrodynamic lubrication and boundary lubrication.

29.2. What do you understand by the term "wetting"? To what extent is it important in lubrication?

29.3. Classify the following substances as hydrophilic or hydrophobic: toluene; *n*-butyl alcohol; *p*-hydroxybenzoic acid; polyethylene; linoleic acid.

29.4. What steps could be taken to render polystyrene water-soluble?

Appendix

SI Units

THE International System of Units has as its foundation seven basic units (Table A.1).

TABLE A.1. BASIC SI UNITS

Physical quantity	Basic unit	Unit symbol
Length	metre	m
Mass	kilogram	kg
Time	second	s
Electric current	ampere	A
Thermodynamic temperature	kelvin	K
Luminous intensity	candela	cd
Amount of substance	mole	mol

From these basic units there are derived other units a selection of which is shown in Table A.2.

TABLE A.2. DERIVED SI UNITS

Physical quantity	Name of unit (unit symbol)	Unit definition	
Force	newton (N)	mass $\times$ acceleration	$= \text{kg m s}^{-2}$
Pressure	pascal (Pa)	force $\div$ area	$= \text{N m}^{-2} = \text{kg m}^{-1}\text{ s}^{-2}$
Energy	joule (J)	force $\times$ distance	$= \text{N m} = \text{kg m}^2\text{ s}^{-2}$
Power	watt (W)	work $\div$ time	$= \text{J s}^{-1} = \text{kg m}^2\text{ s}^{-3}$

Although by no means essential we find it convenient to use prefixes to denote fractions and multiples of basic and derived units. A selection of the most common, recommended prefixes is shown in Table A.3.

TABLE A.3. SOME RECOMMENDED PREFIXES
FOR SI UNITS

Fraction or multiple	Prefix	Symbol
10^6	mega	M
10^3	kilo	k
10^{-1}	centi	c
10^{-2}	deci	d
10^{-3}	milli	m
10^{-6}	micro	μ
10^{-9}	nano	n

Some non-SI units and their SI equivalents are shown in Table A.4.

TABLE A.4. NON-SI UNITS AND THEIR SI EQUIVALENTS

Physical quantity	Name of unit (unit symbol)	SI equivalent
Length	(a) angstrom (Å)	10^{-10} m
	(b) inch (in)	$2 \cdot 54 \times 10^{-2}$ m
Mass	pound (lb)	$0 \cdot 4535$ kg
Volume	(a) litre (l)	10^{-3} m^3 = dm^3
	(b) gallon (gal)	$4 \cdot 546 \times 10^{-3}$ m^3 = $4 \cdot 546$ dm^3
Force	dyne (dyn)	10^{-5} N
Pressure	(a) atmosphere (atm)	$1 \cdot 01325 \times 10^5$ N m^{-2}
	(b) millimetres of mercury (mmHg)	$133 \cdot 322$ N m^{-2}
Energy	(a) erg (erg)	10^{-7} J
	(b) calorie (cal)	$4 \cdot 184$ J
	(c) British thermal unit (Btu)	$1054 \cdot 4$ J
Temperature	degree Celsius (°C)	0°C = 273 K

All of the units in Table A.4 must now be regarded as obsolete and their use will progressively diminish, but old habits die hard and during the time it takes for new habits to be acquired, familiarity with the old and new unit systems will be necessary.

Table of Elements, Symbols, Atomic Numbers, Atomic Weights

Name	Symbol	Atomic number	Atomic weight	Name	Symbol	Atomic number	Atomic weight
Actinium	Ac	89	(227)	Gadolinium	Gd	64	157·25
Aluminium	Al	13	26·98	Gallium	Ga	31	69·72
Americium	Am	95	(243)	Germanium	Ge	32	72·59
Antimony	Sb	51	121·75	Gold	Au	79	196·97
Argon	Ar	18	39·948	Hafnium	Hf	72	178·49
Arsenic	As	33	74·92	Helium	He	2	4·003
Astatine	At	85	(210)	Holmium	Ho	67	164·93
Barium	Ba	56	137·34	Hydrogen	H	1	1·0080
Berkelium	Bk	97	(249)	Indium	In	49	114·82
Beryllium	Be	4	9·012	Iodine	I	53	126·90
Bismuth	Bi	83	208·98	Iridium	Ir	77	192·2
Boron	B	5	10·81	Iron	Fe	26	55·85
Bromine	Br	35	79·909	Krypton	Kr	36	83·80
Cadmium	Cd	48	112·40	Lanthanum	La	57	138·91
Caesium	Cs	55	132·91	Lawrencium	Lw	103	(257)
Calcium	Ca	20	40·08	Lead	Pb	82	207·19
Californium	Cf	98	(251)	Lithium	Li	3	6·939
Carbon	C	6	12·011	Lutetium	Lu	71	174·97
Cerium	Ce	58	140·12	Magnesium	Mg	12	24·312
Chlorine	Cl	17	35·453	Manganese	Mn	25	54·94
Chromium	Cr	24	52·00	Mendelevium	Md	101	(256)
Cobalt	Co	27	58·93	Mercury	Hg	80	200·59
Copper	Cu	29	63·54	Molybdenum	Mo	42	95·94
Curium	Cm	96	(247)	Neodymium	Nd	60	144·24
Dysprosium	Dy	66	162·50	Neon	Ne	10	20·183
Einsteinium	Es	99	(254)	Neptunium	Np	93	(237)
Erbium	Er	68	167·26	Nickel	Ni	28	58·71
Europium	Eu	63	151·96	Niobium	Nb	41	92·91
Fermium	Fm	100	(253)	Nitrogen	N	7	14·007
Fluorine	F	9	19·00	Nobelium	No	102	(253)
Francium	Fr	87	(223)	Osmium	Os	76	190·2

Name	Symbol	Atomic number	Atomic weight	Name	Symbol	Atomic number	Atomic weight
Oxygen	O	8	15·999	Sodium	Na	11	22·9898
Palladium	Pd	46	106·4	Strontium	Sr	38	87·62
Phosphorus	P	15	30·974	Sulphur	S	16	32·064
Platinum	Pt	78	195·09	Tantalum	Ta	73	180·95
Plutonium	Pu	94	(242)	Technetium	Tc	43	(99)
Polonium	Po	84	(210)	Tellurium	Te	52	127·60
Potassium	K	19	39·102	Terbium	Tb	65	158·92
Praseodymium	Pr	59	140·91	Thallium	Tl	81	204·37
Promethium	Pm	61	(147)	Thorium	Th	90	232·04
Protactinium	Pa	91	(231)	Thulium	Tm	69	168·93
Radium	Ra	88	(226)	Tin	Sn	50	118·69
Radon	Rn	86	(222)	Titanium	Ti	22	47·90
Rhenium	Re	75	186·23	Tungsten	W	74	183·85
Rhodium	Rh	45	102·91	Uranium	U	92	238·03
Rubidium	Rb	37	85·47	Vanadium	V	23	50·94
Ruthenium	Ru	44	101·1	Xenon	Xe	54	131·30
Samarium	Sm	62	150·35	Ytterbium	Yb	70	173·04
Scandium	Sc	21	44·96	Yttrium	Y	39	88·91
Selenium	Se	34	78·96	Zinc	Zn	30	65·37
Silicon	Si	14	28·09	Zirconium	Zr	40	91·22
Silver	Ag	47	107·870				

PERIODIC TABLE OF THE ELEMENT

	Representative Elements		Transition Elements—d										Representative Elements						Noble Gas Elements
Group:	IA	IIA	IIIB	IVB	VB	VIB	VIIB	VIII			IB	IIB	IIIA	IVA	VA	VIA	VIIA	0
Valence shell	s^1	s^2	$d^1s^2f^x$	d^2s^2	(d^3s^2)‡	d^5s^1	(d^5s^2)‡	(d^6s^2)‡(d^7s^1)‡(d^8s^2)‡			s^1d^{10}	s^2	s^2p^1	s^2p^2	s^2p^3	s^2p^4	s^2p^5	s^2p^6
Period																		
$n=1$ 1s	1 H 1.00797																	2 He 4.0026
$n=2$ 2s2p	3 Li 6.939	4 Be 9.0122											5 B 10.811	6 C 12.01115	7 N 14.0067	8 O 15.9994	9 F 18.9984	10 Ne 20.183
$n=3$ 3s3p	11 Na 22.9898	12 Mg 24.312											13 Al 26.9815	14 Si 28.086	15 P 30.9738	16 S 32.064	17 Cl 35.453	18 Ar 39.948
$n=4$ 4s3d4p	19 K 39.102	20 Ca 40.08	21 Sc 44.956	22 Ti 47.90	23 V 50.942	24 Cr 51.996	25 Mn 54.9380	26 Fe 55.847	27 Co 58.9332	28 Ni 58.71	29 Cu 63.54	30 Zn 65.37	31 Ga 69.72	32 Ge 72.59	33 As 74.9216	34 Se 78.96	35 Br 79.909	36 Kr 83.80
$n=5$ 5s4d5p	37 Rb 85.47	38 Sr 87.62	39 Y 88.905	40 Zr 91.22	41 Nb 92.906	42 Mo 95.94	43 Tc 99	44 Ru 101.07	45 Rh 102.905	46 Pd 106.4	47 Ag 107.870	48 Cd 112.40	49 In 114.82	50 Sn 118.69	51 Sb 121.75	52 Te 127.60	53 I 126.9044	54 Xe 131.30
$n=6$ 6s4f5d6p	55 Cs 132.905	56 Ba 137.34	57–71 ★	72 Hf 178.49	73 Ta 180.948	74 W 183.85	75 Re 186.2	76 Os 190.2	77 Ir 192.2	78 Pt 195.09	79 Au 196.967	80 Hg 200.59	81 Tl 204.37	82 Pb 207.19	83 Bi 208.980	84 Po 210	85 At 210	86 Rn 222
$n=7$ 7s5f6d7p	87 Fr 223	88 Ra 226	89–103 ★															

Inner Transition Elements—f

★ Lanthanide series	57 La 138.91	58 Ce 140.12	59 Pr 140.907	60 Nd 144.24	61 Pm 145	62 Sm 150.35	63 Eu 151.96	64 Gd 157.25	65 Tb 158.924	66 Dy 162.50	67 Ho 164.930	68 Er 167.26	69 Tm 168.934	70 Yb 173.04	71 Lu 174.97
★ Actinide series	89 Ac 227	90 Th 232.038	91 Pa 231	92 U 238.03	93 Np 237	94 Pu 242	95 Am 243	96 Cm 247	97 Bk 247	98 Cf 249	99 Es 254	100 Fm 253	101 Md 256	102 No 254	103 Lw 257

‡Variable valence shells.

Index